MEDICAL PHYSICS MONOGRAPH NO. 17

COMPUTERS IN MEDICAL PHYSICS

Edited by

Anthony R. Benedetto, Ph.D.
University of Texas Medical Branch
Galveston, Texas

H. K. Huang, D.Sc.
University of California
Los Angeles, California

Don P. Ragan, Ph.D.
Wayne State University
Detroit, Michigan

Manuscripts were originally prepared for the 1988 AAPM Summer School held at St. Edward's University, Austin, Texas, 31 July–5 August 1988

Published for the
American Association of Physicists in Medicine
by the American Institute of Physics

Further copies of this report may be obtained from:

American Institute of Physics
c/o AIDC
64 Depot Road
Colchester, Vermont 05446

(1-800-445-6638)

Library of Congress Catalog Card Number: 90-82398
International Standard Book Number: 0-88318-802-3
International Standard Serial Number: 0163-1802

Published by the American Institute of Physics, Inc.
500 Sunnyside Blvd., Woodbury, NY 11797

Printed in the United States of America

FOREWORD

My primary computational tool when I was an undergraduate during the mid-'60s was my Post Versalog slide rule, a marvel of precision bamboo craftwork. My primary word processing tool was an IBM Selectric typewriter, with either special erasable paper or the pencil-type eraser with a brush on the end. My primary database tool was a file box of 3×5 index cards with alphabetical dividers.

By the mid-'70s, the big decision about a calculational tool was whether to upgrade from the Hewlett-Packard HP35 to the HP45. Minicomputers were being introduced into nuclear medicine, but software wasn't available for my routine work. I was still using a Selectric typewriter, but secretaries had the IBM Magcard and MT-ST memory typewriters for elementary word processing. My database was still on 3×5 cards, but some of our departmental data were kept on the tapes and cards of the IBM memory typewriters.

In 1990, I'm awaiting the installation of my own personal desktop IBM PS/2 Model 70, equipped with an Intel 80386 32-bit microprocessor operating at 25 MHz, a 115 Mbyte hard disk, a laser printer, an Intel 80387 coprocessor, a spectacular color display, an optical disk, and a connection to our divisional network, our departmental network, our campus network, and from there to the rest of the world. I'm already using Word Perfect 5.1, and we'll be installing dBase IV as soon as the hardware arrives. The computational power on my desktop is roughly equivalent to the IBM mainframe computer that was the mainstay of our university's computer center when I started college!

At the other end of the spectrum, supercomputers now operate at billions of operations per second, allowing calculation of π to a ridiculous number of digits. These high speeds also allow rapid calculation of 3-D tomography and real-time display of 3-D reconstructions of PET, CT, and MRI data. Radiation therapy systems allow beam's-eye views of treatment plans in 3-D and color and performance of conformal therapy wherein many parameters are varied during the course of a treatment session.

The 1988 AAPM Summer School explored the modern world of computers, with special emphasis on applications in medical physics. Authors of the articles in this book, who were also presenters at the Summer School, were asked to assume that the attendee-reader was familiar with computers in general but was not a computer scientist or a hacker. The manuscripts were reviewed by at least two experts prior to inclusion in this book. Since not all presenters submitted manuscripts, this volume is not a proceedings in the traditional sense.

The accuracy of information presented in these articles is the responsibility of the authors, not of AAPM, although the extensive review process should have eliminated any egregious errors. Opinions expressed in the articles are those of the authors only and do not constitute official AAPM policy or recommendations or statement of state-of-the-art medical physics practice. Some of the opinions created spirited arguments at the Summer School—in any fledgling discipline, controversy inevitably accompanies technological advances. Only time will tell if some of the material presented here will become accepted practice.

Authors were encouraged to attribute commercial products, trademarks, etc, in their articles. Any oversight by the authors or the editors was purely unintentional.

Special thanks are due to my co-editors (and co-chairmen of the Summer School) Bernie Huang and Don Ragan. Medical physics is such a broad field that no one person can possibly know everyone and everything. Multiple authorship has come under fire recently, but the Summer School and this book couldn't have happened any other way. Special thanks are also due to Lou Levy and Dave Goff, of the Southwest Chapter, who were superb local arrangements chairmen. They and their families and colleagues made the Summer School a pleasant as well as an educational experience.

I hope the reader finds the book educational, interesting, challenging, and perhaps even a little controversial.

Anthony R. Benedetto, Ph.D.
5 February 1990

CONTENTS

AUTHORS/SUMMER SCHOOL FACULTY

Ernst Ammann, Dr.-Ing.
Siemens AG, Medical Engineering Group
Erlangen, West Germany

Steven Balter, Ph.D.
Cornell University Medical College
New York, New York 10021

Gary T. Barnes, Ph.D.
University of Alabama at Birmingham
Birmingham, Alabama 35233

Anthony R. Benedetto, Ph.D.
The University of Texas Medical Branch
Galveston, Texas 77550

James C. Carlson, M.S.
Current Computer Concepts
Greenwood, Indiana 46142

Chee-Wai Cheng, Ph.D.
Joint Center for Radiation Therapy
Boston, Massachusetts 02115

Trevor D. Cradduck, Ph.D.
Victoria Hospital, Univ. of Western Ontario
London, Ontario, Canada N6A 4G5

Benedick A. Fraass, Ph.D.
University of Michigan Medical Center
Ann Arbor, Michigan 48109

L. Stephen Graham, Ph.D.
VA Medical Center
Sepulveda, California 91343

Philip H. Heintz, Ph.D.
N.T. Enloe Memorial Hospital
Chico, California 95926

Bruce Ho, Ph.D.
University of California at Los Angeles
Los Angeles, California 90024

Steven C. Horii, M.D.
Georgetown University Hospital
Washington, D.C. 20007

H.K. Huang, D.Sc.
University of California at Los Angeles
Los Angeles, California 90024

Ira J. Kalet, Ph.D.
University of Washington
Seattle, Washington 98195

Jack L. Lancaster, Ph.D.
The University of Texas Health Science Center
San Antonio, Texas 78284

Tom K. Lewellen, Ph.D.
University of Washington
Seattle, Washington 98195

Jonathan M. Links, Ph.D.
The Johns Hopkins Medical Institutions
Baltimore, Maryland 21205

Nicholas J. Mankovich, Ph.D.
University of California at Los Angeles
Los Angeles, California 90024

Don P. Ragan, Ph.D.
Wayne State University
Detroit, Michigan 48201

Osman Ratib, M.D., Ph.D.
University of California at Los Angeles
Los Angeles, California 90024

George W. Sherouse, M.S.
University of North Carolina
Chapel Hill, North Carolina 27599

I. HARDWARE AND SOFTWARE FUNDAMENTALS

COMPUTER BASICS

Jack L. Lancaster, Ph.D.
University of Texas Health Science Center
San Antonio, TX 78284

Introduction

There are analogous relationships between computer characteristics dealing with input, output, processing, and storage of data and human characteristics relating to seeing, hearing, speaking, cognition, and decision making. I will use such analogies when possible to relate some of the characteristics of current computer technology to our everyday experiences.

A computer system consists of hardware (physical attributes) and software (logical attributes). The hardware is usually broken into functional elements relating to input, storage, processing, and output of data. In a rudimentary fashion these hardware elements have parallels to human physical attributes which provide similar capabilities. For instance, input relates to the human senses and sense organs, storage and processing relate to the brain, and output relates to many forms of human expression. Computer software determines the operating characteristics of the computer system within the physical limits of the hardware. The best analogy between computer software and a human characteristic is probably the catch-all term personality. The software of a computer system determines its personality and the hardware determines its physical capabilities. Just as a broad range of personal achievements are attainable from people with different personalities and physical characteristics a broad range of computer accomplishments are possible due to multifaceted hardware and software. As with people, success is often attributed to the software (personality) and less often to the hardware (physical capability).

The method of input, storage, processing, and output of information by humans is extremely complicated when compared to that of most computers in use today. The bottom line is that a set of rules must be followed to guarantee accurate and concise communication of data. Such a set of rules is the basis for a language.

The Binary Language

To truly understand how the computer functions a good knowledge of the language of computers is necessary. As with humans there are many languages and dialects. However, at the lowest level all computers use a binary (on/off) system of coding to represent language elements. While the specifics vary from system to system it is instructive to learn about the binary language features which are common to all digital computers.

NUMBER SYSTEMS - Several different number systems are used as coding schemes for programming, operator convenience, and to simplify communications. It is essential to learn about these number systems in order to read and understand the technical literature dealing with computer systems. Below is a summary of the most common number and coding systems:

BINARY NUMBERS - 01110101 00011101

Base or Radix - 2
Characters - (0,1)
Bit - Binary digit
Place value = 2^n.

Here n = bit location in number starting with n = 0 for the least significant bit (the one on the right above). An interesting result of the binary representation for odd numbers is that the right most, or least significant bit, is always one. The largest number in binary representation will have 1's in all bit positions.

OCTAL NUMBERS -	01 110 101	00 011 101
	1 6 5	0 3 5

Base or Radix - 8
Characters (0 - 7)
Place value = 8^n.

Note the grouping of 3 binary places (Bits) for 1 octal place. For clarity in printed material octal numbers may have a K or Q at the end. For example 165Q or 165K. Largest octal number has 7's in all places.

HEXADECIMAL NUMBERS -	0111 0101	0001 1101
	7 5	1 D

Base or Radix - 16
Characters (0 - 9, A - F)
Place value = 16^n.

Note the grouping of 4 binary places for 1 hexadecimal place. Again for clarity printed material may have an H at the end, i.e., 75H. Largest number has F's in all places.

DECIMAL NUMBERS - 117 29

Base or Radix - 10
Characters (0 - 9)
Place value = 10^n.

Note that groupings of bits cannot be performed for base 10 because 10 is not an even power of 2. In fact 2^3.3219... = 10. Largest number has 9's in all places.

Note from the examples above:
01110101 = 165Q = 75H = 117
00011101 = 35Q = 1DH = 29

BINARY CODED DECIMAL (BCD) - This is a coding scheme whereby the decimal digits (0-9) are represented using 4 bits per digit. This form of coding is not as compact as straight binary because codes representing decimal 10-15 within the 4 bit code are not used. With some special handling, data coded in this form can be manipulated mathematically.

EXTENDED BINARY CODED DECIMAL INTERCHANGE CODE (EBCDIC) - An 8 bit coding scheme which is an extension of BCD which includes 4 extra bits (or zones) that provide encoding of alphanumeric characters and control codes as well as the decimal numbers. This coding scheme is used on IBM mainframe models and similar machines produced by other manufacturers. It is commonly used for data storage on magnetic media.

AMERICAN STANDARD CODE FOR INFORMATION INTERCHANGE (ASCII) - This coding scheme is widely used in data communications work and to represent character type data internally in personal computers. The ASCII code can be either 7 or 8 bits with the 7 bit version allowing for 128 characters and the 8 bit version 256 characters. In personal computers the 8 bit coding is used for graphics characters, foreign language characters, and other special symbols. The ASCII character set for an IBM PC is organized as follows:

Decimal	Hex	Use
0-31	00-1F	Control Characters
32-47	20-2F	Symbols (punctuation, math, etc.)
48-57	30-39	Numeric Characters (0-9)
58-64	3A-40	Typing (math, etc.)
65-90	41-5A	Upper Case Alphabet
91-96	5B-60	Symbols (punctuation, etc.)
97-122	61-7A	Lower Case Alphabet
123-127	7B-7F	Symbols
128-255	80-FF	Special Symbols

The lower four bits of the ASCII numeric symbols are coded as simple binary numbers and have the same numeric value as the numeric character they represent. Both lower and upper case alphabetic characters are arranged in numerical order which facilitates alphabetizing. The lower case characters are displaced from the upper case characters by decimal 32 which aids in shifting case.

The various number representations such as Hexadecimal facilitate recalling the esoteric sequences of 1's and 0's of binary numbers and provide a more compact representation in printed material.

General Information - Number System

The largest number which can be represented by p places can be calculated for any base as follows:

$\text{base}^p - 1$ = largest number represented by p places

Examples:

3 place binary number	$2^3 - 1 =$	$8 - 1 =$	7
3 place octal number	$8^3 - 1 =$	$512 - 1 =$	511
3 place decimal number	$10^3 - 1 =$	$1000 - 1 =$	999
3 place hexadecimal #	$16^3 - 1 =$	$4096 - 1 =$	4095

The number of distinct values represented by p places is just one larger than the largest number. The difference comes from the fact that zero is counted as a value.

$base^p$ = number of distinct values represented by p places

Examples (binary numbers):
$2^5 = 32$ (These examples are common to
$2^6 = 64$ medical image matrix dimensions)
$2^7 = 128$
$2^8 = 256$
$2^9 = 512$
$2^{10} = 1024$ (Commonly referred to as 1K)

General Information - Binary Number System

BYTE, NIBBLE - These terms are used to refer to groups of bits as commonly used by the computer. A byte is 8 bits and a nibble is just half a byte or 4 bits.

WORD - A logical unit of information. A group of bits, characters, or bytes considered as an entity and capable of being stored in one memory location. The word length is usually 8, 16, or 32 bits.

SIGN - The sign of a binary number (in the computer) is maintained by using one of the bits (the left or most significant bit). Positive numbers are represented by a sign bit = 0 and negative numbers by a sign bit = 1. The largest number representable is reduced by the loss of one place but the total number of values representable remains the same including positive and negative values.

Numeric Data Types

INTEGER - Integers are positive (unsigned) or positive/negative (signed) whole numbers. For example, a 16 bit unsigned number can take on values from 0-65535 whereas a signed 16 bit integer may range from -32768 to +32767.

FIXED POINT - A numeric representation assuming a fixed location of the radix point. An integer is a special case of fixed point representation with no fractional part. The standard form allows for an integer part and a fractional part. The integer place values and sign are handled just as integers. The fractional place values are equal to 2^{-p} where p is the bit position relative to the radix point. The range of numbers which can be represented by fixed point is limited by the number of bits.

FLOATING POINT - An alternate approach to numeric representation which is the binary equivalent of scientific notation. The range of numbers represented in this form is much greater than fixed point. A floating point number is represented as $N \times r^p$ where N is the mantissa, r is the base or radix, and p is the exponent or power to which r is raised. In the computer the radix is 2. The mantissa is normalized for storage such that the highest value nonzero place appears to the right of the radix point.

Ex:	Fixed Point Binary		Floating Point Binary
	101.11	=	0.10111×2^{-2}
	-0.001101	=	-0.1101×2^{-2}

PRECISION OF NUMERIC DATA - Precision is determined by the number of bits utilized to represent the data. Higher precision requires more bits. Precision is a characteristic which is placed on the basic data types so that we can have low and high precision integers, fixed point, and floating point data. This control allows programmers to manage program speed and storage requirements.

Binary Mathematics

Addition - This operation can be performed easily by a digital computer because the rules are well suited for binary logic circuitry.

Rules:
$1 + 1 = 0$ with a carry of 1
$1 + 0 = 1$
$0 + 1 = 1$
$0 + 0 = 0$

Addition of like bits produces a zero result while unlike bits produce a one result. If the like bits are 1's then a 1 carry is generated. The radix point must be adjusted such that the exponents are equal before adding floating point numbers.

Multiplication - This is another operation which is reasonably well suited for computer manipulation because of the simplicity of rules.

Rules:
$1 \times 1 = 1$
$1 \times 0 = 0$
$0 \times 1 = 0$
$0 \times 0 = 0$

If either the multiplicand or multiplier bit is zero the product is zero. Binary multiplication is conceptually a very simple process which involves shifting the binary or radix point and adding results. Shifting comes from the fact that multiplication by 2 is equivalent to moving the binary point one place to the right, multiplication by 4 equivalent to 2 places to the right, etc.

Example of Binary Multiplication:

```
  101    =  5
 x101    = x5
-----      --
  101
 000
101
-----
11001    = 25
```

The computer adds intermediate results after each shift in order to handle the carry produced when 1 + 1 is encountered. Multiplication results generally require more bits for storage than the input numbers. Multiplication of large numbers requires a large number of shifting and adding processes and takes much longer than addition of the same two numbers. Multiplication of two floating point numbers is accomplished by multiplying the mantissas of the two numbers and adding the exponents.

<u>Subtraction</u> - The computer usually performs subtraction by modifying numbers such that they can be manipulated using addition rules. The modification process is called complementing, and the most commonly employed method is referred to as 2's complement. Utilizing this approach the computer does not have to know how to subtract.

Rules for 2's complementing:
1. Change all the 0's to 1's and the 1's to 0's.
2. Add 1.

A second method to derive the 2's complement of a binary number is the following algorithm:

1. Beginning with the least significant bit (LSB) and proceeding to the most significant bit (MSB), copy each bit until a 1 is encountered.
2. Copy the first 1.
2. Copy the complement of the remaining bits.

Example: Subtract 3 from 5.

Form the 2's complement of 3 = 0011 (using 4 bit arithmetic processor).

Binary 3	0011
complement	1100
2's complement	1101

Now add the 2's complement of 3 to 5.

```
 0101    =  5
+1101    = -3
 ----      --
10010              (Drop the carry bit)
 0010       2
```

Note that the result is the same as if 3 had been stored as -3 and addition was performed. For this reason negative numbers are stored in 2's complement form.

The radix point must be adjusted such that the exponents are equal before subtracting floating point numbers.

Division - There are many algorithms for division. One commonly used method is to repetitively subtract the Divisor from the Dividend until a borrow is needed. The Quotient is the number of subtractions with the remainder being what was left in the Dividend.

In terms of speed of these mathematical operations addition is the fastest followed closely by subtraction. Multiplication and division are generally much slower with division being the slowest of the basic mathematical operations performed within the computer. However, there are certain cases where multiplication and division can be fast. The first and most obvious is when either of the numbers is small. Second, when one of the numbers is a multiple of two multiplication is just a shift of the radix point. When the divisor is a multiple of two division is also just a shift of the radix point.

Speed is also determined to a great extent by the precision of the data to be mathematically manipulated. The more bits associated with numeric data the longer it takes for mathematical manipulation.

There are a number of mathematical functions such as trigonometric functions, logarithms, and exponentiation which are also available. These functions are implemented using a power series expansion such that the CPU task is reduced to the more basic operations of add, subtract, multiply, and divide. These scientific functions are sometimes performed by dedicated hardware to improve processing speed.

Special tests such as comparison of numbers and logical operations are also available to round out the mathematical operations in binary arithmetic.

Computer Logic

The central processor (CPU) of a computer is composed of basic decision-making cells or logic elements. The intelligence of the CPU is derived from the interconnection of these logic elements providing the capability to perform more complex functions such as interpretation of codes, control, and mathematical manipulations. A two-element subset of Boolean algebra or switching algebra is the mathematical tool used in the analysis and design of digital circuits. The basic operations of this two-element Boolean algebra are (1) OR, the sum, (2) AND, the product, and (3) NOT, the inversion or complement. These basic operations can be combined to form seven basic combinatorial logic functions. In the computer these combinatorial functions are performed by logical circuits which act as gates to control the state of an output based on the conditions at two inputs. The 7 basic combinatorial logic functions assignable to the two binary inputs and one binary output are listed in the following table:

INPUT		OUTPUT FOR LOGIC TYPE						
A	B	AND	OR	NOTB	NAND	NOR	XOR	XNOR
0	0	0	0	1	1	1	0	1
0	1	0	1	0	1	0	1	0
1	0	0	1	1	1	0	1	0
1	1	1	1	0	0	0	0	1

The fundamental logical building blocks of the computer are these three-element (2 inputs, 1 output) logic gates and the two-element NOT. Further combinations of these logic gates provide functions used for coding, decoding, and the mathematical and logical operations which provide the CPU and associated circuitry with its basic capabilities.

Hardware

In common usage hardware refers to the macroscopic physical components or equipment that make up a computer system, such as a keyboard, disk drive, and visual display. However, all physical components comprising a computer system are hardware.

Architecture - This section relates to the physical structure of the computer's internal operations, including its registers, memory, instruction set, input/output structure, etc.

CENTRAL PROCESSING UNIT (CPU) - This is the component of a computer system with the binary logic circuitry to control the interpretation and execution of instructions. The CPU includes the following:

ARITHMETIC LOGIC UNIT (ALU)
CONTROL UNIT (CU)
GENERAL PURPOSE REGISTERS
SPECIAL PURPOSE REGISTERS

For mainframe and minicomputers the CPU is usually distributed among numerous integrated circuits whereas the CPU for microcomputers is within a single chip called the microprocessor chip.

The CPU is the heart of a computer system since most activities or data manipulation are performed within the CPU. The control unit (CU) is a manager and interpreter of instructions which directs the step-by-step operation of the entire computing system. The arithmetic logic unit (ALU) is where math and logic operations take place. Registers are places in the CPU where data is located for temporary storage or manipulation. A complete description of CPU functions can be found in various manufacturers' publications and will not be pursued further here.

INSTRUCTION SET - The complete command or instruction word vocabulary of a CPU is referred to as its instruction set.

COPROCESSOR - (1) An auxiliary processor that performs time consuming tasks (e.g., numeric processing) to free the central processing unit, thus resulting in faster execution time for the overall system. (2) A CPU that works in tandem with another to increase the computing power of a system. (3) An auxiliary microprocessor that lets one computer imitate a different computer.

Computer Memory.

DATA STORAGE AND RETRIEVAL - User data is stored outside the computer memory on media such as optical platters and magnetic tape or disks. However, to utilize data within a computer program it must first be moved from the storage medium to a location within the main memory. Once inside the memory such data can be rapidly moved into and out of the CPU when needed.

MAIN MEMORY - Main memory consists of the locations at which data can be stored and retrieved directly for use by the CPU. Memory capacity is the number of addresses within the main memory. A computer with 64K memory capacity has 65536 addressable locations within the main memory. The number of bits used for data storage at each memory location is important because it determines the largest number which can be stored at a single address. Most computers use 8, 16, or 32 bits per memory location. It is generally more efficient to use memory with 16 or 32 bits per location. To store data which exceeds the limit of a single memory location more than one memory location is used. For example two eight bit memory locations can store the same size number as a single 16 bit memory location. Most memory currently in use is composed of a number of integrated circuits (ICs) which will maintain their memory content only while electrical power is supplied (referred to as volatile memory). An older type of memory which was non-volatile was made up of tiny ferrite rings and was commonly called CORE memory. Core memory maintained data in terms of magnetization changes and would hold its information even after the electrical power was removed. The most popular type of memory components today are MOS (metal oxide semiconductor) because of speed, power requirements, size, and price.

MEMORY ADDRESSING - The direct addressing capability of a computer system is determined by the number of bits in the address register of the CPU. Many older minicomputer and microcomputer CPU's had only 16 bit addressing capability allowing direct addressing of only $2^{16} = 65536$ locations. This capability has been expanded in some newer microcomputers to as large as 2^{32} memory addresses (greater than 4 billion locations).

VIRTUAL MEMORY - If there is insufficient main memory to handle data and/or programs, the memory address space can be expanded to include storage on fast disks. The memory space external to the main CPU memory is often referred to as virtual memory. Virtual memory is buffered into main memory in large fixed size blocks (usually called pages) to reduce the number of reads and writes from the disks, which operate much slower than main memory.

Another form of virtual memory seen on some microcomputers is RAM which is outside the normal address range of the CPU. This memory is addressed as a small segment of the microprocessor's main memory space. Using a technique called bank switching many different segments (or banks) of memory can be addressed by the CPU using the memory addresses associated with the fixed small segment and the bank address to determine which bank of expanded memory is to be associated with these addresses. This form of memory access requires some additional overhead to manage bank switching so that the process is transparent to the user. One form of the bank-switched memory access for the IBM AT is the Lotus/Intel/Microsoft Expanded Memory Specification.

RAM - Random Access Memory. Name given to memory which can be accessed in a random fashion rather than sequentially. This name has become associated with integrated circuit (IC) read/write memory almost exclusively though other types of memory can be accessed randomly. The two most common types of RAM are dynamic (DRAM) and static (SRAM). DRAM's are less expensive but must be refreshed periodically (dynamically) to keep stored data from decaying away. SRAM is more expensive but operates much faster since no refreshing is needed. RAM type memory is volatile and data is lost when electricity is removed.

ROM - Read Only Memory. Name given to IC type memory which cannot be altered (written into) but can be read. This form of memory is non-volatile. Often crucial system software dealing with hardware related tasks is stored on ROM so that the programs don't have to be read from disk each time they are needed.

PROM - Programmable Read Only Memory. ROM type memory which can be user (or field) programmed. Once programmed it cannot be written to again. This form of memory is also non-volatile. PROM's are used on some medical imagers such as gamma cameras to store correction factors for such things as uniformity, energy, and linearity correction.

EPROM - Erasable Programmable ROM. Allows the user to program the memory into a ROM. Ultraviolet light can be used to erase the EPROM if subsequent reprogramming is desired. The erasure and programming requires that the IC be removed from the socket.

EEPROM - Electronically Erasable Programmable ROM. Similar to EPROM but can be erased electronically in place. This type of ROM requires special programming circuitry.

EAPROM - Electronically Alterable Read Only Memory. Similar to EEPROM, in that memory contents can be altered without removing the chip from its socket. The alteration does not require a full erasure as for EEPROM. Since write times are very slow compared to read times this form of memory is sometimes called Read Mostly Memory (RMM).

RAM DISK - A block of RAM memory allocated to emulate the behavior of standard disks supported by the computer system. This form of disk is usually much faster than a physical disk, but since RAM is volatile important data and modified programs should be stored on a permanent disk before the computer is powered down.

Input/Output (I/O) - The transfer of information into and out of the computer system. The information transferred may take the form of characters, numbers, voice, images, etc.

When data enters the computer system it is often encoded utilizing the ASCII code to represent characters, symbols, and control commands. The ASCII data format is also used for storage of character type data in memory during program execution or on auxiliary storage devices for subsequent retrieval.

Computers get their input from a variety of devices, the most common being the keyboard and the mouse. The range of input devices is increasing with the following being very popular:

KEYBOARD - Textual Information
MOUSE - Pointing, Drawing
JOYSTICK - Pointing, Drawing
TRACKBALL - Pointing, Drawing
BUTTONS - Selecting
TOUCH SCREENS - Pointing
LIGHT PENS - Pointing, Drawing
BARCODE READERS - Specialized Coded Input
MICROPHONE/DIGITIZER - Voice Recognition
SCANNERS - Image/Text Input
FRAME GRABBER - Video Image Input

Keyboards utilize typing as the standard method of data entry. However, most current versions have features far beyond those found on a typewriter. Features such as special function keys, direction keys, separate numeric keypad, and keys which alter the meaning of other keys give the user a broad range of capabilities for data input as well as formatting and control.

The mouse comes as either an optical position sensing device requiring a special grid encoded surface or a mechanical sensing device with a ball which rolls when the mouse is moved on a desk or other flat surface. Motion feedback is provided through some form of indicator on a video screen. The mouse is usually used in selecting options from menus, drawing, or just moving a cursor around. The mouse comes with from one to three buttons which are pressed for selection of pointed items and other options.

Computers also have a wide range of output devices. The ASCII code is used for many common output devices. The most common are character oriented and include printers, video screens, plotters, and MODEMS.

COMMUNICATIONS RATES - The rate at which I/O data is transmitted serially (one bit at a time - one wire for data) is measured in terms of bits/second, which is called the BAUD rate. To estimate how many bytes (or characters) can be sent per second divide the baud rate by 8. A printer or MODEM may receive data at a baud rate of 1200 which is approximately 150 characters/second. Input/output is most often accomplished with 8 bit (byte) transfers which can be either serial or parallel (all 8 bits at once) fashion. Various codes have been developed for the bytes which are sent back and forth in I/O communications, but the ASCII code is the most common for microcomputers.

The rate at which data can be transmitted in a parallel fashion is much higher than that afforded by serial techniques. Parallel communications are used when speed is needed such as for real-time control of external hardware. Personal computers often use the parallel communications I/O port for printers or external disk drives. The high speed is useful for printers which have large data buffers (onboard memory).

Error Checking - While computers are very reliable they are not infallible. If a single bit is lost during data input or output an incorrect character code results. Such an error can be caused by particles on storage media, improper humidity levels, or just electrical noise during data transmission. An error of just one bit in a program or data can produce catastrophic results. Several error detection schemes have been devised to deal with problems associated with errors.

PARITY CHECK - One additional bit is transmitted with the data to make the number of 1's either even (even parity) or odd (odd parity) for each data record (usually a byte) sent. The sending and receiving devices must be set to the same parity type, and if the wrong parity is received in a transfer of data an error will be indicated. This method of error checking is good for detecting an odd number of errors which may have occurred in data transmission. An error occurs whenever a 1 is changed to a 0 or a 0 is changed to a 1 due to noise or other faults in the transmission media. Parity errors will generally halt the operation and the system will request operator intervention.

CYCLIC REDUNDANCY CHECK (CRC) - This error detection method incorporates a check variable which is calculated from the data transmitted and included with the data at the end of the data block. The receiver recalculates the CRC variable from the data and compares it with the transmitted CRC. In some communications protocols a CRC error is used to signal that the data block needs to be retransmitted. For IBM personal computer disk drives a CRC value is calculated and stored on disk for each sector written to the disk. When a sector is read from the disk the CRC value is recalculated from the data and compared with the CRC value read from the disk. If an error is detected the sector is read several times in an attempt to recover the data before the read is reported as a failure.

Auxiliary Storage -

MAGNETIC TAPE - This is a 9 track recording medium with data commonly stored in 8 tracks using an 8-bit EBCDIC code with 1 track reserved for error (parity) checking. Data is stored on the tape in various formats but all tapes have a HEADER block at the beginning. Following the header block are data blocks containing the information as characters separated by gaps with an end-of-file block (EOF) following the last recorded data block. Tapes have a large storage capacity but they are limited by access time (the data being stored sequentially) and read/write speed (much slower than hard disks). Storage capacity depends on reel size, recording density (800, 1600, and

6250 bits/inch), number of tracks and number of inter block gaps. Ex: For a 9-track tape with a recording density of 1600 bits/inch (BPI) in each track, 46 million bytes can be stored on a tape that is 10-1/2" diameter. This diameter tape is approximately 2400 feet in length (nearly 1/2 mile!). Read/write speeds range from 25 to 250 inches per second (IPS) such that a high speed unit can move end-to-end on a 2400 foot tape in less than 3 minutes. For 9-track systems BPI effectively means characters per inch. Tapes packaged in smaller containers called data cartridges are available which provide storage of up to 67 Mbytes on a 1/4" tape just 600 feet long. Magnetic tape is generally used for long term storage of data and for backing up (making extra copies of) data and software files.

DISK - Magnetic disks are the most popular devices for storage and retrieval of moderate volumes of data when rapid, direct access is required. Disks fall into two general categories: (1) Flexible (Floppy) and (2) Hard. Floppy disks rotate at approximately 300 RPM while hard disks rotate at approximately 3600 RPM. A read/write head (or heads) moves in discrete steps in and out along a radial path. Data is written/read using block addresses on the disk determined by the radial position (track #) and circumferential position (sector #) Floppy disks of 3-1/2", 5-1/4", and 8" diameter are in current use. Storage capacity ranges from 360 kbytes on 5-1/4" floppys to several megabytes on 8" floppys depending on the storage format used. Whereas the read/write heads actually contact the flexible surface of a floppy disk, this is not allowable because of the inflexible surface of hard disks. Removable hard disks are usually 14" in diameter and a common storage capacity for a single platter system is 5 megabytes. Storage capacity is often increased for hard disk systems by adding additional platters which rotate on the same spindle. Each new platter increases the storage capacity and keeps access time small compared to increasing the disk diameter. Access times are much smaller and data transfer rates are much higher for the higher RPM hard disks than for floppy disks. If the read/write head happens to impact on the surface of a hard disk damage can occur to both the head and the disk surface. This damaging event is referred to as "head crash" since the heads of high rotational speed hard disks are designed to fly above the surface of the disk platter during normal operation. Once a head has been damaged it should be replaced before another disk is used or the new disk may be ruined also.

Another type of hard disk which has gained popularity is called a "Winchester". The winchester disk and drive are in a sealed environment (nonremovable media) with the read/write head riding closer to the surface than for other hard drives (allows for higher recording density). The heads in some sealed drives fly along at an altitude of only 20 millionths of an inch. Even a small smoke particle (250 millionth inch diameter) can cause damage, therefore these units must be constructed in what is called a "Clean Room" with filters which remove such microscopic particles from the environment. Winchester drives are available with platters of 3-1/2", 5-1/4", 8", or 14" diameters. All platters are hard and rotate at approximately 3600 RPM. Storage capacities range from about 10 million bytes for some 5 1/4" drives to 1 billion bytes (a gigabyte) for multi-platter 14" systems.

Software

Software refers to a group of instructions (a program) that tell a computer what to do. The operational interface between hardware and software is within the CPU where binary coded instructions are interpreted and a predetermined sequence of events executed. This interpretation and execution can be fully implemented in a purely hardwired fashion or by microprogramming combined with a reduced number of hardwired functions. If a microprocessor is microprogrammed, the control necessary to execute normal program instructions (macroinstructions) is generated by a series of microinstructions or microcode. The microcode is stored in a special memory within the microprocessor separate from the CPU's main memory. Therefore software is a group of macroinstructions compatible with the CPU's instruction set. Optionally, software may have some pertinent data imbedded within the group of instructions.

Operating System

While the CPU's instruction set determines the codes and basic operations which can be performed by the computer, the computer system is not very useful without a systematic scheme to manage its operation. This management is provided by operating system (OS) software. For microcomputers such as the IBM PC the lowest OS level is the basic I/O system (BIOS) software. The BIOS is stored on ROM so that the basic I/O routines are available without delay when needed. ROM-BIOS software provides routines to operate various peripherals connected to the computer such as the video screen, keyboard, serial and parallel ports, and disk drives. Additionally ROM-BIOS provides for a real-time clock and a bootstrap routine used to boot the disk portion of the operating system.

The next level of software is called the disk operating system (DOS) partly because the software is stored on disk rather than in ROM. The key routines from DOS are loaded from disk when the computer is turned on. This loading of the critical DOS routines is called booting. For the IBM PC the PCDOS consists of three major programs: (1) IBMBIOS, (2) IBMDOS, and (3) COMMAND.COM. The IBMBIOS contains routines which supervise the use of ROM-BIOS. The advantage of such a supervisor is that newer versions of the operating system may link BIOS routines in a different way to provide improved I/O without changes in the ROM. IBMDOS contains higher level routines for accessing the keyboard, the display, disk drives, and other I/O hardware. The COMMAND.COM is responsible for controlling the overall activities of the operating system. This is the part of DOS that prints the familiar A> prompt and figures out what action to take when you type something. COMMAND.COM provides the computer with a basic form of intelligence. COMMAND.COM contains a "resident" portion and a "transient" portion. When a user program is executed the user program assumes the role of the transient portion and the transient portion may be overwritten if needed. While the user program is executing it has access to all the facilities provided by IBMDOS and IBMDOS just as COMMAND.COM does when it is in charge. When the user program is terminated the transient portion of COMMAND.COM resumes control. If the program was sufficiently large to have overwritten the transient portion of COMMAND.COM it is reloaded from disk.

The highest level in the management and control of computer operation is the user. The user selects and uses software and hardware to perform tasks deemed useful. The organization of PC-DOS has been described in terms of management and control as follows:

The User	Chairman of the Board
COMMAND.COM	Chief Executive Officer (CEO)
User Programs	Acting CEO
IBMDOS	The Management
IBMBIOS	The Foreman
ROM-BIOS	The Workers

System software makes the computer into a computer system and is fortunately somewhat transparent to the user. This software provides the basic lower level characteristics which relate to the "lower level personality" of the computer system. It is the user programs which perform user oriented tasks, rather than computer oriented tasks, that can transform the computer from an operating machine to a useful tool.

Prewritten User Programs

User programs include prewritten software for text manipulation, data analysis, storage and retrieval, and integration of such.

TEXT MANIPULATION - This type software includes wordprocessors, outline processors, desktop publishing, and data communications.

DATA ANALYSIS - This includes spreadsheets; investment, accounting, and statistical analysis; mindware programs; and graphics packages.

STORAGE AND RETRIEVAL - This includes data base management packages, project management, and many memory resident programs.

Computer systems which have a highly standardized user interface can generally transfer data between programs. An example is the Apple Macintosh. Other systems such as the IBM personal computers and clones require special software or integrated software packages to share data between programs. Newer computers and operating systems are heading toward a more standard means to share data between users' application programs.

User Written Programs

Software which allows users to develop their own programs is called a programming language. The purpose of such software is to allow a source program written in a human-readable language to be transformed into an object program which contains only the instructions allowed by the CPU. Before a program can be run it must be converted to the machine language or binary level. A programming language consists of symbols, characters, and usage rules that permit programmers to communicate with computers. Most programming languages have instructions that fall into (1) input/output, (2) text & numeric manipulation/calculation, (3) logic/comparison, and (4) storage/retrieval categories. Though each programming language has these features, the symbols, characters, and syntax vary drastically between languages.

Machine and Assembly Languages

The lowest level language is referred to as machine language because it must be programmed using the basic instruction set of the CPU (the machine). To ease the burden of programming with purely binary codes a language referred to as assembly language was developed in the early 1950s which allowed the programmer to replace crude binary code with mnemonic code. Assembly language source code is converted to machine language object code by a program called an assembler. The mnemonic codes are specific for each CPU and assembly source code written for one CPU cannot be directly assembled into correct object code for another. Programs called cross-assemblers have been written to convert assembly language source code from one CPU's format to another's.

High-Level Languages

Programming languages which are considered to be higher level than assemblers have evolved which are oriented toward a particular class of processing problems. Higher level languages provide the following advantages over assembly language:

1. Source code may be used with many different makes of computer regardless of CPU type.
2. Easier to learn.
3. Requires less time to write programs.
4. Programs are easier to maintain.
5. Documentation is better.

Generally high level source programs are translated into a machine-usable code by a program called a compiler. The compiler may produce assembly code which is subsequently converted by an assembler into machine code or bypass the assembly code stage and directly produce machine code. The final step is to link together from a library of machine coded routines those routines which were called by the high level program. This final step is called linking and is performed by a program called the linker. The output from the linker is directly executable. An alternative translation scheme is an interpreter. Instead of saving a permanent copy of the executable code, the interpreter translates the high-level source code to executable object code on a line-by-line basis during execution.

Major High-Level Languages

The major high level languages used for scientific programming are BASIC, FORTRAN, PASCAL, and C.

BASIC - Beginners All purpose Symbolic Instruction Code.

BASIC is a very interactive language which is often implemented as an interpreter, allowing direct communication between the user and the computer system during preparation and use of programs. A problem-solver with little or no knowledge of programming can learn to write BASIC programs easily. Most personal computers offer BASIC for this reason. BASIC source code is fairly portable in that code written on one machine generally works on another machine. BASIC has also been implemented as a compiler with products such as Microsoft's Quick Basic V4.0. The compiler version of BASIC operates much faster than the interpreted version because no interpretation is required with compiler produced programs.

FORTRAN - FORmula TRANslator.
FORTRAN is a scientific-mathematical language noted for the ease with which it can express mathematical equations. FORTRAN has the advantage that huge libraries of engineering and scientific programs have been developed which are available to scientists and engineers. Because there are well established standards for FORTRAN, programs written on one machine are usually easily converted for use on another. However, FORTRAN does not offer the I/O capabilities (especially for communications and graphics) available in other high level programming languages.

Pascal - Named in honor of Blaise Pascal, a 17th century French mathematician. Pascal was the first major high level programming language created to incorporate the concepts of structured programming. Pascal can be used for both scientific and business applications. A growing number of high school and college instructors are using it to teach programming using structured techniques to their students. Pascal is one of the major languages running on personal computers.

C Language
The C language was developed at Bell Labs in the early 1970s. Although C is a general-purpose high level language, it allows programmers to write low level machine instructions almost as compactly as those written in assembly language. C is thus a favorite language among professional programmers.

Summary

The basics of computers were presented in a form which covered the binary language, binary mathematics, computer logic, hardware, and software. The material should be useful for reference purposes. Analogous relationships between computer and human characteristics were used when possible as an aid to understanding. The data presented here should prove useful to those who are asked to teach others the basics of computers.

References and Suggested Reading

1. Webster's New World Dictionary of Computer Terms, Third Edition, Websters New World, New York, 1988.

2. Krutz, Ronald L., Microprocessors and Logic Design, John Wiley & Sons, New York, 1980.

3. Norton, Peter, Inside the IBM PC, Revised and Enlarged, Prentice Hall Press, New York, 1986.

4. Sanders, Donald H., Computers Today, Third Edition, McGraw-Hill, New York, 1988.

5. Lafore, Robert., Assembly Language Primer for the IBM PC & XT, Plume/Waite, New York, 1984.

Classical vonNeumann Architecture; Special Processors

Tom K. Lewellen, Ph.D.
Department of Radiology
University of Washington
Seattle, Washington 98195

ABSTRACT

This paper reviews the basics of classic vonNeumann computer systems. After a brief introduction and review of the history of development of such machines, the various subsystems in a typical system are discussed. Finally, comments about different design approaches in modern computer systems (complex and reduced instruction set machines) and auxiliary processors are included.

INTROUDCTION

The first goal is to define the term "vonNeumann architecture". VonNeumann described an environment that utilizes a central memory store and that intermixes instructions and data (figure 1). Many uses of systems have come to use the term "vonNeumann" machine to refer to the sequential execution of instructions. It is more proper to regard a vonNeumann machine as a system where the difference between instructions and data is determined only by the CPU logic that decides where the next instruction is to be found in memory. The original vonNeumann machine concept consisted of a central memory with all input being passed through an arithmetic logic unit (ALU) that translated input into machine readable form and machine readable output into human readable form. A control unit cycled the machine hardware through the various steps required to fetch and decode an instruction and to send the proper control signals to the rest of the machine components. As we shall see, this basic system is still used in the typical machines we deal with today.

```
 -------------------control------------------
 |                    |   |                 |
 |                    |   |                 |
input               memory               output
 |                    |   |                 |
 |                    |   |                 |
 -------------------A.L.U.------------------
```

Figure 1: Diagram of the basic vonNeumann machine

It is of some interest to note a bit of history in the development of these machines. As indicated in table 1, the early machines were given generation numbers, a practice that is no longer feasible in the rapid development of computers in recent years. The first machines were tube devices and were the reason that the term "bug" was coined. As the story goes, a moth was attracted by the glowing tubes in one of the earliest machine and flew into the contacts of one of the relays. The resulting malfunction was finally traced to the relay and the discoverer noted that there was a bug in the computer. Now a days, the bugs we find are not due to insects, but to normal human errors in the design and implementation of hardware and software.

ENIAC Generation 0 1946

18,000+ tubes, 1500 relays main function= ballistics
20 registers, 312 word storage
5000 additions/second all hard wired

EDSAC 1949 first system with stored programs

1024 word memory, 4600 word magnetic drum

IAS machine (Princeton Institute for Advanced Studies)

definition of the vonNeumann machine
first version: 4096 40 bits word memory
64 opcodes

Univac I (1951) Generation 1
Magnetic tapes,
followed by Univac II (1957) with core memory
followed by Univac 1100 series and IBM 704, REAL computers

IBM 7094 (1962-64) representative of Generation 2 machines
32 k core memory, 185 opcodes, 7 registers, transistors

Table 1: A short list of some of the early computer systems.

In the very complex world of modern day computing, the time needed in good engineering (hardware and software) can never be underestimated and needs to be invested. At a recent conference at the University of Tennessee, it was noted that in modern day computers the speed of development is so rapid that if it works, it must be obsolete. So, the user needs to understand the requirements of the computing task to be done and select the hardware and software environments accordingly. If new software is being developed, considerable time should be spent in careful definition and design of the software prior to coding. In any event, careful testing of the computer system to assure that the calculations or other tasks to be performed meet the specifications must be undertaken.

I like to talk about computer system architecture by dividing up a typical computer system into five major subsystems:

1) The central processor unit (CPU).
2) The system memory (usually random access memory - RAM).
3) The input/output (I/O) devices.
4) The memory bus (the address and data paths connecting the CPU to the RAM)
5) The I/O bus (the address and data paths connecting the CPU and RAM to the I/O devices)

Before getting into these various systems, we should define some terms that often get confusing in discussing computer systems. Figure 2 indicates the terms that are normally used to refer to the various ways bits of information can be combined to provide storage of data or instructions in a computer. The convention to use "word" to indicate 16 bits is standard in the world of micro and mini computers. It should be noted that in larger systems (main frames, super computers, etc) the term word is often used to refer to a larger collection of bits (e.g. a 64 bit word). However, in the systems currently in use in medical imaging, the terms in figure 2 are the standard definitions.

The differences between byte, word and double word lead to another area of confusion over the last few years. A computer is often described as a 8, 16 or 32 bit system. The questions is what the bit width is being used to describe. It can be used to refer to registers within the CPU, the way memory is organized to store data, or the number of parallel data paths connecting the CPU to the RAM (e.g., how many bits are transferred during one read/write cycle to memory). In the past, the description of the bit depth of a computer has often been used to describe the bit width of the main CPU registers. With the arrival of some microcomputers with 32 bit registers but 16 bit buses as well as some systems with 16 bit registers and 8 bit buses, the classification of the CPU register width could give a false impression of the full computing power of a system. More recently, system specifiers seem to be using bit width to describe the bus width, assuming it is less than or equal to the CPU registers.

When looking at the potential compute power of a system, it is very important not to get too enamored with the various specifications such as bit width, millions of instructions per second (MIPS), number of floating point operations per second (FLOPS), or other standard speed descriptors. Any system can have very fast components in part of the machine, but can completely or partially negate that speed with other parts of the system (e.g., the CPU is fast, but the disk drives are slow and can not move data into the CPU fast enough for processing of large data files). In any application, it is important to select new systems based on the performance of carefully selected benchmarks of the applications to be run on that system. The standard benchmarks offered by the various computer vendors can only be used as rough guides and should be treated with healthy skepticism. Often the speed of an application will be far more dependent on the quality of compilers and the care and skill in software engineering of the application software than on the speed of the hardware.

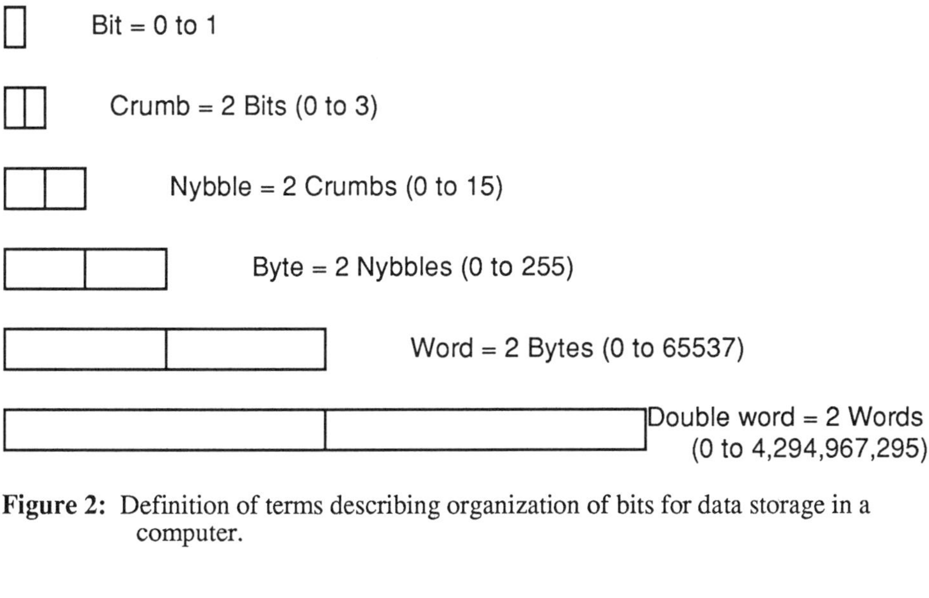

Figure 2: Definition of terms describing organization of bits for data storage in a computer.

MEMORY AND I/O BUS

With the general comments out of the way, let us start looking at some of the details of the five basic systems defined above. A good place to start is with the two principal bus systems. There are several methods used to configure the memory and I/O bus systems. In some machines (e.g., DEC Unibus and Qbus machines, VME bus systems), the memory and I/O devices are connected by the same set of wires - a common bus. With a common memory and I/O bus, the I/O devices appear as sets of memory locations which can be used to send data and commands to/from the I/O device. An alternative approach (e.g. Data General computers, mainframe computers such as the Cray XMP) is to use separate buses for memory and I/O.

Additional pathways may well exist in a machine to allow very rapid movement of data between memory and I/O devices. Most machines support some form of direct memory access (DMA) that allows an I/O device to move data to/from memory without using the machine's CPU. Typically, such a pathway is used by the CPU (executing instructions in a program) sending information to the I/O device over the normal pathway that defines how much and where data is to be moved. Then the I/O device uses the DMA channel to do the data move and then informs the CPU when the task is done. The reason for such an arrangement is speed (both in not having to move data through the CPU and in letting the CPU do other tasks while the data is being moved).

Typically, DMA systems can move large amounts of data, but the moves are still made one data element at a time. In such an arrangement, the DMA I/O device sends an address to memory, then does a read or write of one data element. As demand for speed has increased, many manufacturers have implemented another form of DMA, often referred to as a block move channel (BMC). A BMC device is again sent information on the data to be moved by the CPU, but after it sends an address to memory, it then moves a block of data without issuing any further memory addresses. This arrangement greatly increases throughput, usually by factors of 2 to 4, but requires that both the I/O device and the memory system to have additional hardware handle the automatic increment of the memory address.

The main memory usually consists of solid state random access devices, normally dynamic devices (as opposed to static RAMs). The address bus width determines the maximum amount of physical memory that can be accessed by the machine. Over the years, several schemes have developed to allow a machine to provide the user with more memory space than the CPU can directly address. The first approach used, introduced in minicomputers, was to include additional memory in the machine and issue special instructions to bank switch the memory into part of the memory address space. Initially, these extended memory machines used the additional memory for storage of data only - essentially what we would now call a RAM disk. As the operating systems became more sophisticated, the bank switching became more or less transparent to the user software and allowed both instructions and data to be stored in the memory banks. The same basic approach has been used in some of the more popular microcomputers, including IBM PCs and Apple II series systems.

In larger multi-tasking, multi-user systems, the approach of dividing memory into a series of pages was introduced, very early in the development of computers. With a paged memory system, the users memory space (the logical memory) could be scattered through physical memory in a series of pages (typically 512 to 2048 bytes per page). Special hardware in the machine mapped a given user's logical space to the series of pages his software was really using in memory. In this way, the available memory could be more efficiently used as various tasks ran or finished resulting in a continuous change in the amount of memory available for new tasks and the location of tasks within the memory. As the memory configuration changed during run time, a new task that would require a logical space of memory that was larger than any contiguous space of physical memory available could still run as long as there were enough pages of memory scattered throughout the physical memory to support the application's requirements.

The presence of paged memory systems then lead to the introduction of virtual memory systems. In these systems, the user has a logical memory space available that is larger than the physical memory space available. The computer loads in as much of the software or data as there are pages available for the user. When the logical memory addresses a part of the program or data that is not currently in physical memory, the computer copies some of the least used pages of physical memory allocated to the task to disk and then loads the needed program or data into the "freed" pages of physical memory. In 32 bit minicomputers and high end microcomputers, special hardware is included to provide memory protection of the various pages being used by a given task and to speed up the process of identifying page faults (requesting data not currently in physical memory), mapping the user space into the physical memory space, and moving data to and from the disk. The result is that the user can run very large programs without regard to the physical memory - in terms of being able to run the program. However, the process of page faulting and disk swapping is still time consuming, and the software must still be carefully designed to minimize such faults. In some cases, there is no way to avoid operations such as accessing large matrices, and adequate physical memory must be available to hold the arrays or the performance of the computer may be unacceptable. For example, we had an image reconstruction algorithm that was run on a 32 bit super-minicomputer with 2 megabytes of memory available and the program took 16 hours to run. When 2 more megabytes of memory were made available, the same program ran in 20 minutes.

CENTRAL PROCESSOR

The CPU is the part of the system that decodes instructions and performs manipulation of data. The basic concept is that the machine retrieves (or fetches) an instruction from memory, executes that instruction, then fetches the next instruction and so on. In modern systems, the CPU is often designed so that as one instruction is being executed, the next instruction is also being fetched. In some systems, several instructions that take multiple clock cycles of the CPU can be executed at once by splitting the instructions into several parts that use different logical units of the CPU. This process is sometimes referred to as pipelining the processor. The more common form of pipelining in modern CPUs is that of interleaving the process of fetching, decoding and executing instructions. An example is illustrated in figure 3 and is not uncommon in modern machines. During the first CPU clock cycle, an instruction is fetched from memory. During the next cycle, the first instruction is decoded while a new instruction is fetched. By the third cycle the decoded instruction is used to transfer control to the logic unit that controls the rest of the CPU components, the second instruction is decoded, and a new instruction is fetched. Finally, at the fourth clock cycle, the first instruction is executed, the second transferred, the third decoded and a fourth instruction is fetched. However, the basic concept is still that of fetch, execute, and fetch again.

There are several major variations in processor design, including the use of cache memory (that fetches large portions of a program into very high speed memory within the CPU), the use of pipelined processors, the number of bits used within a CPU word, and the use of memory management hardware. One major difference between different design approaches is that of the use of stacks or many registers for storing intermediate results within the CPU. Essentially all systems have some mixture of registers (high speed storage of one word per register within the CPU) and the use of stacks (special areas of main memory that are serviced by special instructions for rapid pushing and pulling of data from memory). Examples of machines that use many registers include the DEC VAX line of systems and the 68000 series of microprocessors. Some machines, such as the Perkin Elmer systems, have multiple banks of registers, with each bank assigned to a specific program to allow rapid switching between processes in a multi-program environment. Examples of stack based machines include the DG MV line of systems and the 6502 series of microprocessors.

To provide an example of how a CPU is organized, we will look in more detail at a Data General MV10000 minicomputer system (figure 4). As can be seen in Figure 4, the internal structure of a computer can become quite complex. A common feature of modern CPUs is the microinstruction sets. Typically, the instruction set of a computer (the assembly level instructions a programmer or high level compiler generates) are used as an

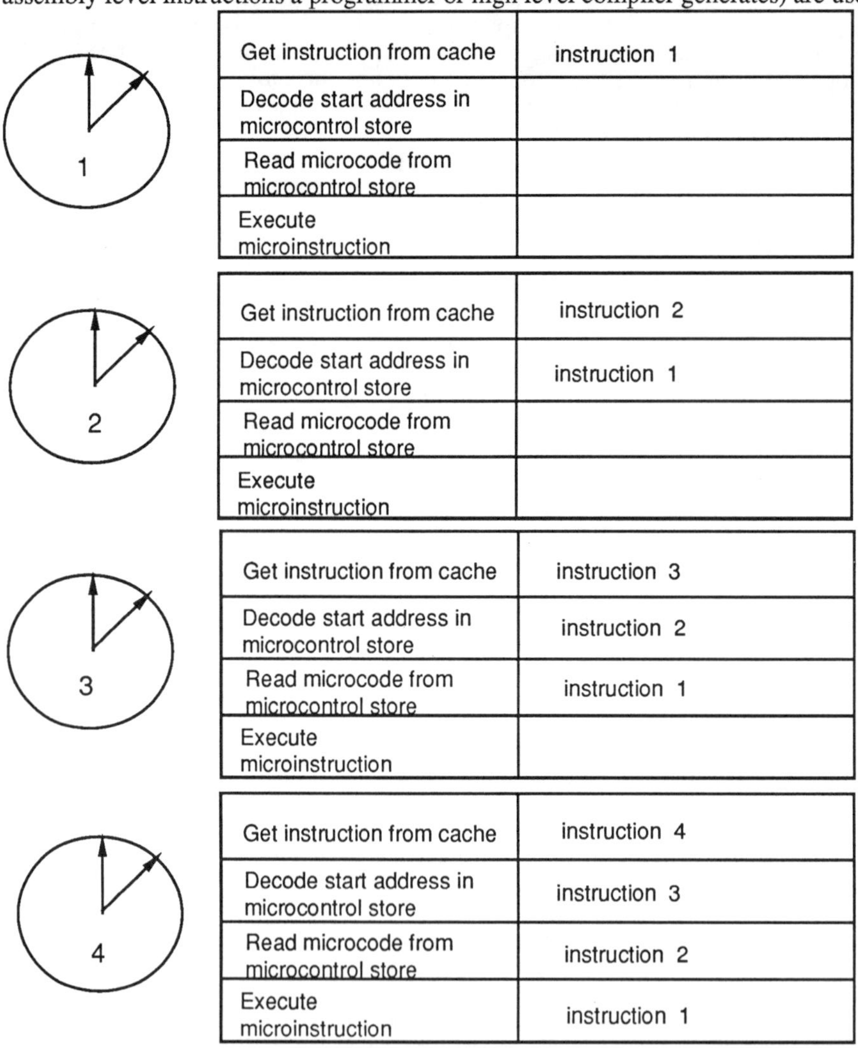

Figure 3: The pipeline in the instruction fetch and execute logic of a Data General MV10000 mini computer.

index into a more detailed set of instructions, the microinstructions. These microinstructions consist of collection of bits that turn control lines in the various parts of the CPU on and off. These control lines control the transfer of data to and from registers and the several data/address buses that service the CPU, control adders, bit shifters, comparators, and all of the various discrete logic devices that implement the functions of the machine. Often these microinstructions use very large word sizes (on the Data General MV10000, the microinstructions are 104 bits wide). The microinstructions are either stored in read only memory (ROM) or in very high speed RAM. In microprocessors, the storage media is usually ROM, while the RAM stores are used in the larger minicomputer systems. An advantage to the use of RAM is that the instruction set of the machine can be upgraded or expanded by simply putting new microcode on the system disk that is loaded into the microinstruction RAM when the system is started up. The disadvantage is cost of the high speed RAM.

The CPU normally consists of several major sections. In the MV10000 processor, these major systems consist of:

a. Arithmetic logic unit: This part of the CPU does the adding, subtracting, and bit manipulations of integer data. Higher order functions such as divide and multiply are implemented as multiple passes thru this unit. It is driven by the microcode bus, the CPU memory bus, and the CPU data bus. This device is quite complex including a set of 32 bit registers, a bit shifter, a 256 x 32 bit very high speed memory scratchpad and data paths for 16 and 32 bit data manipulation.

b. Floating point processor: The FPU implements floating point operations. In the MV10000, it is actually part of the ALU instead of a separate device. This part of the ALU use 64 bit wide data with 72 bit data paths (to avoid roundoff errors), scratch pad registers, and shifters.

c. Microsequencer: This device reads the current decoded machine instruction from the program and reads in the sequence of basic control bit manipulations for the control lines in the CPU (microcode) and executes the microcode operations.

d. Instruction processor: This device decodes the machine instruction from the program and sends the starting address of the microcode instructions to be executed to the microsequencer. This device is pipelined in that it fetches the next instruction from memory while it is decoding the previous instruction and sending the starting address to the microsequencer for the instruction before that (figure 3).

e. Address generator: This unit produces logical addresses for use in reading/writing data from memory and is controlled by data sent from the microsequencer or from the system main memory. It also calculates effective addresses when jump or offset address instructions are executed.

f. Address translation unit: This device takes the logical address from the address generator and determines the physical address to be accessed in the system main memory. This type of device is often separate from the main CPU and is termed a memory management unit. It is needed when the program memory may not be contiguous (as if often the case in minicomputers with multi-user, multi-tasking operating systems).

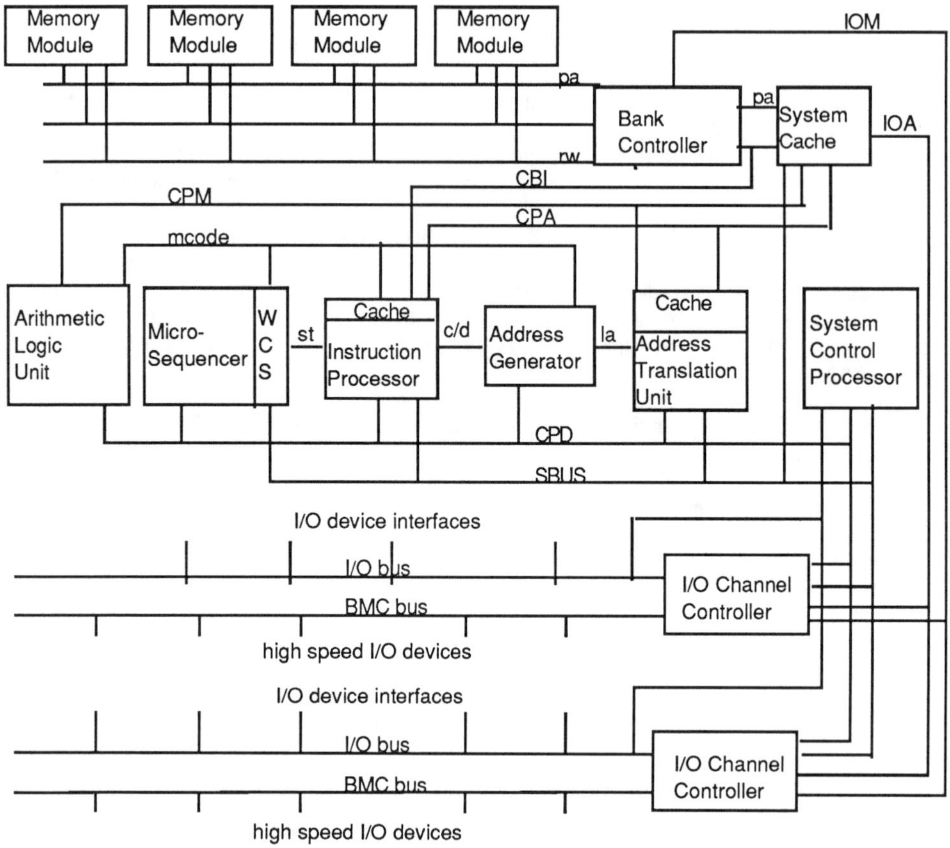

pa = physical address and control bus CPD = CPU data SBUS = diagnositc scan
r/w = read/write data code = microcode bus IOM = I/O memory data
CBI = data (instruction) bus st = starting mcode address IOA = I/O memory address
CPM= CPU/memory data c/d = address control I/O bus = data channel
CPA = CPU/memory address la = logical address BMC = burst multiplexor bus

Figure 4: Block diagram of a Data General MV10000 computer

INSTRUCTION SETS

Two basic styles of processors have arisen and are differentiated by the nature of the processor instruction sets. The "classical" systems have evolved ever increasing numbers of instructions that often do very complex operations with a single assembly level instruction. These machines are now termed complex instruction set computers (CISC) and include the majority of computers utilized in medical imaging (e.g. Digital VAX, Data General MV, Motorola and Intel microprocessors in the 68xxx and 80xxx family, etc.). Normally the computers implement these CISC systems with a microsequencer that executes a complex series of microcoded instructions for each assembly level instruction used. These microsequences are stored in very high speed RAM or read only memory (ROM), resulting in a computer within a computer with the accompanying complexity and resources, such as the MV10000 system described above. The alternate approach is to use a very small number of instructions optimized to what the designer feels are the most used instructions in a computer and then implement the instructions in hardware, not using microcode. Such systems are termed reduced instruction set computers (RISC). CISC systems provide hundreds of instructions that can be executed, often doing very complex operations in a single instruction. This style of machine offers the primary advantage that each instruction is highly optimized for its purpose allowing efficient processing for a wide variety of tasks. The principal disadvantage of a CISC machine is that it is difficult to select the most efficient sequence of instructions for optimized performance of a program. At the assembly level it is difficult for the programmer to keep track of the hundreds of instructions available. The high level language compilers need to be very sophisticated in selecting the instructions to use. The good news is that many systems have had hundreds of man years of effort put into developing such sophisticated compilers and produce very efficient code indeed.

In RISC systems, the basic approach is to use a few instructions that are implemented in hardware for very fast execution. These machines offer the advantages that the few instructions are executed very fast and it is less confusing to select the best sequence of instructions. The main disadvantage is that the limited set of instructions can result in a much longer program to perform a task than a CISC processor, partially negating the faster execution speed (in terms of millions of instructions per second). If the application being programmed results in a large expansion of the program size, then the system resources in terms of memory can be quite large. Also, since in some cases the number of RISC instructions needed to implement the equivalent of some CISC instructions may be quite large, many more machine cycles are needed to get the task done.

It is worth restating that in any system, the speed of the hardware is no excuse for not doing good software engineering. Seemingly small things (e.g., aligning array manipulations to the page fault size in a virtual memory machine, compressing loops, etc) can make order of magnitude differences in execution speeds. The quality of the compilers one is using, especially the sophistication of optimizing compilers, may be far more important for speed of a program than the hardware the application is running on.

AUXILIARY PROCESSORS

Many computers have reasonable execution speed for integer operations but do not include instructions for floating point operations. The result is that the floating point operations are often implemented in a higher level language and suffer from relatively slow execution speed. A floating point processor unit (FPU) is often an option and is a special ALU that performs floating point operations. Normally, the floating point library for the high level languages is replaced with one that uses the FPU, or it has routines that can check if a FPU is present and use it accordingly. Usually, the main CPU waits for the FPU to finish its task before it processes the next instruction. The FPU often uses a series of pipelined devices to implement the algorithms (usually series expansions) for many of the operations.

That brings us to the topic of pipelined processors and array processors. The term pipelined processor refers to a series of logical processing devices (e.g. adders, shifters, etc.) that accept data from the preceding device in the pipeline and sends their data to the next device in the pipeline. The connection between these devices (the pipeline) is fixed and cannot be altered under software control. The result is a very fast processor dedicated to a specific task. Examples include types of image processors that have pipeline devices for performing image rotations, convolutions, and other operations. Pipelining is also used in the "micro architecture" of many CPUs that execute one instruction while another is being fetched from memory.

Array processors are special processors designed to do a restricted set of operations very fast. The general layout of the system consists of a control processor, memory, one or more arithmetic processors, and one or more I/O processors. The system is then interfaced to the host computer. The key to the system are the arithmetic processors that are generally a series of pipelined processors designed for specific types of operations. A typical configuration is to use a pipeline consisting of:

1. a format convertor to convert the data from the host into the internal representations.

2. A set of high speed registers

3. A floating point multiplier

4. A second set of high speed registers

5. A set of floating point adders

6. Return data paths that provide the digital equivalent of a feedback loop from the second set of registers to the first.

The distinction between array processors and parallel processors has become increasingly vague. Modern array processors can have several ALUs and are programmable. Basically they have become an option to increase the compute speed of existing host computers with minimal investment in additional general purpose computing equipment. Large scale minicomputers are themselves distributed processing systems, often incorporating many of the features that have been the province of array processors. The trend is to now offer many type of add-on processing engines that increase the compute speed of host systems for specific applications. Some systems are optimized for image processing, some for image reconstructions in tomographic applications, and many other processing problems. For the user, the evolution of modern computers and auxiliary processors has resulted in increasingly lower cost systems with large compute capacity.

Parallel Architecture

Tom K. Lewellen, Ph.D.,
Department of Radiology
University of Washington
Seattle, Washington 98195

ABSTRACT

The use of parallel systems in the design and implementation of computer systems is reviewed. Examples of the different approaches to use parallel systems are presented. The architecture of a Cray XMP and an Alliant FX-80 are presented along with some examples of how a simple Fortran routine can be adapted to a parallel system.

INTROUDCTION

The term parallel computing can be applied to many systems. In some respects, many "common" computers have parallel computing with parallel processors. For example, many minicomputers utilize "intelligent" interfaces that are computers in their own right, including CPU and memory with programs running independently from the main system CPU. Such interfaces often include serial interfaces, disk interfaces, graphic display devices and a host of other systems. Often, programs can have parallel tasks that execute when there is data or a task to be performed. Such tasks can be thought of running in parallel, although in the conventional computer system with one main CPU, the various tasks still execute on the CPU one at a time. In networked computer systems, the various tasks of a large program may run on different processors, sending messages to each other. In fact, the operating system on most machines could be considered a parallel environment in the sense that the user program is being serviced by the operating system which handles tasks in parallel with the user program - at least from the programmer's point of view if not in terms of the CPU usage.

PARALLEL PROCESSORS AND PROCESSING

How then do we define what is meant by parallel computing? I like to divide the topic into parallel processors and parallel processing. I use this division since a machine can have multiple (parallel) processors while using only one processor for a user program - what I like to call parallel processors as opposed to parallel processing. Such a system often consists of an environment where several CPUs share common memory and peripherals and are interconnected on some sort of high speed bus inside a single computer. The CPUs can then be assigned different tasks and run in "parallel", providing a cost effective way of increasing throughput for a multi-tasking system without having to duplicate all portions of a computer system.

For example, several minicomputers offer two or more processors. However, in most systems one of the CPUs serves as the master processor and is also used for most I/O operations and all processes driven by interrupts. The operating system usually watches the loads on the various processors and assigns tasks to the processors to even the computing load. The different CPUs run different programs or tasks. Usually, each CPU is running programs that do not use the other processors in the system (except for I/O operations provided by the operating system). Many of the super computers and mini-super computers utilize multiple processor architectures with very fast CPUs. Again, having multiple processors does not automatically mean that one is doing parallel processing.

For this discussion, I will define parallel processing to mean a computational problem that is split up into tasks that can be run in parallel on multiple compute engines. The simplest approach is to set up the problem as a set of parallel tasks that do not need to interchange information. An example would be doing a Monte Carlo calculation of photon transport in the human body. Since each photon is independent, each processor could track the history of one photon and the the results can be combined in a history file at the end of each calculation run. Perhaps the more interesting approach is to break up the tasks on the various processors, but also interchange information between the processors. This approach has the problem of efficiently mapping the programs onto the processors and establishing efficient methods to exchange the information between the processors.

So why bother setting up machines with many processors and go to the trouble to split up tasks on these parallel compute engines? Parallel processors running separate programs can get higher throughput for each program by efficient sharing of the computer resources. Further, the overall compute power for the user can be increased without duplicating many of the normal system resources (e.g. disk drives, main memory, etc), resulting in an effective cost savings.

Parallel processing, where a single program is executing on multiple processors, can result in enormous increases in computing speed. If a problem has a huge number of computations, then running on parallel, fast processors can achieve major speed increases since more computing operations per unit time can be performed. Such speed increases can mean the differences between some computational approaches being strictly research or made available for routine use (e.g., advanced image processing in Radiology), or it may mean that some problems that were previously unapproachable can be run (e.g., large simulations).

CRAY XMP SYSTEM

Let us consider one of the most famous names in super computing and define several terms that describe many of the standard ways that programs can take advantage of the high speed computing engines in these machines. Figure 1 is a simplified diagram of a Cray XMP super computer. The system contains one to four CPUs, each with multiple registers sets (64 and 24 registers with some registers having 64 64-bit elements), vector ALUs, floating point ALUs, scalar function ALUs, and address ALUs. Each CPU has four direct ports to the central memory (up to 16 million 64 bit words). Each CPU has a 8.5 ns clock cycle and is essentially a RISC processor.. Auxiliary bulk memory systems (up to 4096 mbytes of memory) and high speed disk and I/O channels (100 megabytes/second) are standard accessories. The system includes a separate CPU communication channel for interprocessor communications.

It should be noted that the machine can be set up with only one CPU and yet provide major increases in execution speed over more conventional systems. The speed is partially due to the very fast clock, but the most impressive speed increases come with utilization of the vector registers and the general implementation of many of the basic instructions in very fast hardware instead of microcoded, general purpose ALUs (a RISC processor).

Vector registers or compute engines are very useful for operations that work on arrays of numbers. Basically, the vector system allows groups of numbers to be transferred to the registers and then the same operation performed on the all of the numbers in the registers. Typically, there are two sets of input registers (call then A and B) and one set of output registers (C). The software issues a single instruction to use the contents of A and

B and a manipulation (e.g. add, subtract, multiply, etc) and put the result in C. The key here is that the register operations are done in parallel, so that with 64 registers in A, B, and C, one machine cycle in the vector set is roughly equivalent to 64 cycles in a conventional, non vector CPU.

With a multiple processor system, each processor can execute its own set of programs, or a single program can be split among the processors and different segments run in parallel. The different processors communicate via shared memory and the special interprocessor communication channel. The difficult part is writing the programs to take advantage of the vector and multiple processor facilities. Fortunately, tremendous advances have been made in providing compilers that can both vectorize code and split the program into multiple tasks that can be run on parallel processors.

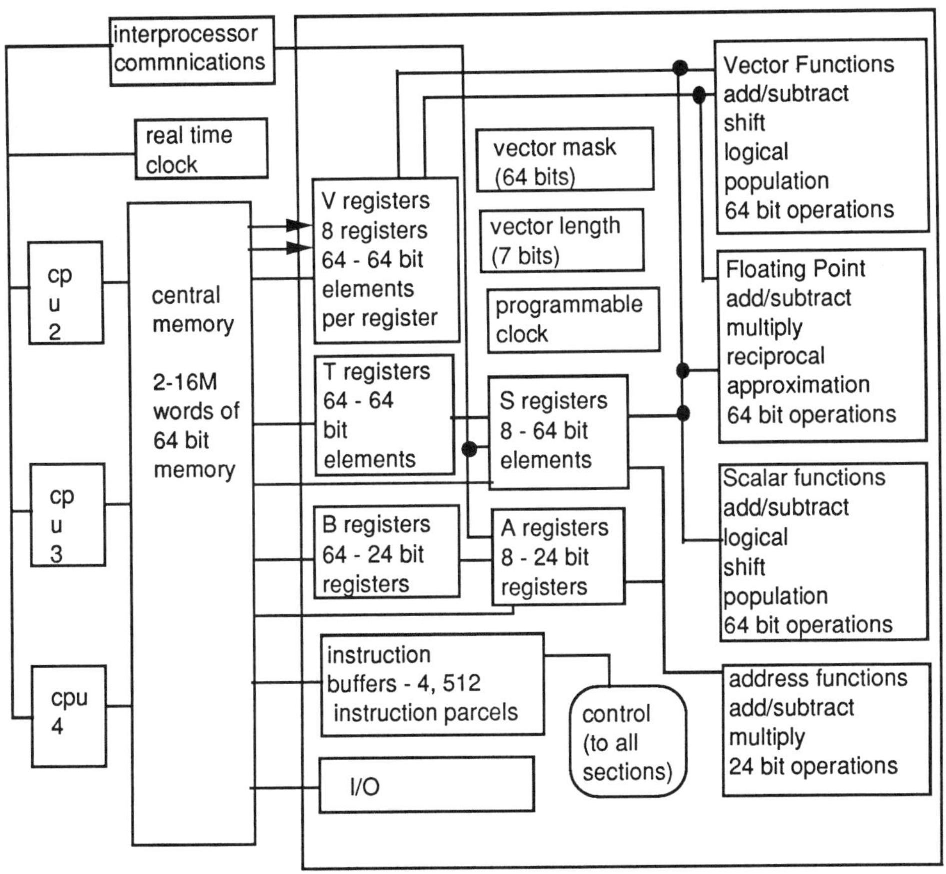

Figure 1: Simplified block diagram of a Cray XMP super computer. Note that the detail of CPU 1 is shown on the large block on the right of the diagram. CPUs 2,3 and 4 are identical to CPU 1.

COMBINED PARALLEL AND VECTOR PROCESSING

To illustrate the many ways of exploiting the vector and parallel processors now found in parallel systems, let us take a look at a "mini super" computer, the Alliant FX series.

Figure 2 is a simplified diagram of the FX system. It has some features in common with the Cray, but it uses more processors, each running at a slower speed that the Cray. With the slower processors, many of the other system components can be made less expensive since the speed requirements are less (e.g., memory, I/O channel speeds, etc.).

The central part of the system is the main memory and the memory bus that is connected to a series of cache memories. One to 12 I/O processors (68012 CPU with 512 kbytes of RAM and a multibus interface for external devices) are connected to 4 of the 6 cache memories. One to eight of the main computing CPUs are connected to 2 of the cache memories. The operating system can execute programs on the compute CPUs (termed CEs) or the I/O processors (termed IPs). The system can allocate the processors to single programs (the parallel processor environment discussed earlier) or can allocate several CEs for one job (a program set up to run with parallel tasks) providing a parallel processing environment. In addition, each CE has a vector engine as part of the CPU architecture. At run time, the system can allocate the 8 CEs and 12 IPs so that each one is executing a separate task or it can allocate groups of processors to programs that have been compiled to take advantage of the parallel CEs.

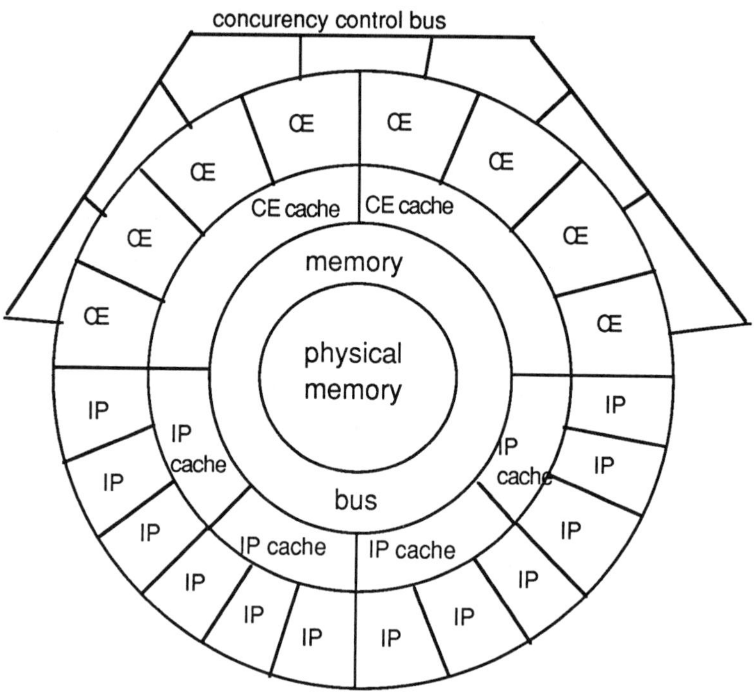

Figure 2: Block diagram of an Alliant FC/8 mini-super computer. IP stands for I/O processor that connects the machine to peripherals and communication systems. CE stands for Compute Element which is the high speed compute engine including vector processing.

VECTOR, CONCURRNET, VECTOR CONCURRENT

To illustrate the possible ways to use these processors, we will consider three ways to map a simple array manipulation among these various CEs. Figure 3 indicates the speed up in compute time if we use a single CE with its vector engine as compared to a traditional CPU such as a 68000 series microprocessor. In the example, the arrays have 256 elements and the vector processor has 64 element vector registers. The vector processor can do the calculation in 4 "machine cycles" while the conventional system requires 256 cycles (where the term cycle is used to cover all the steps in a single addition). Thus, the vector processor can achieve a large increase in speed. In the "real" world, there is some additional setup time in moving the data into and out of the the vector registers, so the speedup indicated by figure 3 is a bit optimistic, but it is still true that the vector processor can achieve very large increases in speed when dealing with array operations. This type of system is usually referred to as vector processing.

Machine cycle	Scalar Execution	Vector execution
1	C[1] = A[1]+B[1]	C[1:12] =A[1:64]+B[1:64]
2	C[2] = A[2]+B[2]	C[65:128]=A[65:128]+B[65:128]
3	C[3] = A[3]+B[3]	C[129:192]=A[129:192]+B[129:192]
4	C[4] = A[4]+B[4]	C[193:256]=A[193:256]+B[193:256]
5	C[5] = A[5]=B[5]	
.		
.		
.		
256	C[256] = A[256]=B[256]	

FIGURE 3: Example of the difference in the number of CPU cycles between a conventional computer (scalar execution) and a system with a vector processor in adding the elements of a pair of arrays. In this example, the vector processor has three sets of 64 element registers.

In a parallel processor system, different parts of the program can be shared among several processors. A possible arrangement is indicated in figure 4 where the same array problem of figure 3 is shared among four processors instead of one. In this case, none of the processors is using vector engines and the use of multiple CPUs to execute different parts of the same code is often termed scalar concurrent processing. The speedup is about a factor of four in the example, although there is some overhead in coordinating the processors so that the actual time will be somewhat less. If vector engines are added to the parallel processors, than an even greater speedup is achieved as indicated in figure 5. In this last example, the type of processing is termed vector concurrent.

Cycle	Scalar Execution	Processor 1	Processor 2	Processor 3	Processor 4
1	C[1] = A[1]*2	C[1]=A[1]*2	C[2]=A[2]*2	C[3]=A[3]*2	C[4]=A[4]*2
2	C[2] = A[2]+B[2]	C[5]=A[5]*2	C[6]=A[6]*2	C[7]=A[7]*2	C[8]=A[8]*2
.					
.					
.					
256	C[256] = A[256]=B[256]				

FIGURE 4: Example of the difference in the number of CPU cycles between a conventional computer (scalar execution) and a system with four parallel processors in adding the elements of a pair of arrays (scalar concurrent processing).

Cycle	Scalar Execution	Processor 1	-----------------------	Processor 4
1	C[1] = A[1]*2	C[1:64]=A[1:64]*2	....	C[193:256]=A[193:256]*2
2	C[2] = A[2]+B[2]			
3	C[3] = A[3]+B[3]			
4	C[4] = A[4]+B[4]			
•				
•				
•				
256	C[256] = A[256]=B[256]			

FIGURE 5: Example of the difference in the number of CPU cycles between a conventional computer (scalar execution) and a system with four parallel processors, each with a vector engine, in adding the elements of a pair of arrays (vector concurrent processing). In this case, the parallel/vector machine would do the entire array calculation in one "cycle", as compared to 256 cycles in the conventional system.

PROGRAMMING PARALLEL SYSTEMS

The use of parallel processors (as opposed to parallel processing) where each processor runs a single job that is not connected to the jobs running on other processors is relatively easy. In most cases, the operating system does all of the allocation and mapping for the user. In some instances, the user can specify which processor is to run the job. In parallel processing, where a single job is split among several processors, it is more complex. If the tasks are independent (do not need to send messages to each other), then it is usually easy to instruct the operating system to create the tasks on separate processors. If the tasks need to interchange messages, then life is much more complicated and care must be taken to efficiently setup the tasks and the means of task communication. In some machines, this is still done as a set of multi-tasks that send messages as in a single processor system using standard interprocess communication facilities or pipes. In more sophisticated systems, the code is not setup as separate tasks in the conventional sense, but the code is mapped to different CPUs and special instructions tell each CPU where it has to wait for values or information from another CPU.

The mapping of a program onto multiple processors has been the major challenge in parallel processing. The task of directly instructing a compiler on how to allocate code segments to different processors can be a most difficult job. With the rapid development of min-super computers, the necessity of developing sophisticated compilers that can vectorize and map code onto multiple processors has become of major importance in order to develop a large enough market for such machines. The efforts in compiler development are becoming successful and many vendors offer compilers that can do an amazingly good job of mapping programs onto a few (less than 9) processors.

To illustrate how a compiler can map a program to multiple processors, I will use a code fragment produced by an Alliant compiler mapping a program onto three processors. The standard code is a DO loop. The fragment in Table 1 is what the programmer would provide to the compiler. This code segment will be split and mapped so that 3 CPUs run different parts of the do loop and provide a major saving in compute time (Table 2). Note that the splitting of the do loop into several concurrent calculations and the mapping onto the three processors is done by the compiler and the operating system, not by the programmer! The key is that the CPUs

```
scalar code
.
.
DO 24 i = 1,6
x = FT(i) * FLOAT(N)
X2 = 2. * X
X21 = x2 - 1.
DF1 = X21 / X2
DF2 = DF1 * DF1
FF = F(i) + DF2
F(i + 1) = FF
AF = ABS( DF2 / F(i + 1))
IF (AF.LE.EPS) goto 25
24 continue
```

Table 1: Code fragment of a Fortran loop on a single processor machine.

```
CE1                         CE2                           CE3

serial code
.
.
DO 24 i = 1,6
x = FT(1) * FLOAT(N)        x = FT(2) * FLOAT(N)          x = FT(3) * FLOAT(N)
X2 = 2. * X                 X2 = 2. * X                   X2 = 2. * X
X21 = x2 - 1.               X21 = x2 - 1.                 X21 = x2 - 1.
DF1 = X21 / X2              DF1 = X21 / X2                DF1 = X21 / X2
DF2 = DF1 * DF1             DF2 = DF1 * DF1                   DF2 = DF1 * DF1
FF = F(1) + DF2             WAIT                          WAIT
F(1 + 1) = FF
AF = ABS( DF2 / F(1+ 1))    FF = F(2) + DF2
IF (AF.LE.EPS) goto 25      F(2 + 1) = FF
24 CONTINUE                 AF = ABS( DF2 / F(2 + 1))     FF = F(3) + DF2
x = FT(4) * FLOAT(N)        IF (AF.LE.EPS) goto 25        F(3 + 1) = FF
X2 = 2. * X                 24 CONTINUE                   AF = ABS( DF2 / F(3 + 1))
X21 = x2 - 1.               x = FT(5) * FLOAT(N)          IF (AF.LE.EPS) goto 25
DF1 = X21 / X2              X2 = 2. * X                   24 CONTINUE
DF2 = DF1 * DF1             X21 = x2 - 1.                 x = FT(6) * FLOAT(N)
FF = F(4) + DF2             DF1 = X21 / X2                X2 = 2. * X
F(4 + 1) = FF               FF = F(5) + DF2               X21 = x2 - 1.
AF = ABS( DF2 / F(4+ 1))    F(5 + 1) = FF                 DF1 = X21 / X2
IF (AF.LE.EPS) goto 25      AF = ABS( DF2 / F(5+ 1))      DF2 = DF1 * DF1
24 CONTINUE                 IF (AF.LE.EPS) goto 25        FF = F(6) + DF2
                            24 CONTINUE                   F(6 + 1) = FF
                            AF = ABS( DF2 / F(6+ 1))
                            IF (AF.LE.EPS) goto 25
                            24 CONTINUE

serial code continues
```

Table 2: Code fragment of the same code in Table 1 mapped to a three processor system by a parallel compiler.

support a special set of instructions that allow a processor to wait for a data message from another CPU via the special interprocessor communication facilities. The three processors all start the code within the do loop with CE1 starting with the index set to 1, CE2 with the index set to 2, and CE3 with the index set to 3. CE2 goes into a wait loop until CE1 calculates the value of F(2) and CE3 waits for CE2 to calculate F(3). The three processors continue on, going into a wait loop whenever a value is needed from a different processor.

When more processors are involved, the problems become much harder to map and usually require a great deal of programmer intervention or the use of very sophisticated communication methods to implement algorithms of very large parallel systems (e.g., large numbers of processors, sometimes called massively parallel systems). When modeling the behavior of systems such as the flow of fluids through capillaries, the massively parallel systems can be set up such that each processor represents a specific portion of the physical model. In such a setup, the flow can be started at one end of the many processors (such as the Connection Machine with thousands of processors) and the machine will mimic the flow in the capillary system. The investigator can interrogate individual processors to get the flow data at that point in the system as a function of time. Considerable progress has been made in compilers that support such machines, and in the next few years, it may be fairly easy to develop programs for system analysis with such processors. The success of such devices will be highly dependent on the communication channels between processors, and the amount of overhead spent in communications.

In systems with a few processors, it is now fairly straightforward to program the systems since the compilers do much of the difficult mapping job. It is my understanding that systems with up to 8 processors can be mapped quite well by these new compilers. For the massively parallel systems, it will probably be necessary to have dedicated software analysts and programmers to implement new algorithms on the systems for the foreseeable future.

APPLICATIONS

When should parallel systems be used? When the problem being computed can not be done fast enough on the non parallel systems. When a system is not responsive to multiple users and a cost effective way must be adopted to improve response time and yet share large resources (such as large memories and disk subsystems). Some obvious examples are to be found in image reconstruction and image processing.

1. In time-of-flight PET, there is a problem in long reconstruction times. Reconstruction times of 20 to 30 minutes per image are not uncommon. At the Washington University Biomedical Computing Laboratory (BCL), an NCR system with 2300 1 bit processors can reconstruct an image in less than one second. The cost of the system is on the order of a standard large minicomputer ($150,000).

2. In image reconstructions using iterative algorithms the compute time can be excessive, particularly if corrections are being made for effects such as scatter and detector resolution. Again, a system such as used at the BCL laboratory should be able to do such reconstructions in a few minutes.

3. Image processing systems are using many parallel systems for rapid image manipulation. The differences between pipelined logic devices and parallel computing systems is becoming hard to separate and many modern image processing systems utilize parallel systems for individual pixel manipulations. In

3D display systems, the use of parallel systems to provide rapid rotation and splitting of 3D shaded surface volume data sets are being exploited.

4. In analyzing metabolic image data, the use of parallel systems to solve kinetic models based on distributed models is most promising and some work is underway for such applications.

Many more applications can be stated. As the cost of these parallel systems continues to decrease (e.g., the mini-super computer workstation), and the compilers do better and better jobs of mapping problems onto the multiple processors, the range of applications in medicine will grow at an enormous pace.

Computer Bus Architecture

Bruce Ho, Ph.D.

Division of Medical Imaging
Department of Radiological Sciences
UCLA
Los Angeles, CA 90024

ABSTRACT

Standardized computer buses are a key factor which lead to the prevalence of low cost, high performance computing systems. The basic designs and some special features of the few popular buses are described. Their implications in imaging applications are also discussed.

INTRODUCTION

A computer bus provides an efficient and organized way of interboard communication. It makes possible a partitioned system design approach which greatly enhances the flexibility and cost-effectiveness offered to system integrators and end users. An established bus convention allows system vendors to market at board level rather than (closed and complete) system level. This has largely been responsible for the success in high volume commercialization of application-specific microcomputers.

Survey of Common Buses in the Market Place

New Generation of 32-Bit Buses

The class of buses most relevant to the imaging field due to their high bandwidth is the new, 32 bit generation. The chief contenders in this class are:

1) VME Bus.
2) Multibus II.
3) Nubus (Mac II).
4) Microchannel (PS/2, closed).
5) EISA bus (only recently announced).

Of the buses listed above, only IBM's Microchannel is legally and technically closed to third party board level support. The EISA (Extended Industry Standard Architecture) Bus supported by over 60 competitor companies was recently made public to combat IBM's effort to dominate the next generation personal computers with its proprietary bus. The Nubus, heavily supported by Apple and Texas Instruments, is the first 32-bit bus made available for personal computers. Both VME and Multibus II are aimed at the workstation market with VME taking the dominant market share.

Other Common Buses

Digital Equipment Corporation has three bus names under its trademark: Unibus, Q bus and the latest BI bus, all of which are proprietary to DEC. Low cost and wide functionality are therefore not likely to become the traits for DEC compatible products. STD and STE are physically rugged buses preferred for rough, noisy industrial environments. In terms of the one bus which has had the most prominent presence in the imaging market, PC/AT bus unquestionably stands as the champion. Many small companies have successfully exploited the open architecture of the AT bus and the low price tag induced by massive cloning of AT's to market board level solutions for a wide range of imaging applications. Data Translation, Matrox and Data Cube are a few of the most notable ones. The one parameter which most clearily underscores the merit of a given bus is undoubtedly its bandwidth, measured in megabytes per second. Bandwidths of the previously mentioned bus standards are charted in figure 1 for comparison.

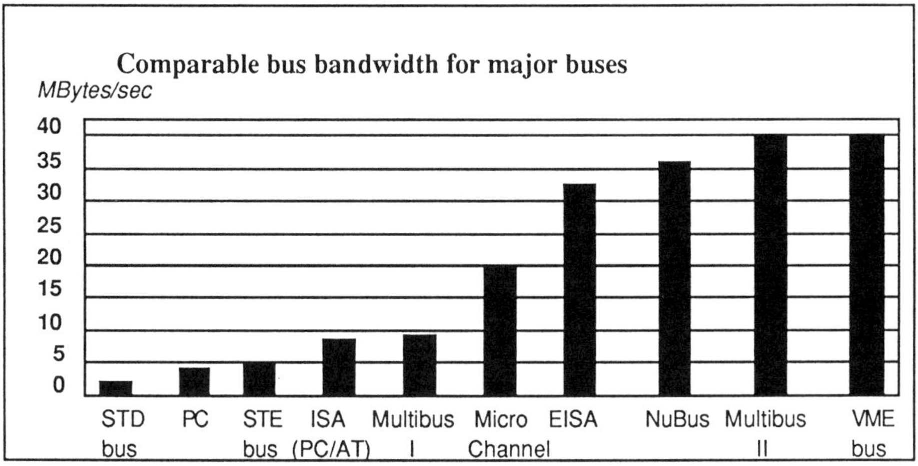

Figure 1. Bandwidths of major buses

Philosophy and Practicalities of Bus Standards

Because of the grave economical implications in a bus standard, establishment of one is elaborate and costly, and it only comes to life after strategic planning and commitment from major corporations. Each major sponsor designs into its bus standard features which are competitive for the existing technology and nourishes user loyalty by providing both board level and chip level support. The corporate sponsors of Multibus II, VME bus, and Nubus, as well as the buses' salient features, are tablulated in table 1. Common features among them are transfer speeds of approximately 40 MBytes/sec, 4 GBytes addressable space as implied by the 32-bit address pins, 96-pin DIN type connector(s) (as opposed to the less reliable card edge type used on the AT) and a comparable size form factor which is an indicator for packing density per board. The author's experience has been that component overcrowding is a serious problem for all but the simplest type of design. Many board vendors have turned to surface mounted technology (SMT) to overcome the board real estate shortage. Deviants do exist as in the case of Sun Microsystems, who designs component boards that adhere to the VME protocol but circumvent the packing problem by using a much larger size format. The issue behind synchronous vs asynchronous protocol remains a debated item. Proponents of synchronous buses claim a superior performance in multi-master systems. This point remains largely an unproven theory. On the other hand, asynchronous buses users are going for the simplicity in interface design.

Table 1

New buses for 32-bit multiprocessor systems				
		Multibus II	VMEbus	NuBus
Source	Originator (backers)	Intel (AMD, Zilog)	Motorola (Signetics, Mostek)	MIT and TI
Performance	Bandwidth	32 bits at 10 MHz	32 bits at 12 MHz (20 MHz on VMXbus)	32 bits at 10 MHz
	Data-transfer rate	40 Mbytes/s	48 Mbytes/s (max) 24 Mbytes/s (typ)	37.5 Mbytes/s for block data transfers
	Memory addressed	64 terabytes	More than 4 Gbytes	More than 4 Gbytes
Architecture	Bus components (exclusive of supply buses)	1. Multiplexed main data and address bus serving 4 Gbytes 2. Local data and address bus serving 64 Mbytes of memory 3. Two-line serial message bus 4. DMA bus 5. I/O expansion bus	1. Nonmultiplexed main data and address bus 2. Expansion bus (32-bit data and 24-bit address lines) 3. Serial message bus	Single multiplexed data and address bus
	Communication between bus elements	Synchronous	Asynchronous	Synchronous
	Connectors	One 96-pin DIN type	Two 96-pin DIN types	One 96-pin DIN type
	Form factor	9.19 by 8.66 in.	9.19 by 6.3 in.	Not available

Table 2

Corporations Behind the Buses

Commited support from a major chip/or system corporation is essential for the survival of a bus standard

VME	Motorola	open
Multibus	Intel	$500 license fee
Nubus	MIT, TI, Apple	ID # assignment
Microchannel	IBM	closed

The degree of "openness" is also a practical matter of concern to system integrators facing a bus choice decision. The battle between IBM and Apple at the onset of the PC revolution has demonstrated that a proprietary bus in a closed system may do more to harm a product's appeal than to ensure the producer's prosperity. On the other hand, DEC's long standing success in defending its turf against invasions of copycats represents a winning strategy based on monopolizing its own design. In today's bus war, we see a range of protection measures used by corporate participants as shown in table 2. Motorola clearly leads in "openness" with its VME offering.

In Search of a Standard Bus for Imaging

A bus standard benefits the users in various ways from both technical and economical points of view. When computers sell in the volume of lettuces, supermarket price tag will also emerge. This concept is partially brought to realization in the case of IBM PC. By the same token, one looks for an imaging bus standard to stir competition among manufacturers with the result of driving quantity up, price down, and broadening selection of functionalities. Rising demand for chip level support also provides incentive for VLSI designs which are unaffordable at low quantity.

Market reports currently show the VME bus as the leader with a rosy outlook for the future (see tables 3 & 4). Functionality selection (see table 5) which ranges from single board computers to communication controllers also underscores the maturity of this standard. In terms of chip level support, Motorola conscientiously designs their processor chips to match the bus in signals and protocols as shown in figure 4. In addition, at least half a dozen companies are offering interface chips to the VME bus to simplify the designer's task and free up precious board space.

<u>Table 3</u>

Market Size (year end 1987)

VMEbus	\$320M
MB-I	\$260M
MB-II	\$27M
Qbus	\$210M
IBM PCbus	\$110M
Other	\$190M

Overall Market World-Wide: \$1.1B

Table 4

1990 Board Market

Bus Type	Revenues	Share	Character
VMEbus	$680M	34%	mature growth
Multibus-I	250M	13	slow decline
Multibus-II	290M	14	rapid growth
Qbus	200M	10	decline
other	630M	29	very fragmented
total	$2.050B	100%	

Table 5

Available Categories of VME Boards

Single Board Computers (SBC), 68030, RISC

Memory, 32MB

Disk Controllers, SCSI, SMD

Graphics/Display

I/O, printer, terminals

Acquisition, A/D

Communication, ethernet , MAP, DR11-W

Image Processor, TMS34010, DSP56001

The VME Bus Architecture

A description of the VME bus basics will be given in general terms here. As such, many features mentioned will be applicable to buses in general. The VME bus has four functional parts: 1) data transfer, 2) arbitration, 3) interrupt and 4) utility. The data transfer cycle starts with an address broadcasting phase in which the master board selects a slave board sharing the bus. The data transfer phase follows in either read or write mode with Data Strobes and Data Acknowledge signals handling interboard synchronization. Thirty-two bit buses carry the built-in burden of having to be upward compatible with 8-bit and 16-bit boards. Replication of strobe signals is used for such purposes.

Table 6

	Addr.Str.	Data Str.	Interrupt	Bus Arbitr.
VME	AS*	DS0*,DS1*	IRQn* IACKin*/out*	BREQn*,BBSY BGnin*/out*
68020	AS*	DS*	IPLn*	BR*,BGACK* BG*
68450	AS*	UDS*,LDS*	IRQ* IACK*	BR*,BGACK* BG*
56001	HEN*		HREQ* HACK*	BR* BG*

Crucial to Motorola's philosophy of the VME architecture is the provision of an express bus which links resources on a local group of boards. It is named the VSB (VME Subsystem Bus). Design of express buses leaves much room for innovation. Sun Microsystems, for example, proposed their own express bus standard called the Sun P2 bus. Both achieve physical compatibility with the VME bus by placing express bus signal pins on the unused rows of the VME bus connector J2. The VSB has two noteworthy specialties: 1) dynamic bus sizing, which allows the master to handle slaves of different data width (32, 16 , 8, etc.) without the user's prior knowledge, and 2) participating slaves which allows data transfers to multiple destinations in each cycle. The Sun P2 bus, on the other hand makes use of multiple strobes in a way which allows for multiple data widths as well as simplify interface designs for DRAM (Dynamic Random Access Memory) arrays.

300 ns /transfer, horizontal or vertical

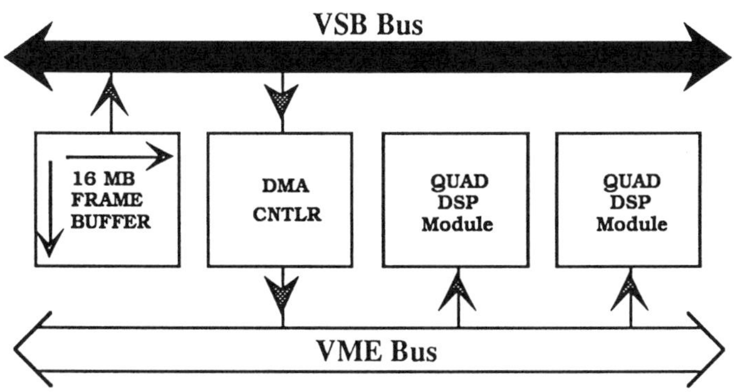

Custom DMA controller simultaneously broadcasts addresses over VME and VSB buses for single cycle transfers

Figure 2. DMA transfer over two buses

Doing DMA over Two Buses

The key factors in measuring the performance of any imaging system are data transfer speed and data processing speed. Obviously the virtue of a good bus design and efficient use of such a bus is in how much data transfer bandwidth it yields for a given system. Direct Memory Access (DMA) is technique for achieving fast data transfer as old as computer technology itself. The normal DMA mode for memory to memory transfer is a two cycle process involving a data read and then a data write. At UCLA, an innovative design takes advantage of the dual bus structure in a VME system and performs a one-cycle transfer to speed performance by a factor of 2. As figure 2 illustrates, addresses for both source (frame buffer) and destination (quad DSP module) are simultaneously broadcast but over different buses, making single-cycle transfer possible.

When Will VME Bus Run Out of Steam?

As technology progresses, memories are faster and denser, processors go to higher precision, packaging density multiplies, and as we are seeing, even new bus architectures are invented to keep pace with system requirements. Before we sink deeply into one bus standard, let's think about the day such a bus will face obsolescence, and with it, much of the hardware investments we are making today. The VME buse is electrically capable of 12 Mhz transfer (83 ns). This limit will not be reached anytime soon. Both memory speed and processor speed will have to advance for some generations to reach comparable speed (see figure 3). The latest announcement of RISC (Reduced Instruction Set Chip)

When will VME bus run out of steam?

Not for a while!

12 MHz VME bandwidth	⟶	**83 ns/transfer**
available DRAM boards	⟶	**270 ns/transfer**
SRAM access time	⟶	**100 ns**
20 MHz 68020	⟶	**100 ns/transfer**

Bottleneck at the backplane may occur due to contention between masters, but not due to bus transfer speed. Special imaging buses aim at resolving bus contention situations.

Figure 3. Chip speed in current technology

claims 40 MHz clock rate which is rapidly approaching this bus limit. Short of this case, we should not find that bus transfer speed is the culprit for system performance loss. A much more critical factor to consider is bottleneck created by multi-master contention (more than one master tries to get the bus at one time). Special imaging buses have been proposed by various manufacturers to get around bus-contention. Examples of this are given in figure 4.

Special Imaging Buses

Pixar — **Sun host**

Data Translation

Data Cube

} **PC/AT host**

Figure 4. Some special imaging buses

As a concluding remark, we will also give a special look at the PC/AT bus as a reverent gesture to its omnipresence. Some key features are listed in figure 5. Of particular importance is its maximum data transfer rate of 700 KByte/sec. IBM's solution for an upgrade is the Microchannel with 17 MBytes/sec, while the EISA is designed for 33 MBytes/sec. For fairness, these should be compared to Apple's Mac II, which according to Nubus specification, can handle up to 37.5 MBytes/sec.

PC/AT Bus

24 bit addr (16MB)

8 or 16 bit data access

Interrupts IRQ9-12,14,15, 3-7

DMA channels DRQ0-3 (8 bit)
DRQ5-7 (16 bit)

I/O wait-state generation

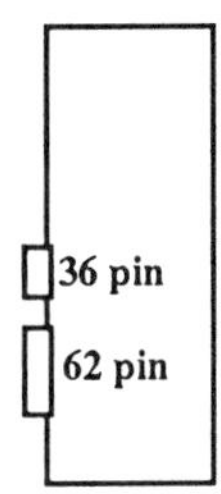

Max. DMA rate = 700KB/sec

Max. disk access rate = 500KB/sec

Figure 5. Salient features of the AT bus

CONCLUSION

In this paper, we discussed some aspects of bus architectures from the point of view of an imaging system integrator. VME bus is shown to have many advantages in terms of market dominance and chip level support, whereas its technical merits are comparable to Multibus II and Nubus. A point is made that the VME bus standard is mature enough to ensure survival in the market and yet fresh enough to last through many more generations of chip technology upgrade.

DIGITAL STORAGE TECHNOLOGY FOR MEDICAL IMAGING

Nicholas J. Mankovich, Ph.D.
Department of Radiological Sciences
University of California
Los Angeles, CA 90024

ABSTRACT

Medical imaging requires tremendous storage resources. This tutorial reviews digital magnetic and optical storage with an emphasis on the underlying technologies and how they apply to digital imaging. Magnetic tape allows sequential access to data at an extremely low cost while storing up to 5,000 megabytes (MB) in a single tape cartridge. Although useful for backups and short-term storage, tape's sequential-only access limits its use in radiology. Magnetic disks allow high-speed access regardless of sequential position of the data and now store over 1,000 megabytes (MB) on a single drive. Where high transfer speed is crucial, parallel storage systems maintain data on one or more disk drives to provide nearly instantaneous image display. Optical disk storage, in contrast, provides slower access from single platter disk drives housed in multi-disk, robot controlled systems (jukeboxes) storing nearly 1,000,000 megabytes on either write-once or erasable disks. Parallel disk architectures are now being used in optical systems to increase access and transfer speed. Medical imaging systems can take advantage of this diversity of storage hardware to allow fully integrated digital image libraries configured to minimize cost while maximizing performance.

INTRODUCTION

Radiology departments traditionally provide three basic services to primary care physicians, including (1) image acquisition, (2) interpretive consultation, and (3) image record keeping. Radiology departments are continuing the trend toward increased use of digital imaging modalities and are maintaining these images in their original digital form. This discussion focusses on the technology available to solve the problem of storing digital radiological images.

As with conventional film libraries, the main operational concerns in digital image storage include the speed of image recall, the space required for storing the images, and the cost of library operation. There are many measures of the performance of digital storage technologies including the total amount of storage, the cost of the storage, the time it takes to get to the start of the information, and the actual transfer speed of the information, once located. Digital data storage is measured in megabytes (abbreviated MB) corresponding to 2^{20} bytes (just over 1 million bytes). Because a single digital radiographic image can range in size from 0.01 MB to 16 MB, we must also use gigabytes (1 GB = 2^{30} bytes) or terabytes (1 TB = 2^{40} bytes) when considering the storage needs of an entire department. Multi-image patient exams will contain from 0.16-25.0 MB.[1] A medium-to-large hospital produces hundreds of megabytes of data per day and therefore a few terabytes per year.[2]

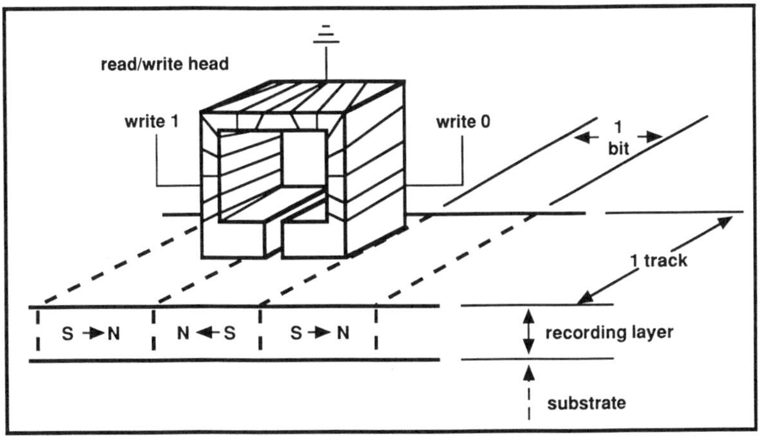

Figure 1. Schematic diagram of the basic components of the magnetic storage process.

The cost of storage is measured by calculating the total dollars per megabyte for either the entire storage device (interface, drive, and media) or the storage media alone. The average access time measures the speed of data access in seconds for large archival storage and in milliseconds for on-line data storage. Data transfer speed is measured in megabytes per second (MB/s).

Magnetic storage is based on the induction and detection of local magnetic dipoles in a recording layer (Figure 1). There are thousands of dipoles per inch on the surface of the recording media. A small solenoidal magnetic head on or above the surface of the media provides both the magnetic field for writing and the induction device for reading. A single head reads or writes a single track of data in the form of serial binary integers (0 or 1 bits) along the recording path either down the length of a tape or around the circumference of a disk.

MAGNETIC TAPE

Magnetic tape is one of the oldest media for the storage of digital information. It has traditionally relied on the sequential streaming of data to a reel of polyester-based tape coated with a ferro-magnetic material. Data are encoded onto the magnetic layer by several electromagnets embedded in a single read/write head resulting in parallel data tracks across the tape, hence the name multi-track tape. The three types of tape tracking technologies are distinguished according to their pattern of reading and writing bits: longitudinal, serpentine, and helical (Figure 2).

Longitudinal multitrack tape typically records data with a nine element head that writes a single byte (8 bits of data, 1 bit parity) across the width of a 0.5 inch tape wrapped onto a 10.5 inch diameter open reel. Data are written at 800/1600/6250 bits per inch (bpi) along the tape at 60-125 inches per second (ips). Longitudinal magnetic tapes offer one of the most inexpensive recording technologies with a media cost at about $0.06/MB. The disadvantages of this type of storage lie in the time it takes to retrieve a single file and the general problem of media degradation due to excessive wear.

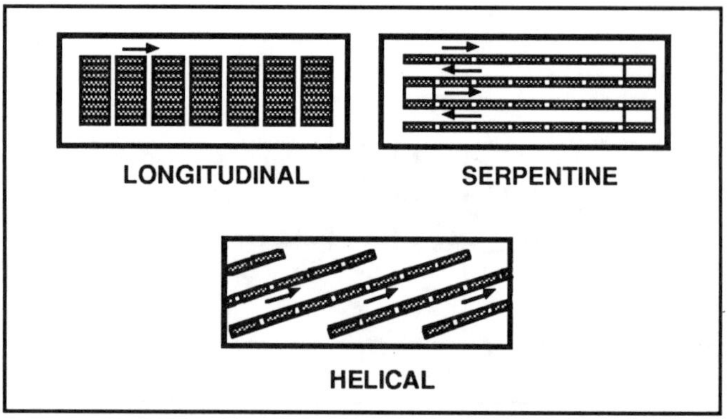

Figure 2. Organization of information on the three types of tape recording technologies: longitudinal records data from start to end, serpentine streams data back and forth along the tape, and helical writes diagonal stripes of data from tape start to end.

Serpentine track tape drives, unlike longitudinal drives that read the tape end to end once to obtain all information, first read one to four tracks on the forward pass then shift the head, reverse the tape, and read the tape backward for the next set of tracks. This back and forth reading continues until all tracks are passed. This reduces the amount of tape that must be passed to get to any particular piece of data and thus reduces the average access time, but significantly increases the tape wear.

Serpentine tapes contain 20-30 tracks and store 200-750 megabytes (MB) on a single cartridge. The most popular size and encoding format for this type of tape is the Quarter Inch Cartridge (QIC) standard. The most recently proposed QIC standard is QIC1350 for 1,350 MB tape drives operating at 597 kilobytes (KB) per second on a 30 track tape. Prototype 0.25 inch cartridge drives are expected to allow up to 2,000 MB per cartridge running at 3 MB/s and the introduction of data compression into the QIC standard is likely to double storage capacity in the near future.

One of the most exciting advances in storage technology over the past three years is the commercial availability of helical scan tape recording. This technology relies on 0.5 inch, 8 mm, or 4 mm magnetic tape packaged in cartridges. Like longitudinal and serpentine tape, a read/write head encodes data by direct contact with a magnetic coating on a polyester based tape. However, helical scan tape writes data in distinct data blocks or tracks recorded as a slanting stripe. This tilted-write technology developed from videotape recorders that use single track read and write heads rotating within a tape transport drum. The magnetic tape is partially wrapped around this drum in such a way that the intersection of the head and the tape describes a 6 degree swipe across the tape surface. One of the major benefits of this combination of moving tape and moving head technology is that the tape can be transported slowly (e.g., at 0.32 ips) while the spinning head gives a relatively high tape to head speed comparable to longitudinal tape systems (123 ips). The slower transport eliminates the tape retensioning problem that can lead to reduced data integrity. Digital helical tapes come in 0.5 inch, 8 mm, and 4 mm formats. Many of the 8 mm and the 4 mm systems use the new Digital Audio Tape encoding.[3, 4]

Because of cartridge tape's use as a backup device for large disks, tape drive manufacturers and system integrators have combined tape drives with automatic tape storage and loading devices. The simplest device is a tape stacker that allows 5-10 tapes to be loaded into the device. Data archiving software then automatically switches to the next tape during a backup or restore operation, allowing ten gigabytes to be written without operator intervention. A more general and sophisticated device for unattended tape operation is the magnetic tape jukebox that allows automatic access to hundreds of gigabytes of data within minutes of a request. One design uses a 54 cartridge carousel storing 124,000 MB in a standard 19-inch computer rack. When fully loaded, two racks (about 12 square feet) can store over a terabyte of data.

Overall, tape technology has shown great improvement over the past five years. The cost per megabyte is unbeaten by any other storage technology. The entry of helical scan formats and new tape technologies has minimized the longevity problems of magnetic tape that traditionally include (1) data print-through, (2) self-erasure, and (3) binder failure. Current helical scan tapes can expect to last 10 to 15 years and store in the 5,000 MB range with moderate data access times. With a best average access time of 20 seconds, however, tapes will not see widespread use as a primary storage device. Tapes will remain useful for (1) data backup, (2) software distribution, and (3) data interchange.

MAGNETIC DISK

Computer disks rely on a movable, single-track, read/write head either on or close to the surface of a magnetically coated disk of metal, plastic, or glass. The same principle of local N-S orientation of magnetic fields is used with magnetic disks, but instead of linear tracks along a tape, *disk tracks*, are formed of concentric annular rings on the surface of each disk. The tracks are further divided into short segments called *sectors*. which contain disk address information, user data, and error detection and recovery information. In general, all tracks on a disk have the same number of sectors.

Disk performance is measured in terms of the *average seek time*, indicating the average time taken to position the read/write head over the desired track, and the *average latency time*, recording the average time taken for the requested sector to rotate under the head (usually one half of a disk rotation). These two numbers are combined into the most widespread figure of merit on disk drive performance, the *average access time* (equal to average seek plus average latency).

One of the most popular types of magnetic disk is the *floppy disk*. This technology uses the same plastic substrate as magnetic tape but uses a 3.25-inch to 8-inch diameter disk of this bendable material enclosed in a hard or flexible plastic housing. The most widespread floppy disks store from 1-2 MB on a two-sided, high density disk. Modern floppy disk drives have read/write heads on both sides of the disk, allowing the user to consider the two-sided disk as a single storage surface. Floppy disk heads are in constant contact with the disk throughout the read/write process. For this reason, the disks generally spin at 300-500 RPM, and the average access times are slower than for rigid disks. Commercial floppy disks for the personal computer market now allow the storage of up to 10 MB on a 5.25-inch disk, although these have not reached the popularity of the 1-2 MB 3.25-inch disks. More unconventional approaches push storage on a 3.5- or 5.25-inch floppy disk up to 40 MB. Methods for increased storage capacity range from disks with optical components for precision tracking to carefully designed aerodynamic systems that allow the read/write head to fly above the surface of the floppy, providing performance near to that of a rigid disk.

The most appropriate technology for medical image storage is the *rigid magnetic disk*, composed of a disk assembly and a computer system interface board. The disk assembly contains multiple disk platters, each with two disk surfaces (Figure 3). Over each disk surface is a read/write head. The 2-10 disk platters with diameters from 2.5 to 14 inches are all mounted to a single spindle and all spin in concert, usually at about 3600 rpm[5, 6]. The head assembly is composed of a pivoting arm with a motor or actuator device on one end, and a disk read/write head on the other. The arm usually contains one read/write head per disk platter surface, with all of the heads moving together. Thus, in an eight disk system, there are 16 heads moving at once under commands from the drive controller and computer interface. Either a stepper motor or a voice coil assembly positions the heads over a single disk track. The small read/write head actually flies a few microinches above the recording surface, relying on the carefully designed aerodynamics within the spinning disk enclosure. The term *Winchester disk* was coined by IBM to designate the current generation of rigid disks constructed with a sealed head/coil/platter assembly.

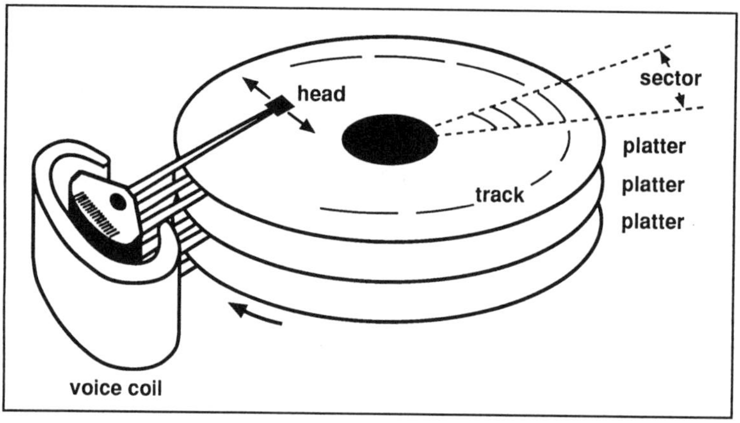

Figure 3. Multi-platter winchester disk components.

The large capacity and performance demands of a medical image workstation require higher data transfer speeds than is available from conventional disk systems. In pursuit of higher speeds, disk system manufacturers have developed multi-disk arrays and parallel transfer disks[7, 8, 9]. Conventional disks stream data byte-by-byte out to the disk drive through one disk head at a time. Data are stored sequentially on the tracks within the sectors. *Parallel transfer disks* simultaneously distribute the data bytes across the disk surfaces by writing the data in parallel on many heads. Typically, computer bus and memory speeds limit the data transfer rate on parallel transfer disks. In imaging, these disks are usually directly interfaced to an image processor memory, where they can transfer data in excess of video data rates (>7.2 MB/s) and effect a 2000x2000x12-bit image display in 1-2 seconds.

The second high-speed technology used to increase disk transfer rates takes advantage of the prevalence and low cost of small (100-500 MB) disk drives. Rather than operating the disk heads in parallel, this technique uses a shared disk interface to operate many disks in parallel so that data is spread across three or five disks. Formerly known as _Striped Subsystems_, these _Redundant Arrays of Inexpensive Disks (RAID)_ are configured to be somewhat fault-tolerant, with a typical five-disk system able to fully recover from the loss of a single disk drive[10]. This ability to sustain high transfer rates (>10 MB/s) and insure the integrity of data has made these schemes part of commercial mainframe systems introduced in the last year.[8]

Just as parallel disks push the transfer rate limits of disk storage, many techniques are on the horizon for pushing the absolute bit density of single disk units. Disks are now available with zoned recording technology[5] that allows disks to have more data stored on the outer than the inner tracks, thus increasing overall storage by 30%. New recording media and heads also promise to increase disk capacity through higher bit densities. Recently introduced systems use perpendicular magnetic recording to align the magnetic dipole perpendicular to the disk surface yielding a two- to three-fold increase in capacity. A recent review of magnetic disk drive technology predicts that the price of 5.25-inch drives will continue to fall 30 percent per year over the next few years and be accompanied by a tripling of data transfer rates[11].

OPTICAL SYSTEMS

The demands of medical imaging and the general problem of large databases in seismology, astronomy, and other sciences has led to the development of optical storage systems. While the emphasis in magnetic systems is frequently on the average access speed, the emphasis in optical systems is the total system capacity and recording longevity. Optical disk systems are divided into write-once, read-many (WORM) and erasable optical (EO) technologies. The WORM systems are well designed to be the ultimate repository of digital information[12, 13]. They offer large volume storage at moderate speeds[14]. The more recent EO systems are designed as the low-cost alternative to magnetic disk technology, where some speed is compromised for increased storage capacity. In general, the WORM systems are used as an add-on to a magnetic disk system where the EO systems are actually used to replace magnetic disks systems.

CD-ROM (Compact Disk-Read Only Memory) disks represent a special type of WORM disk that is amenable to the production of many copies of a digital database. Although important for its potential in radiology education, this technology does not lend itself to the maintenance of an image archive composed of a single copy of each image.

WORM technology relies on a digitally controlled laser diode to permanently change the optical characteristics of a sensitive substrate. This is done through substrate pitting (ablative pitting), bubble formation, or chemical change. Laser light is focused onto the recording layer through a protective housing (Figure 4). Information so encoded is then read back with a lower powered laser. Typical optical disk drives have a single read/write head that operates on only one side of the disk at a time. Total storage on optical disks is 800 MB on a 5.25-inch disk, 6,600 MB on a 12-inch disk, and 8.2 GB on a 14-inch disk. WORM disks are, in general, much slower than magnetic disks, although recent developments have lowered the average access times from over 100 ms to as low as 28 ms. WORM media is rated as having a lifespan of from 10 to 100 years.

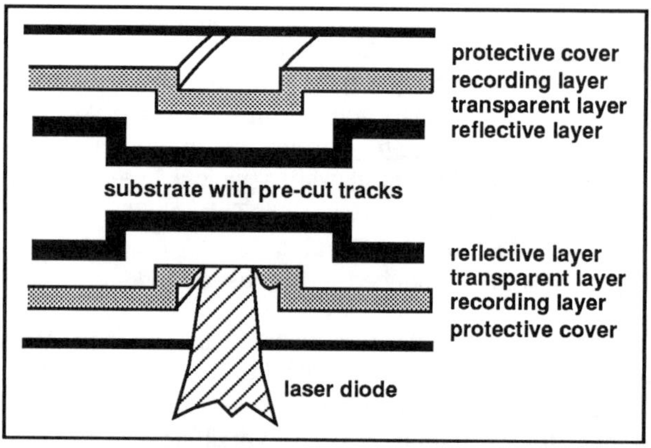

Figure 4. Components of a write-once, read many (WORM) optical disk platter using ablative pitting for writing and substrate reflectance for reading.

WORM disks are single platter devices housed in an openable or removable cartridge. The recording substrate is deposited on a glass, plastic, or metal substrate. The recording layer is within a protective layer, insuring against damage. Unlike magnetic disks, the read/write laser head rides many millimeters above the disk leaving no possibility of mechanical damage due to a disk head crash.

Optical disks' use of a single platter in a lightweight package makes this technology amenable to multi-platter jukebox configurations. Jukeboxes are usually configured with 2-4 optical disk drives, a computer controlled robot arm, and an internal disk shelf space. Typical 5.25-inch jukeboxes store 20-60 disks and the larger 12-14-inch disk jukeboxes store 32-200 disks. The larger jukeboxes can store in excess of 1,000 gigabytes of data with 5-20 second retrieval times.

Erasable optical disks offer an alternative to the permanent storage in WORM technology[15]. EO technology can be based on chemical, phase, or magneto-optical processes, although the latter is the most common. Magneto-optical disks rely on a recording substrate that will differentially polarize light according to the orientation of local magnetic dipoles. The dipole orientation can only be switched when the material is heated above a certain temperature. Magneto-optical disks are written by the simultaneous action of substrate heating via laser and magnetic field induction via bias coil (Figure 5a). To read the data, low-powered polarized laser light illuminates the substrate and a detector notes the direction of rotation of the reflected beam and thus determines the 0/1 content of the bit (Figure 5b).

As sold today, EO disks systems offer relatively low average access speeds at about 80 ms. Systems are now available that minimize this by either using disk striping or the integration of solid state or magnetic disk cache memories within the EO disk subsystem. These systems are frequently configured to resemble more conventional disk cluster hardware to facilitate integration into existing mini- and mainframe computer systems. Gardner[15] predicts that the use of lighter head assemblies and newer multi-beam technologies will allow the 5.25-inch disk to have average access times of 15-25 ms on 6-7000 MB disks.

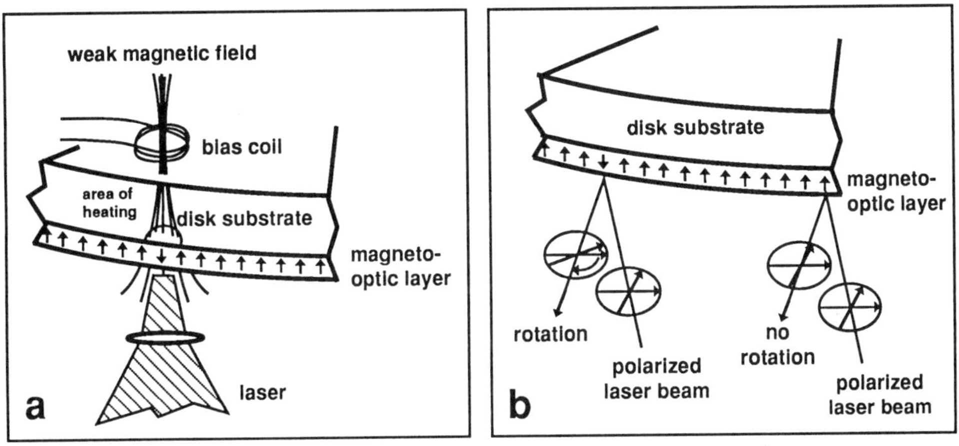

Figure 5. (a) Writing to an erasable optical disk uses a laser to heat a small region of a recording substrate in the presence of a magnetic field. (b) Data bits are read from the surface of an erasable optical disk by noting the direction of rotation of reflected polarized light.

Because of their lower-than-WORM capacity and the ability to rewrite data, the EO disks are being marketed as replacements for the typical office filing cabinet, although jukeboxes handling from 30-100 disks are now being sold for both WORM or EO drives. Because the 5.25-inch package is similar for each, the same robotic mechanism can be used.

During the decade in which storage media have been available, some non-disk systems have been proposed and developed. Most notable has been the introduction of working optical card systems that can store 2-4 MB on a single credit card sized package. The recording media used on these cards is actually an optical coating on a flexible plastic base similar to magnetic tape. This tape technology has been used to develop optical tape systems capable of storing over two terabytes on a single optical tape. These systems carry with them the low access speed problem inherent in any sequential media, but they promise an extremely low-cost storage alternative.

DISCUSSION

Storage technology is not at the point where commercial systems can be purchased that can cost effectively handle the large storage demands of radiology departments. Table 1 contains a comparison of the dominant digital storage technologies based on recently published list prices and specifications. Fortunately, the challenge for medical image storage is not in the basic technology but in the integration of technologies to store and retrieve images in a timely and cost-effective manner. The solution will involve a layered or hierarchical system composed of appropriately sized and priced components with high-speed systems placed where demand is highest and high capacity systems placed centrally. A few simple scenarios should provide an appreciation of such an integration.

Type of Storage Device	Media Physical Size	Device Size (MB)	Average Access Time	Transfer Speed (MB/s)	Cost of Media ($/MB)	Cost of Drive ($/MB)
MAGNETIC TAPE						
Longitudinal	9 track reel	140	90 s	0.20	0.06	100.00
Serpentine	1/2" cartridge	750	44 s	0.17	0.20	20.30
Serpentine	1/4" cartridge	320	20 s	0.24	0.09	19.00
Helical	1/2" cartridge-VHS	2,500	45 s	2.00	0.004	3.00
Helical	8 mm cartridge	5,000	67 s	0.50	0.02	5.70
Helical	4 mm cartridge	1,300	20-30 s	0.25	0.02	5.00
Helical	8 mm Jukebox	580,000	84 s	0.50	0.02	0.36
MAGNETIC DISK						
Winchester	2.5"	40	28 ms	-	-	15.00
Winchester	3.5"	320	20 ms	1.80	-	5.00
Winchester	5.25"	1,600	22 ms	2.80	-	3.00
Winchester	9"	3,100	12 ms	4.50	-	3.00
Winchester	8" parallel disk array	-	-	10.00	-	15.00
Winchester	14" parallel heads	825	30 ms	18.60	-	60.60
OPTICAL DISK						
WORM	5.25" cartridge	800	28-200 ms	0.2-1.5	0.17	4.80
WORM	12" cartridge	6,600	28-120 ms	.5-1.5	0.06	2.90
WORM	14" cartridge	6,800	190 ms	.5-1.5	0.11	6.90
Erasable	5.25" cartridge	600	35-120 ms	.2-1.4	0.25	10.33
WORM/Erase	5.25" jukebox	90,000	3-8.5 s	.2-1.4	.17-.25	0.95
WORM	12" jukebox	372,500	6 s	.5-1.5	0.07	0.81
WORM	14" jukebox	1,020,000	6 s	1.50	0.11	0.26

Table 1. Comparison of the characteristics of current digital storage technologies.

MR/CT image archive. Most moderate-sized hospitals now have both MR and CT scanners on site. Assuming in our scenario hospital that there is one MR scanner and one CT scanner, the total volume of image information might be:

Modality	Patients per day	Images per patient	Images per day	Storage (MB/day)
CT	15	20	300	150
MR	10	25	250	31
			TOTAL	181

Existing MR and CT systems are probably using magnetic tape to store images, requiring both operator time and a significant storage space. Because our scenario hospital just purchased their MR and CT systems from one of the manufacturers who now offer the ability to share archives between MR and CT, the scanning suites are configured with a single optical disk drive to handle all disk archiving on a 12-inch WORM drive storing about 2,000 megabytes/optical cartridge. With this volume, the MR/CT facility purchases an optical disk cartridge every two weeks and relies on the technologists to manually load a stored disk cartridge when archived data is recalled. Hospital policy designates a ten-year retention period, so the entire CT and MR archive will fit on 225 disks taking up about ten linear feet of shelf space, easily stored in 10 square feet. Retrieval of any patient image set older than two weeks would require about five minutes of technologist time. If image review and retrieval patterns make it feasible, the system will be augmented by a WORM jukebox and independent display console allowing on-line access to any study done within the last four years within 20 seconds.

Sectional PACS Module. Although not yet broadly available, the full power of digital imaging in Radiology is found in Picture Archiving and Communication Systems (PACS), in which all phases of image handling would be performed by computer including: acquisition, display for diagnosis, display for review, and archival storage. Within all of these PACS functions there is a need for image storage devices. The subsystems responsible for these functions include acquisition nodes, display workstations, and digital image libraries.

Because departmental PACS systems are likely to grow out of PACS modules like the CT/MR system above, it is useful to consider a more typical PACS module where images are acquired by computed radiography (CR), digital fluoroscopy, digital ultrasound, CT, or MR. In such a module, each modality has an interface unit designed around the particular method and speed of image acquisition. The CR, MR, and CT systems all have conventional magnetic disks in the acquisition systems where images are held pending transmission to the display station and archive. The fluoroscopy and the ultrasound subsystems each might have a high-speed parallel magnetic disk capable of capturing the data at video rates or better (>7.2 MB/s).

The image display station would have high speed local disk storage capable of holding a few days of patient data for rapid (<2 s) recall. Stations designed for real-time image viewing such as digital fluoroscopy or ultrasound might have a parallel head magnetic disk or a disk array (RAID) capable of recalling the images in real-time. This disk technology might also be employed on a CR diagnosis workstation where the typical 2000x2000x12-bit image requires image display in under 1 second (>8 MB/s).

Once diagnosis is rendered, the images would be routed to the sectional (or departmental) archive. This multi-jukebox storage facility would employ WORM disks in a large jukebox configuration, probably with an integrated magnetic disk to allow the queuing of images to and from the slower optical archive. In addition to the central optical archive, a display station might be fitted with a small (600-1000 MB) erasable optical disk system for radiologists to use in their private image collection for reference or research.

An entire department might be made up of 5-15 of these modules, each requiring full internal operation as well as file exchange for case or historical comparison. If the department produced two terabytes of image data per year, then total yearly storage would fit on 300 14-inch WORM cartridges in three optical disk jukeboxes. The total space required for these three units is about 100 square feet.

CONCLUSION

Digital image storage technology can now accommodate the difficult volume and speed requirements of medical imaging. The brief scenarios presented represent the extremes of storage demands in the digital imaging environment in radiology. The introduction of such systems and storage components is likely wherever some small operational benefit can be derived from new technology and thus be cost justified. Complete systems are probably still too expensive and new in concept to be able to turn entire departments over to all digital storage. However, digital image storage is now an integral part of image acquisition. Long term storage systems are commercially available that integrate

MR and CT images. The first commercial systems that integrate on a wider scale are being tested in a few institutions. The future of digital image storage will continue to improve as the cost of digital storage decreases and the performance of systems increases.

REFERENCES

1. Cox, G., S. Dwyer and A. Templeton. 1986. Computer Networks for image management in radiology: An overview. CRC Critical Reviews in Diagnostic Imaging. 25:333-371.
2. Jost, R. G. and N. J. Mankovich. 1988. Digital archiving requirements and technology. Investigative Radiology. 23(11):803-809.
3. Lion, K. 1989. DATA/DAT format provides random access tape option. Computer Technology Review. 9(14):63-67.
4. Vermeulen, B. and C. Watts. 1989. DDS format enhances digital audio tape technology. Computer Technology Review. 9(14):57-61.
5. Spencer, C. E. 1989. High-capacity hard disks. Macworld. November:161-169.
6. Devlin, J., E. Berk and S. Sostack. 1989. New disk subsystems pass the 1-gigabyte mark. PCWeek. 6(5):89-90.
7. Aston, R. 1989. Parallel transfer disk arrays. Advanced Imaging. 4(6):28-58.
8. Cox, J. 1989. Disk-striping speeds I/O. Digital News. November 27:1.
9. Gamerl, M. J. 1989. Parallel transfer drives meet the workstation challenge. Computer Technology Review. 9(14):49-51.
10. Allan, I. D. 1989. RAID arrays making a slow start. Computer Technology Review. 9(11):22.
11. Janson, J. L. and K. Sorensen. 1989. Local storage, file servers tap into system costs. Digital Review. 6(45):47-52.
12. Rosch, W. L. 1989. WORM disks survive in a maturing optical industry. PCWeek. January 2:77-79.
13. Sullivan, K. B. 1989. Kodak eyes gigabyte storage levels with four new systems. PCWeek. 6(45):24.
14. Lawton, S. 1989. Optical disk challenges traditional magnetic tape. Digital News. November 13:56-61.
15. Gardner, R. N. 1989. Integrating erasable optical subsystems into the mainstream. Computer Technology Review. 9(14):39-46.

ADC and the Hardware Interface

Bruce Ho, Ph.D.

Division of Medical Imaging
Department of Radiological Sciences
UCLA
Los Angeles, CA 90024

ABSTRACT

In digital image processing, there are many interface problems to deal with. Three classes of electronic interfaces are discussed: image digitization interface, communication interface and interprocessor interface. A practical example of application of interface techniques in radiological imaging is then given.

INTRODUCTION

Nature is inherently analog. However there is little doubt that the best computational tools made by men are digital. It becomes worthwhile, therefore, to learn how to best represent real world occurrences digitally. In imaging, one is concerned with how to preserve the accuracy of the informational content as well as being able to reproduce an image or a sequence of images in a fashion which closely resembles their original, natural occurrence. This calls for high resolution and high speed analog to digital converters (ADC). Some form of this device exists at the front end of every digital imaging station.

Analog to Digital Converter as an Image Acquisition Device

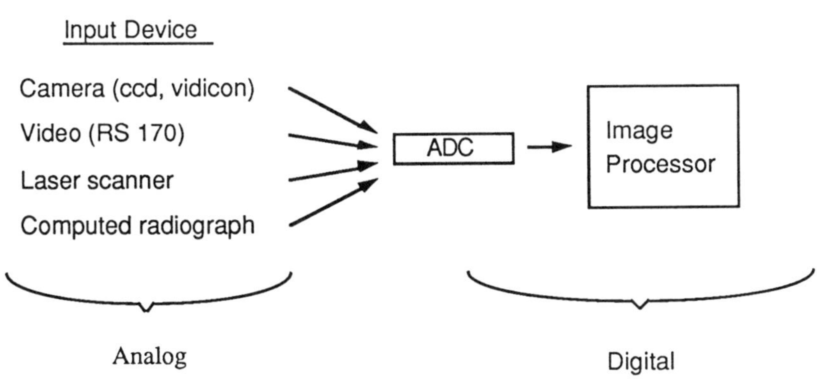

Figure 1. Image acquisitions using ADC

ADC-the Image Acquisition Device

Of equal importance is the problem of transferring image data from one station to another. Distance, transfer speed, complexity of interlinked nodes, connectivity, and cost divides the problem into many classifications, with just as many solutions. Finally, certain design techniques are essential to link together processors with different I/O throughputs. Proven examples of all these techniques will be presented, focusing on hardware aspects.

An analog to digital converter is an electronic device which efficiently samples and digitizes a measurable analog signal and feeds the result into a digital system (i.e., an image processor). As such it is used as an interface between the real world and a digital machine. Many examples of ADC in imaging systems are easily found. In figure 1 is a list of common digitizing devices in radiology, including cameras, video level outputs, laser scanners and computed radiographs. Figures 2 to 5 graphically illustrate the role of ADC in each of these devices.

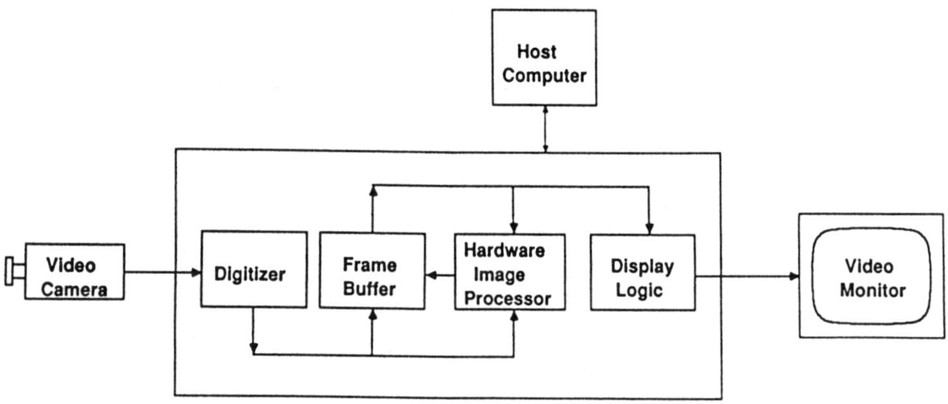

Figure 2. Video camera interface to an image processing system

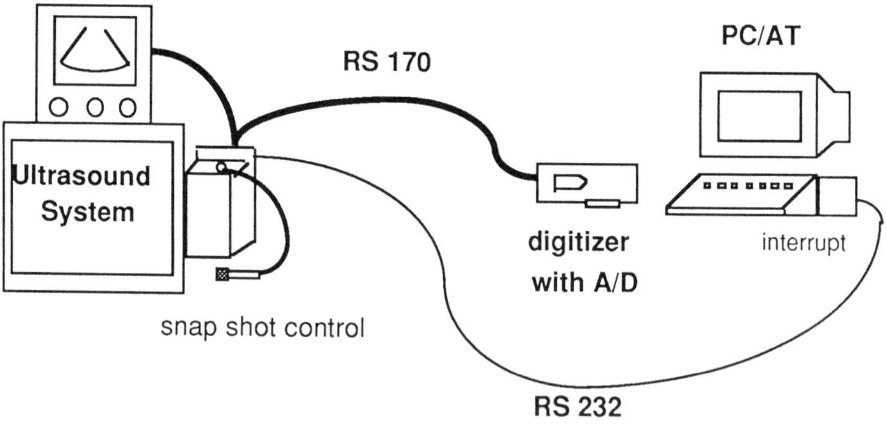

Figure 3. Ultrasound system intercepts images from video output of TV monitor

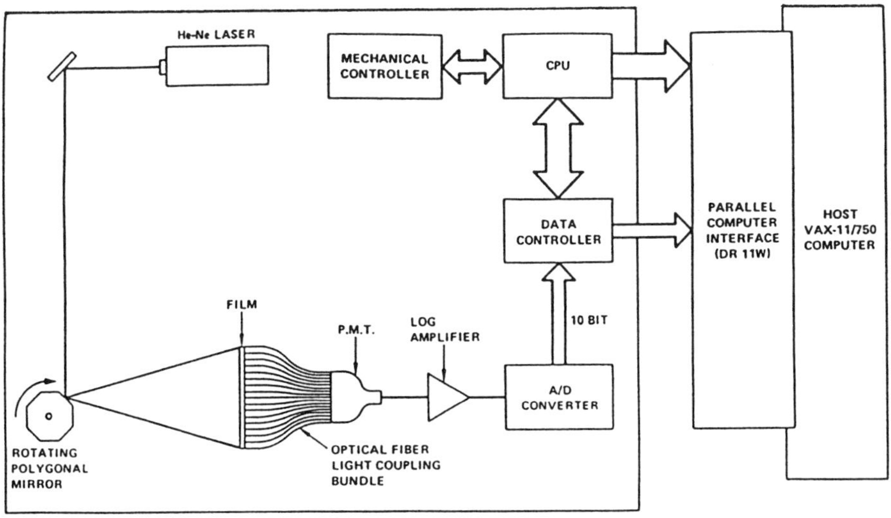

Figure 4. ADC incorporated into a laser scanner design

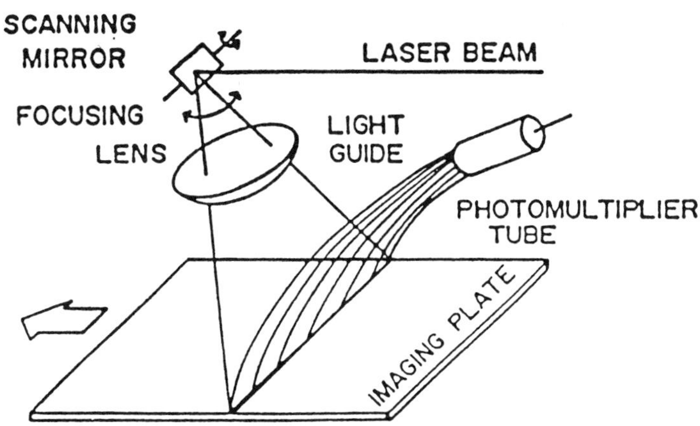

Figure 5. Scanner section of a computed radiograph system. Photomultiplier output goes to an ADC device

There are two major types of ADC as listed in table 1, differing in design approaches and application needs. The flash ADC is the faster of the two with a conversion time of about 100 ns, but as a trade-off provides only 8 bits of accuracy. A diagram of a flash ADC is shown in figure 6. A voltage comparison is made at every resistive node which increments by one-eighth the supply voltage. In a single comparison and encoding cycle, the output is generated. One can easily visualize from this three bit example how quickly complexity multiplies as output bits increase (the number of resistive dividers and comparators go up as a power of 2).

Table 1

Major types of ADC

	conversion time	# of bits of accuracy
Flash	100 nsec	8
Successive approximation	1 usec	10-16
Ramp	ms - sec	10 - 18

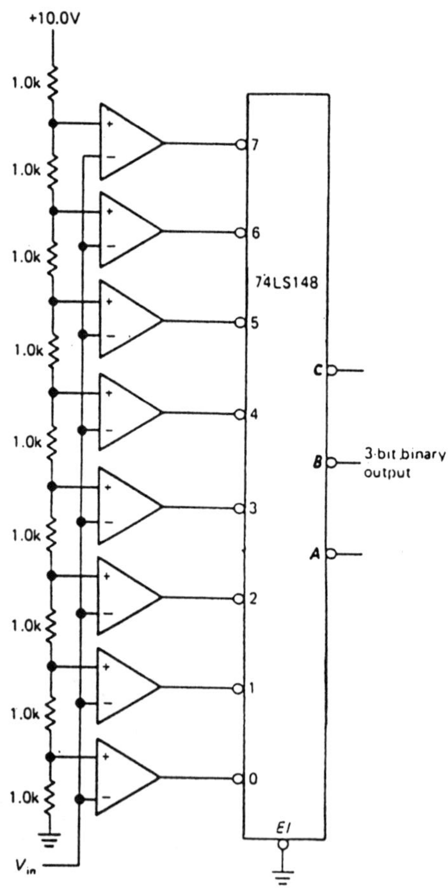

Figure 6. Parallel-encoded flash type ADC

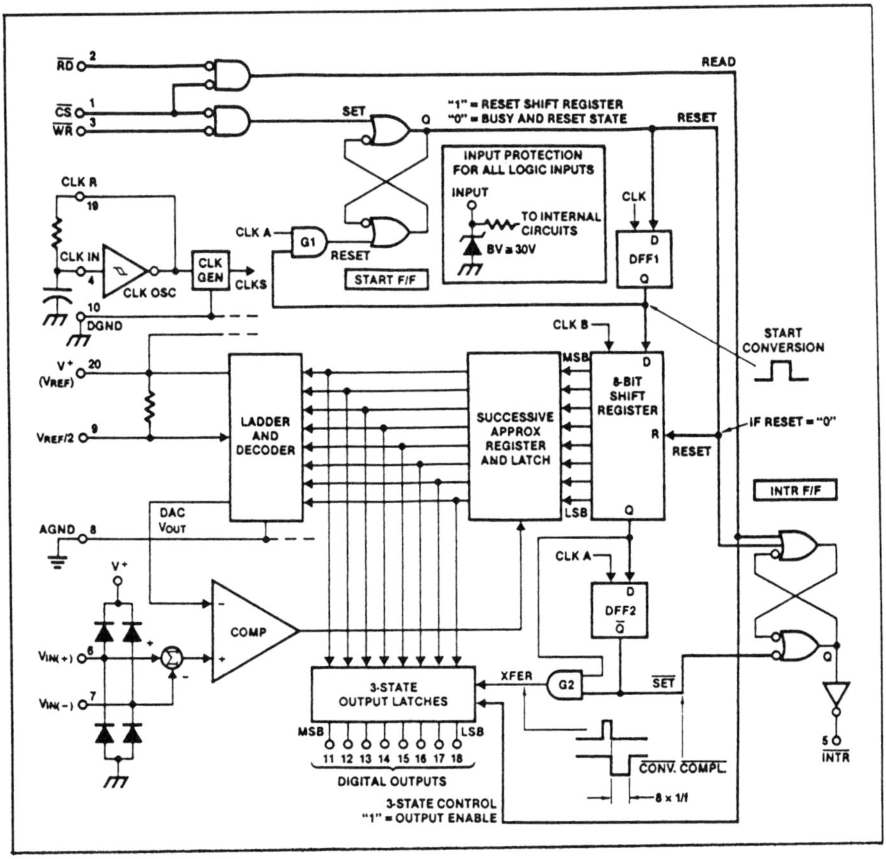

Figure 7. Successive approximation type ADC

The successive approximation type is about one order of magnitude slower in response but can serve applications which require up to 12 bits of accuracy. One such device is illustrated in figure 7. In the beginning of operation a "1" bit is shifted into the 8 bit register from the MSB (most significant bit) side. The ladder and decoder section produces an analog equivalent output which is compared to the input level being sampled. Depending on the result of the comparison, the "1" bit is either retained or reset to "0" at the MSB. Then "1" is shifted into the next most significant bit and the process continues. Finally, after eight cycles of comparison the digitized result is stored in the successive approximation register. As one can deduce, 12 bits of accuracy will require 12 cycles of comparison, which explains the one order of magnitude increase in conversion time. Several new ADC products are listed in table 2. From the rule of thumb just given, one should be able to guess what type of design Sipex used for its product.

Table 2

company	*Part*	*bits of accuracy*	*time*	*A/D type*
IDT	IDT75C58	8	33 ns	flash
Datel	ADC-511MC	12	1 us	flash (two passes)
Precision	ADC-908	8	6 us	successive appr.
Monolithic	ADC-9012	12	12 us	successive appr.
Sipex	SP674A	12	12 us	not given
prema	ADC 5601	14	10 ms	multiple ramp
		25	20 s	

Communication Interface Standards

One essential element of a PACS (Picture Archival and Communication System) is the ability to perform image transfers between image stations within a network. Data transfer speed, distance covered and serial vs parallel data path are the key factors in determining hardware requirements for a particular data link. Various conventions have been established. Their driving and receiving stages are now handbook items (McNamara, 1988). A summary of the most common conventions is listed in table 3.

Table 3

Communication Interface Standards

TTL	board to board	<10Mb/s	5V	74AS240 74AS14
Coaxial	video, broadband	100kb/s-20Mb/s	5V	8T23 8T24
RS 232	terminal, modem	110-19,200b/s	12V	ICL232
RS 422	parallel, <4000 ft	<10Mb/s	5V	26LS31 26LS32
Current loop	parallel, <2000 ft	<10Mb/s	5V	74AS38 6N136
Ethernet	serial, baseband	1Mb/s		

The TTL interface is capable of handling board to board communications for frequencies up to 10 Mbits/sec. Ribbon cables serve as the connecting medium. The simplicity of this design is being exploited for proprietary imaging buses by various PC (personal computer) based imaging companies. For longer distances (>4 inch) or faster signals, it may be necessary to implement termination resistors as shown in figure 8. The coaxial cable is generally used for carrying video or broadband signals at a rate ranging from 100Kbits/sec to 20 Mbits/sec. A 50 Ohm characteristic impedance is typical. A stage of translation to and from TTL is shown in figure 9.

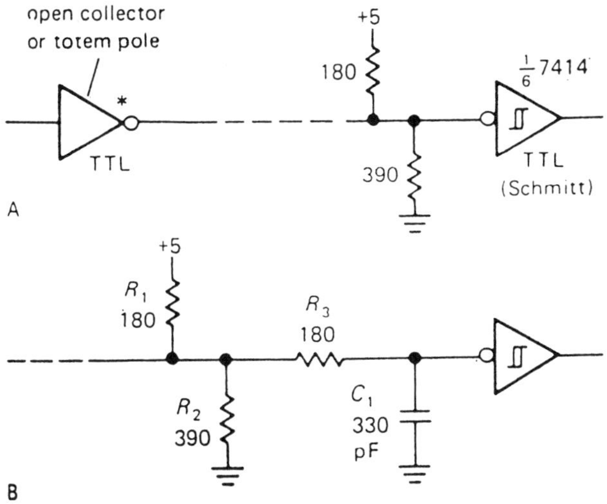

Figure 8. Termination resistors used in TTL interface

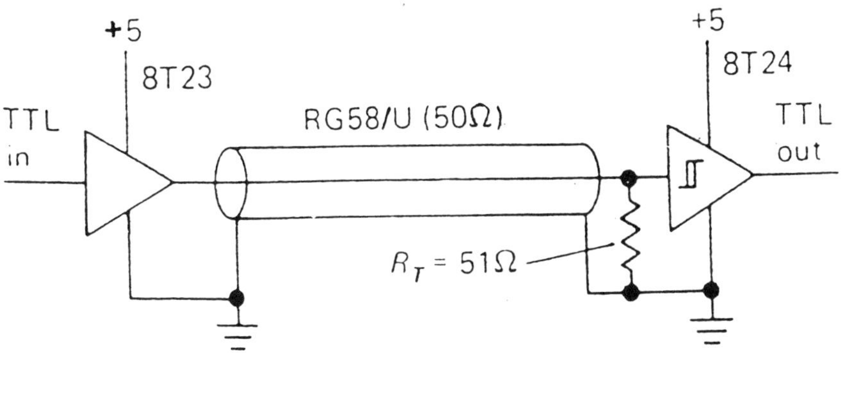

Figure 9. Coaxial interface circuitry

The most prevalent type of interface standard seen around the office environment is the RS232, entailing 12 voltage drivers and receivers and thin cables. There are chips available now, such as the ICL232, which generate positive and negative 12 volts internally from a single 5 volt power supply for easy interface to RS232 from a TTL logic circuit. The RS232 convention has 25 pins in the complete set. On IBM PC's, one sees a stripped down 9 pin subset implemented. More commonly, a still simpler 3 pin subset is seen on almost all computer terminal connections, including only the "transmit", "receive", and "ground" lines. This convention was originally created for linking terminals (DTE) to modems (DCE) in a communication system as configured in figure 10. As such the signal names clearly identify the function and directionality of each pin; pin numbers on two ends of the connection simply match. Confusion could arise, however, in the case of terminal to VAX connection, in which the computer as well as the terminal is configured as a "DTE" port.

In such cases, it is necessary to swap the pins 2 and 3 as the wires go from terminal to computer. The RS422 (shown in figure 11) is the interface of choice when one has many data lines (say 32 bits from 68020) going long distances (say 1000 feet). The driver and receiver chips work off a single 5 volt power supply. However, twisted pair wires are required, and in case of noisy environment, ground shield around them recommended. These requirements on top of parallel data path easily make the cable cost an important issue. Two dollars per foot is a conservative estimate. A performance guideline for RS422 running over 24 gauge twisted pair cables is given in figure 12. The wide range of speed and distance spells out the attractiveness of this convention as an interface solution.

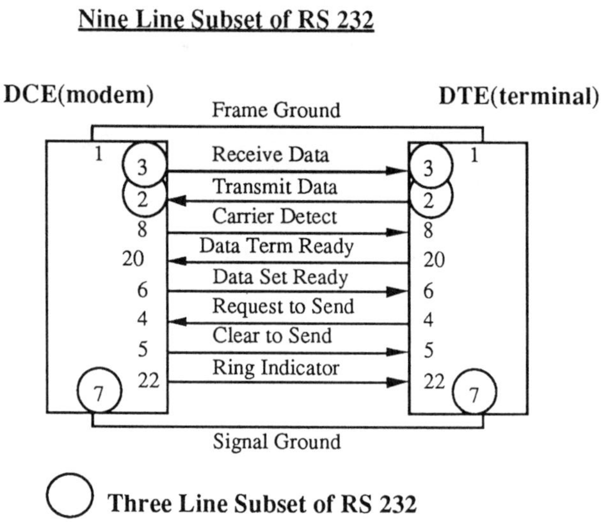

Figure 10. RS232 signals

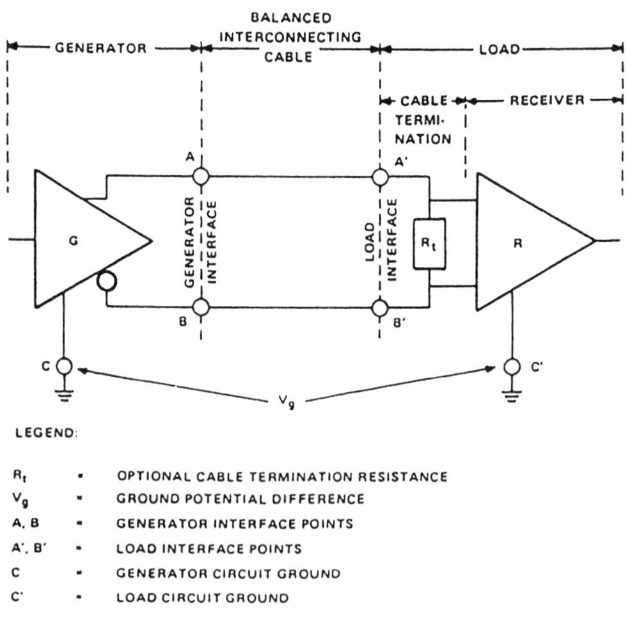

Figure 11. RS422 interface circuitry

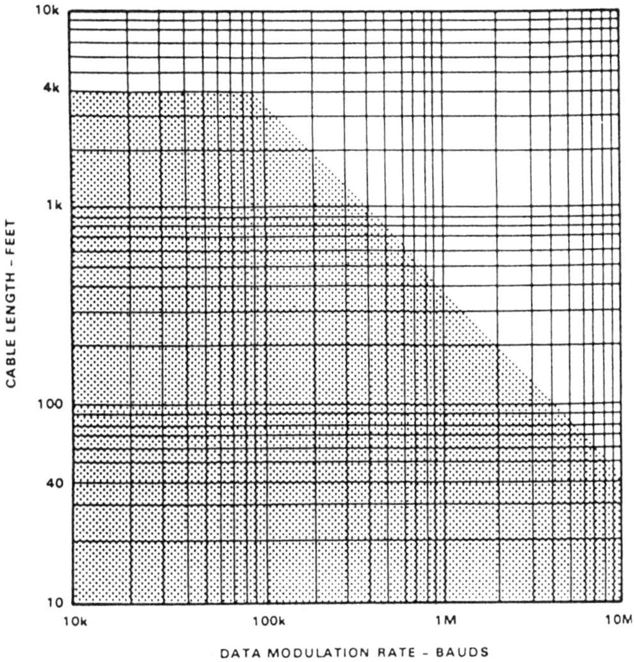

Figure 12. Performance specification for RS422

Current loop is another way for long distance, high speed datalink which uses twisted pair cables. The gruesome looking driver and receiver circuitry and limited lifetime of optical coupler chips (shown in figure 13) makes current loop a less friendly standard for designers. However, it does offer a better protection against ground loop problems due to optical isolation between driving and receiving voltage levels. Finally, coming on as a household name is the Ethernet standard, offering 10 Mbit/sec in specification, but is reportedly good for up to only 1 Mbit/sec in practice. It is a proven technique for connecting large numbers of computer systems for medium range transmission speed. The investment in both hardware and software to implement the Ethernet connection is in a much higher category than any of others mentioned, and the complexity of its protocol makes the design task of this interface standard an item for the corporate level developers.

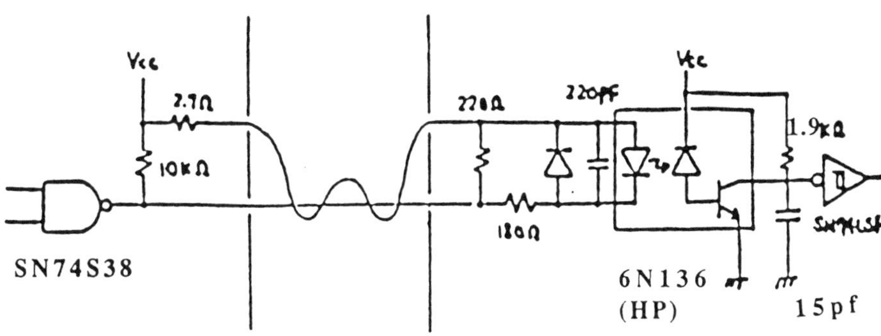

Figure 13. Current loop interface circuitry

Interprocessor Interface Designs

Once the electronic implementation is over, timing and protocol become the main concern. How does the system designer synchronize two systems with drastically different I/O throughputs? Two methods, both simple in concept, work very well. The ping-pong buffers illustrated in figure 14 swap between two buffers

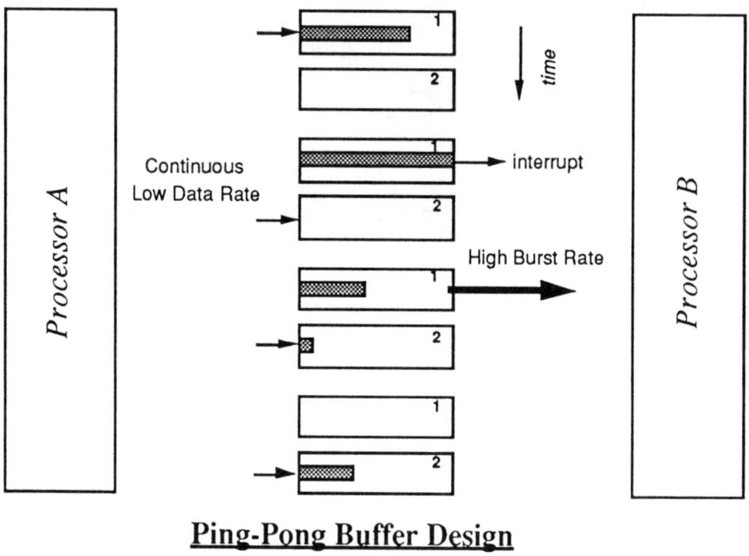

Figure 14. Ping-pong buffer design for interprocessor interface

for sending a data stream to one processor and receiving from another, so that the two interface processes become independent and each runs at the rate suitable to the processor involved. In the second scheme, illustrated in figure 15, a FIFO buffer does the same job by having its input and output ports functioning independently. An interrupt is send to the faster processor by a "half-full" flag so that processor B has enough time to start clearing the FIFO before it overflows. A functional block diagram of a FIFO (shown in figure 16) illustrates how it handles input and output data with separate pointers, each of which can increment by its own clock.

When one is shopping for a packaged solution for interprocessor communication, a very useful product not to be overlooked is the DR11-W, originated by DEC, but which now stands on its own backed by multiple vendor sources. Both processor-to-processor and process-to-device link can be implemented as shown in figures 17 and 18. Vendors of these boards generally provide software which, with slight modification, can be integrated into computers ranging from the VAX and VME systems, to the PC/AT.

The FCR to VAX Interface

An good example of applications of the many interface techniques mentioned is the FCR (Philips Computed Radiography) to VAX interface project, which aimed at acquiring digital images generated in the FCR for archival and display in CRIS (Clinical Radiological Imaging System), which is a VAX based mini-PACS module at UCLA. The diagram in figure 19 shows the data path for image data and electronic interfaces used. Within the FCR, module to module communication is based on current loop design. Once intercepted, data are transferred as RS422 signals. The ping-pong buffer circuitry temporarily stores one section of an image, and interrupts the VAX for data acquisition via the DR11-W board. The interface module has been released for clinical operation for over two years.

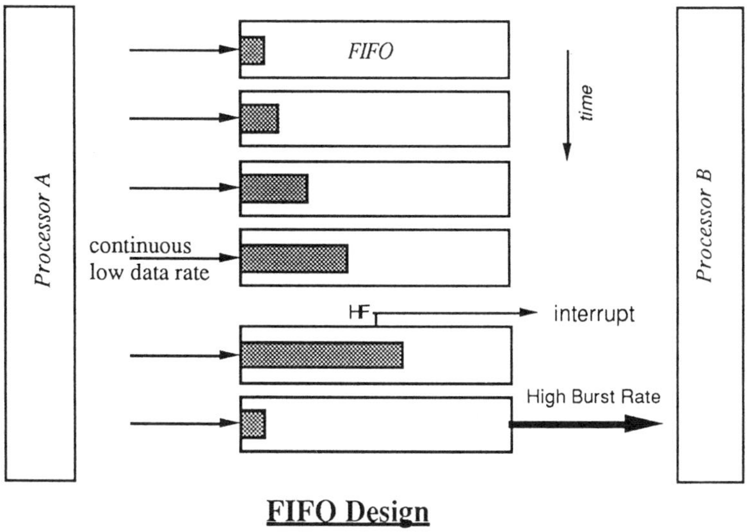

Figure 15. Alternative design using FIFO

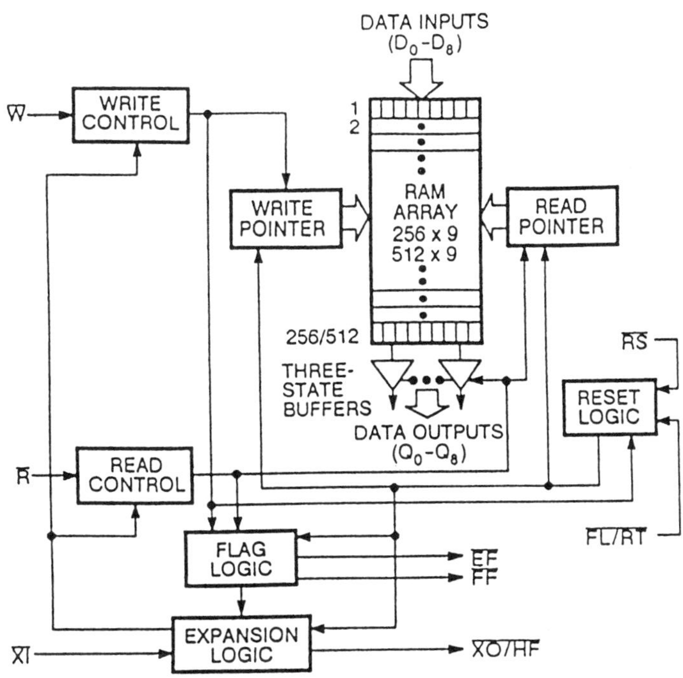

Figure 16. Block diagram of a FIFO component

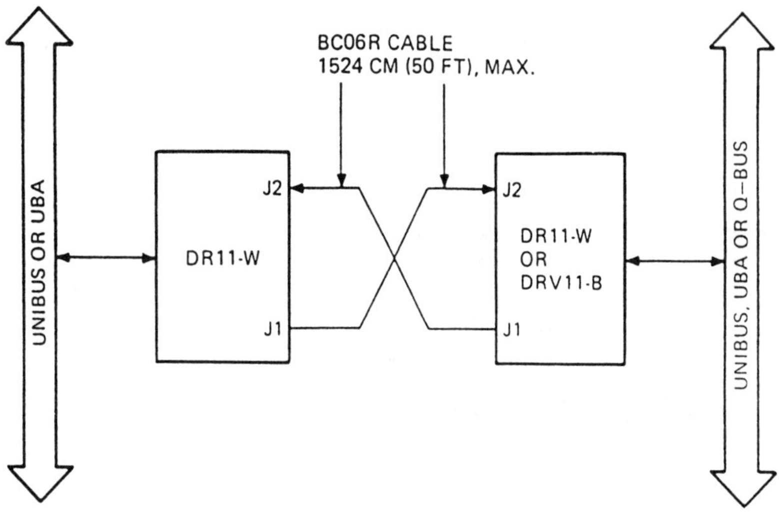

Figure 17. Interprocessor link block Diagram

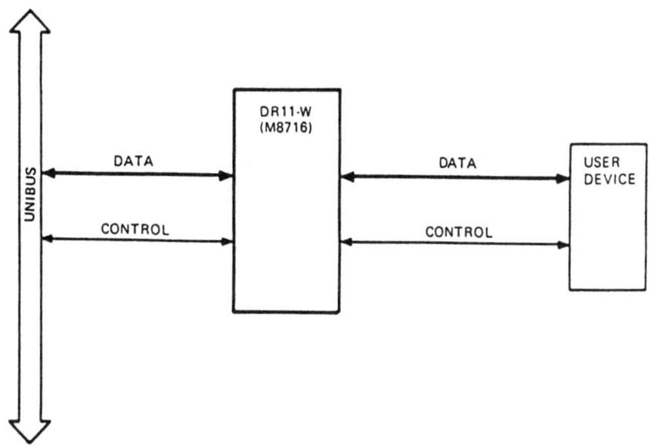

Figure 18. Simplified block diagram of DR11-W used as interface for UNIBUS and user device

CONCLUSION

We have presented some basic concepts in computer interface. Detail circuit diagrams and specification charts have been provided for convenient references.

REFERENCES

John McNamara, Technical Aspects of Data Communication. 1988, Digital Equipment Corporation.

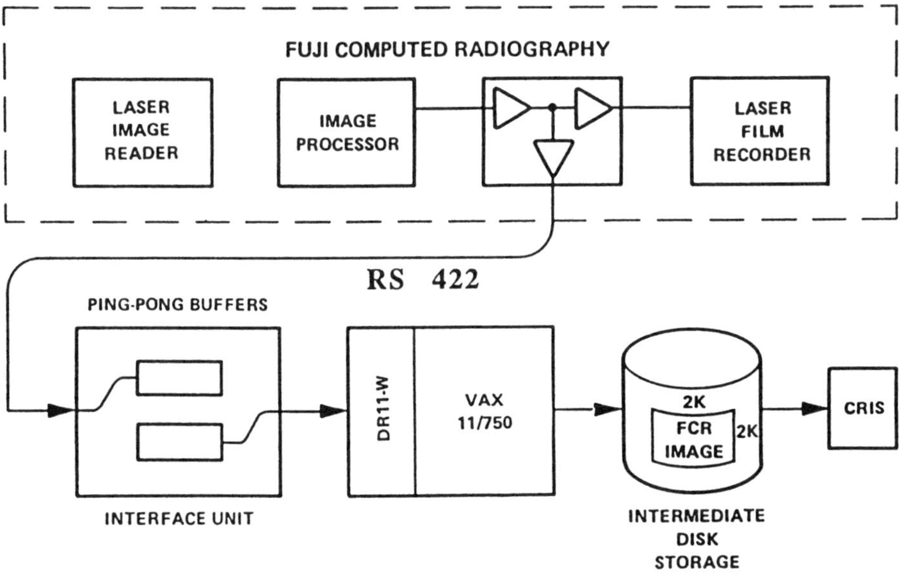

Figure 19. The FCR to VAX interface

OPERATING SYSTEMS

Ira J. Kalet, Ph.D.*
Radiation Oncology Department RC-08
University of Washington
Seattle, Washington 98195

Abstract

Users of computer software take for granted that any computer they might use will have available for it an operating system, a set of programs and program fragments that handle the tasks that any computer user needs done. These tasks include control of access to the system's resources, managing non-volatile data in files and executing basic user commands. The typical contents of an operating system, the kernel, device drivers, the file system and the utility programs, are described here. We discuss the general principles that operating systems follow in order to solve the problems of data management, resource allocation, and multi-tasking. We also include some discussion of the issue of program and operating system portability. While this is not intended to be a catalog or "buyer's guide" to commercially available operating system software, we provide some guidelines that may be used in analysing computer system needs.

1 Introduction

An operating system is a collection of programs or program fragments, available for general use. The programs or program fragments provide certain generic services to a computer user, not specific application functions. The actual services or functions of an operating system have evolved over the last two decades, so that no simple definition is possible. Operating systems began as simple loader programs that allowed "batch" jobs to run unattended in sequence from a punched card reader to elaborate systems by which many users share a collection of computer resources as if each had sole use [1]. The range of software that one usually thinks of as "the operating system" includes the lowest level of programs that transfer bits and bytes to and from devices like disk drives, and the highest level of complex data management programs. In between are programs for creating simple data files and file directories, programs for allocating memory and central processor resources among multiple tasks (programs) waiting to run, command processors that read input from a user, display screen manager programs, language translators (e.g., compilers for Pascal, C, other languages).

The following is a description of the more central components of the software we usually consider part of an operating system. This includes the kernel (or monitor), the file system, command interpreters, and some of the generic utilities that one expects in any system. We do not describe language compilers, network software or database systems, as they are described elsewhere [2,3,4].

*This work was partially supported by NIH grant no. LM04174 from the National Library of Medicine and contract no. CM97282 from the National Cancer Institute

1.1 Why have an operating system?

There are some obvious reasons why people invented operating systems:

- So that multi-tasking would be possible. With an operating system kernel, it is possible to have the computer operate on more than one task at a time, with each task getting some resources in turn (the alternative is that each task has total control of the computer until it completes).
- So a person wanting to use the computer does not have to write his own version of the kernel, etc., when a generic system will serve for all users.
- So computer centers could operate time-sharing computer systems without operators, or with minimal operations staff. The user can be more in control of his/her work and use of the system, without being an operator or systems programmer.

Any user of a computer system must provide the computer with some commands (usually in the form of a program) that define the job to be done. Not long ago, people who used computers for scientific computing would learn a programming language (usually FORTRAN) and create a deck of Hollerith punched cards containing the program statements and the data. The program and data represented input information, and the program produced some output either on punched cards or on a line printer. Even in the early days, much of the work of basic input and output and memory allocation was already done, and available in a standard form so that the programmer did not have to be concerned with these details.

When all computers were large and expensive, a typical installation had a computer center, and a staff including one or more operators. Sometimes a programmer could have the machine to himself, but more often, programmers submitted decks of punched cards at a job entry window, and an operator would feed them into the card reader, running each job by pressing the right combination of buttons or switches at the computer operator's console. The first operating systems were just small programs that did this automatically so that the operator did not have to do anything except keep the card reader hopper full of cards.

Later operating systems provided time-sharing in which several programs (jobs) could run at once [5,6]. Although the computer could only execute one instruction stream at any time, it would allocate a small amout of time for each job in turn, so that it appeared as if all jobs were running simultaneously. When it became possible to provide input from a teletype or CRT terminal keyboard and see rapidly displayed output on a CRT or teletype printer, each user could approach the computer as if it were a "virtual machine", where he/she had the machine to him/herself even though actually the user was only getting a small slice of the machine's running time.

Two factors have contributed to a trend towards a user having a computer to himself again. One is the availability of small cheap but powerful personal workstation computers, and the other is the fact that eventually many time-sharing systems became overloaded. However, the need for time and memory management is still present even on a personal workstation, and the same concepts apply as for larger systems [7,8]. This is particularly important as we move towards workstation networks and distributed computing.

1.2 Early and modern operating systems

In this article we refer to several commercial and proprietary operating systems and other computer equipment and software. We frequently use examples of Digital Equipment Corporation products, primarily because of the widespread use of DEC systems in medical physics applications, and because the examples illustrate well the concepts we describe. The following are trademarks of Digital Equipment Corporation: DEC, RSX, RSX-11, VMS, VAX, PDP, PDP-7, PDP-11, RT-11, DCL, EDT, and Ultrix. Other trademarks (and their owners) mentioned include: UNIX (AT&T), OS/360, JCL, and OS/2 (IBM), CP/M (Digital Research, Inc.), MS-DOS (Microsoft).

Some examples of early and modern operating systems include:

- OS/360 - lessons in software engineering
- MULTICS - an early time-sharing system
- RSX-11 - real time multitasking
- UNIX - small is beautiful
- VMS - a return to "all things for all people"
- CP/M,MS-DOS,OS/2 - the micro world

OS/360 was the operating system that was developed for the IBM 360 series computers. In its initial version OS/360 had over a million lines of assembly language code, and in later versions (by 1970) had grown to several times that size. Fred Brooks, who was the project manager for the IBM System 360 hardware and operating system development, has written a series of entertaining and provocative essays on software engineering [9] that are highly recommended for anyone contemplating a large software project.

One of the first time-sharing operating systems, MULTICS was implemented for Honeywell mainframe computers around 1970. It was a joint development by MIT and Bell Laboratories. Like OS/360 it was a large system, representing (by 1975) more than 20 million lines of assembly language code.

The Digital Equipment RSX operating system started out as a small system oriented to real time applications, but also became popular as a time-sharing (multiprogramming) system. RSX had limitations that made it not optimal for time-sharing, and it is also limited by the PDP-11 architecture, with its 16 bit address space.

The UNIX time-sharing system [10] was developed by programmers at Bell Laboratories who were not getting much response from their overloaded MULTICS time-sharing system, and thought that they could use an old PDP-7 that had been in storage. UNIX originally was written as a simple terse system that would provide good support for programming, particularly programming experiments. The name UNIX is a pun on MULTICS.

When the Digital Equipment PDP-11 computer came out, UNIX was adapted to it, and the PDP-11 implementation pretty quickly became a de facto standard. UNIX provided many features that RSX lacked, most notably a hierarchical file system and a rich collection of software tools. Later UNIX was adapted to the Digital Equipment VAX computer. It has grown and been adapted in various forms to many other computer systems, including everything from small desktop workstations to the largest supercomputers. It is particularly important because of this ubiquity, and today represents the only serious candidate for formulation of an industry standard for operating system specifications.

When Digital Equipment Corp. developed the VAX 32 bit computer, the DEC engineers also undertook development of a new operating system for it, that would (they hoped) incorporate the best features of earlier DEC operating systems, while adding the new capabilities of virtual memory (see Section 5.3 below) and a new system interconnect to get higher rates of input and output. This operating system, VMS, has become the DEC standard proprietary operating system. It supports time-sharing, real time data acquisition and control, batch processing, interactive processing, large scale systems and small systems. While it is very efficient in some ways, it is cumbersome and limited in others. There are significant trade-offs in trying to have one operating system that does everything. This is not peculiar to VMS, but is also a property of the present implementations of UNIX. While both started out small, they are now huge and difficult to manage.

Microcomputer operating systems such as CP/M, MS-DOS, and others are small systems for single user, single task applications. Some people argue that these are not really even operating systems, but simply program loaders. However, for the user of a small system they provide the basic functions you need, and are essential to making the personal computer a useful tool, rather than just a hobby or a toy. It is a further indication of the influence of early DEC operating systems that the file name syntax and command syntax of CP/M is similar to DEC systems such as RT-11.

2 Device Drivers

Some data and programs are still stored on punched cards, but this is now very unusual, and most familiar computer systems provide disk drives that read and write on removable disks (usually flexible or so-called "floppy" disks), cartridge tapes and other magnetic media for storage of data and programs away from the computer itself. In order to use a program in a computer, the program must be copied from the disk to the computer's memory, and then the program will typically send data to the user's display screen, and read data from the keyboard. The program may also read and write data from/to the disk drive.

Most computer systems also include disk drives with the disk(s) permanently mounted in the computer system box. These may also be used for storage of programs and data. They are faster and have larger capacity than most removable media.

For the computer to transfer characters to or from a device like a disk drive to the computer memory requires a standard set of very low level instructions in the computer's machine language. Because every program that reads input from the disk (or prints on a printer or any other device) uses the same instructions for that device, these were written once and them made available to any other program that needs them. It would be a waste of time for each programmer to re-invent the instruction sequence for input and output (I/O).

These sets of instructions for transferring data to and from a particular device are called "device drivers". They are program fragments that transfer user (programmer) supplied data between the computer's short-term (working) memory and external devices.

Examples of devices for which typical operating systems may provide device drivers include:

- Punched cards
- Printers
- Magnetic tape
- Magnetic disks
- Paper tape
- Analog input/output circuits
- Digital relays or control circuits
- Video displays
- Keyboards

A device driver is necessary for several reasons. First, the detailed way in which each device connects to the computer system is different from device to device. Some devices look to the computer like extra memory areas. Others have device registers that the device driver will read and write; the device driver program may set or clear bits and test bits to see what state the device is in. Instead of having to include all these specialized details in each program, the device driver provides a uniform standard interface or way in which programs get data into and out of a device. This interface is available to other operating system components as well as user programs so it makes the rest of the operating system easier to write as well. Then programs need not include any detail regarding which device they transfer data to or from. Reading from and writing to all devices will look about the same if they are generally the same kind of device. For example, a program can read data from a file, and write results to a file, and need not have any specific information about which disk drive it is on.

Most operating systems have not only device drivers for standard devices such as terminals, printers and disk drives, but also high level language support. This means that the language itself provides READ and WRITE statements so that a programmer need not worry about the details of the device driver program. In larger systems, these input and output operations can be device-independent, so that it doesn't matter if the input stream of data comes from a terminal keyboard, a disk file, a tape file or even a network connection to a remote computer. The same applies to the output stream. The READ and WRITE statements can refer to "standard input" and "standard output", or to FORTRAN logical unit numbers, whose actual assignment to data sources and sinks is determined at the time a program runs, and may be different for different runs, without changing the program.

Second, some devices are capable of transferring data on their own, once they are started. Then there must be some way to set up the device to transfer information, start the transfer, and (optionally) proceed with other tasks until the transfer of data is complete. When the transfer of data is complete there must be some way to resume the processing of the data.

Device drivers usually have program fragments that run when a device produces an interrupt, a signal for the computer central processor (CPU) to stop its current activity and instead start executing a different set of instructions. A device driver therefore can be very complex. Typically device drivers are written in the assembly language of the computer system on which they are to be used, and naturally they can be the most hardware dependent part of an operating system.

Finally, it is necessary to insure that at most one program has made a request for input or output. The device driver software must prevent more than one process from setting or clearing bits in the device register because the register will soon contain inconsistent data if this happens, and the system will no longer be able to function.

In order to get data into or out of some devices (for example, analog input converters) it may be necessary to write in your program a subroutine call or procedure call to use the device driver program directly. Each operating system has a method for doing this, and programs that utilize these techniques then become operating system and hardware dependent. On the other hand, it may be unavoidable, as you may need to have a program that works for a specific system anyway.

Writing your own device driver is a very demanding and difficult project. It should not be considered lightly. If there is an error in a user program in a time-sharing system, the program may terminate abnormally but the system as a whole will continue running. If on the other hand there is an error in a device driver, the whole operating system may stop working, or massive amounts of data corruption could occur. Therefore it is a task best left to experienced system programmers. Buy a device driver if you need one, rather than attempting to write your own.

3 File systems

A file system is a collection of programs and program fragments that organize data on long term storage devices (e.g., disk, tape) in the form of files. A file has a name, whose syntax depends on the particular operating system that creates it. The data contained in the file may be simply text information, binary numeric data, other (binary) encoded data, or stored computer programs in binary form. The data belonging to a file occupies a defined area on the storage device; this area can be located by looking up the file name in a special file called a directory. In general, most commands and programs that operate on files use file names and a directory to find or store data. A given device (e.g., a disk or tape) may have only one directory that lists all the files, or a hierarchical arrangement of directories, or no directory at all (this is common for magnetic tape, where keeping a directory updated would be difficult). If there is no directory at all, files are generally read sequentially until the program finds the file it is looking for. Most operating systems come with utilities or commands that provide listings of the contents of a directory, and provide for creation, copy, delete, renaming and other operations that may modify both the file and its directory.

In a system with a large disk, or disks, you might accumulate many files, perhaps hundreds. Even if the disk had a directory it would be difficult to keep track of all the files. Sophisticated operating systems provide hierarchical directories to help organize your files into subgroups.

3.1 Text files

There are several ways that text is represented in files and programs, the most common being the American Standard Code for Information Interchange (ASCII). The ASCII code assigns one byte (eight binary bits) to each character. Most computer files are organized into bytes, but some are word oriented, and the word size may then be different from a byte, the most common being 16 bits (two bytes). Generally text files are divided up into "lines", sometimes also called records, although the latter is not a useful term, because it refers to other things as well. In the text file, the end of a line is marked somehow, so that a program can read a line at a time as well as a character at a time.

An important characteristic of text files is that they are easily manipulated by many operating system utilities, and can be easily read and written by programs in almost any language. Text files can be printed directly on a printer, displayed (typed) on a terminal screen, edited with a text editor (regardless of which editor program you use, any other editor program can edit the same file). Text files can be mailed to other users in a local or wide area network, without any special processing. Programs that read and write text files generally do so in an operating system independent way, so that the program and its data as well can be moved to another operating system or computer with no change. This is particularly important with respect to data files, since conversion of data file formats is an otherwise difficult, time consuming and expensive task.

3.2 Other data files

Files that are not text files represent data in some other (binary) form than streams of characters. Such binary data files cannot in general be read, printed, edited, etc., without the use of specialized programs that depend on knowledge of the content of particular files. Thus these files may not be as portable as text files. They nevertheless are important for several reasons. First, the data being stored (for example, arrays representing dots in an image) may not have an obvious representation as text. Second, even if there were a simple text representation, in general, the data are manipulated in their binary form, not the text form, so there is a lot of conversion overhead. Finally, to represent integers or decimal numbers by characters takes more space, so that for large data sets the binary representation saves storage space.

A data file may be viewed at the lowest level as simply a sequence of bytes. This is the model of a file that is provided by UNIX. It is then up to each program or programming language to provide more abstract layers or more structured layers. Other operating systems provide file organization at the lowest level. In the VMS operating system, for example, a file can be a sequence of bytes, or a sequence of records (collections of bytes). The records may be fixed length (all the same) or variable length, and may be indexed by some part of the content of each record. All these mechanisms are provided by the file system programs. An advantage of the simpler organization is its inherent flexibility. An advantage of the more complex system is that if its capabilities are needed, they are already provided with high efficiency.

Indexed files are useful when you need to look up records in a large file, and you know the value of the key(s) that make up the index. This lookup is much faster than searching the file sequentially for the right entry when the index is much smaller than the data file itself. If the key field takes up much of the data record, it is probably not worth using indexed files since the time saved will not be significant.

Some programming languages provide a more abstract view of a file. In these programming languages (Pascal, for example) it is simply a kind of data structure that one might use in a program, like an array or a scalar variable or other program variables. The file (in this view) is a sequence of components, each of which has the same type or structure (each may be a Pascal record variable type, or a scalar type, or an array). The differences between a file and an array are:

- An array has fixed (predefined) length while a file can grow indefinitely large, and its size is not predefined.
- An array component or element can be accessed or referenced by specifying an index, while a file can only be read or written sequentially.

The operating system and the language implementation then provide some mechanism for connecting a file as a program variable to an external entity, a disk file with an actual name, for example.

Although this abstract view has its limitations, it has the advantage that reading and writing the file is simple and efficient. The organization of the data is more easily understood. This contrasts with the old-fashioned practice of assigning data elements to files in an ad hoc way (e.g., word number 54 is an integer containing the number of slices, word numbers 76-77 are floating and represent the pixel size, bytes 235-258 are the patient's name, etc.).

3.3 Directories

A directory is a file containing the names, locations and other information about a collection of other files. It also has a name, and the way in which the user refers to a directory is also operating system dependent. A single disk or tape volume (a single piece of media) may have just one directory, many directories, or no directories. The importance of directories is that they provide a simple way to get access to a file, find out what files are available, and create a new file without worrying about where on the disk or tape it needs to be created.

In systems which have multiple directories on a disk, the directories are organized into a top level directory (sometimes called the Master File Directory or the root directory) and subdirectories to multiple levels. This is useful when you have so many files that to enter their names in one directory would create an unwieldy list. A directory may have files listed in it, and it may also have subdirectories listed in it. So you can organize your files into groups of related programs and data, just as you might have a file cabinet organized into drawers and sections. Naming of files may vary considerably from one system to another even when both systems provide similar arrangements (e.g., hierarchical directory structures). Nevertheless, the different naming conventions all express the same idea, namely to be able to start from the file name and a path name, without being concerned with tracks and sectors and other highly detailed device dependent stuff. Here are a few examples.

In most Digital Equipment proprietary operating systems, the naming of devices, directories and files is done by distinguishing among these three types of entities. A device name (a string of characters) includes a device type code (usually two letters, e.g., DU for certain types of disk drives), a controller designator (a single letter, A for the first, B for the second, etc.), and a unit number (one or more digits). The device name is followed by a colon. The directory path is enclosed in brackets and the directory names (top, sub, sub-sub, etc.) are separated by periods. The file name itself has parts, the name being a character string, then a period, then the extension or "file type", then (for some operating systems) a semicolon and a version number. Here is an example, for a VMS file:

```
DUA2:[RADONC.OFFICE.MEMOS]DEPT_MEETING.TEX;24
```

In the UNIX operating system, devices are made to appear as simply a special kind of file, as are directories, and the names are separated by "slash" characters. A name beginning with a slash refers to an entry in the root directory, and a name without a slash refers to a file (or directory) in the current (default) directory. So, if the disk drive above were the current mounted file system, the equivalent UNIX path name might be something like this:

```
/radonc/office/memos/dept_meeting.tex
```

The UNIX file system does not provide version numbers, but it does allow periods in file names. It is worth noting also that in the UNIX system file names are case-sensitive, so "memos", "Memos", "MEmoS", "MeMOs", and "memoS" are all different file or directory names. In many other systems all file names you type are converted to upper case so in the "memos" example above, the names would all refer to the same file. In this and other details, you will have to read the manuals describing the operating system you are using to learn the specifics of your system.

In the MS-DOS operating system your disks can be organized into hierarchical directories, and the path name convention is similar to UNIX, except that the "backslash" character is used in place of the "slash" character. Also MS-DOS file names are not case sensitive. On the other hand the disk drives are simply referred to by a letter and a colon, rather like a very simplified DEC VMS device name. Here is an example:

```
C:\radonc\office\memos\depmtg.tex
```

3.4 Space allocation

On a magnetic tape where data are (usually) written sequentially there is no special problem or process of space allocation on the tape. Data on a magnetic tape are usually written in blocks of some number of bytes with short gaps between. A tape may be organized into files simply by a program writing a tape mark, a special sequence of data that indicates end-of-file. There is a more complex schema for writing files on tapes, defined by ANSI standard X3.27-1978. This standard provides for blocks of information called volume labels and file labels. The information in the file labels includes a file name and other information that usually is in a directory, so by reading through the tape sequentially a program can create a kind of directory-like listing of the files on the tape. However, in general, programs can not delete or rewrite files, but can only add a new file at the end of the information area on the tape (after the last file).

On a disk drive, any physical area of the disk can be read or written, so the file system programs and program fragments provide a means of organizing the information area on the disk. Space is allocated on the disk for a table of allocated and free blocks, so the file system does not overwrite existing data or read garbage that is not part of a file.

On systems like MS-DOS, UNIX and VMS, the disk blocks (a block may be 512 bytes, or some other fixed size) are allocated in groups. Each group is contiguous, i.e., the blocks are adjacent, but a file may be made up of several non-adjacent groups. Therefore the disk can become fragmented, which means that in order to read data from a file, the disk drive read head has to move many times over the drive surface to get the data. This is less efficient than having the files be contiguous. On the other hand, it is more efficient in space utilization, since it may be that after creating and deleting a lot of files, there is no contiguous space large enough to create the next file. Most operating systems include one or more utility programs that can be used to relocate all the files on a disk so that they are contiguous.

3.5 File Protection

In a multi-user system, there may be many files, some belonging to the operating system, and many belonging to each user. It should be possible to insure that each user's files cannot be altered or deleted by other users, unless the user who owns that file explicitly allows it. This is done by providing as part of the file system a protection mask for each file. The protection mask says what kind of access (if any) each category of user may have to a file. In the UNIX and VMS operating systems users are assigned User ID Codes (UIC's) and the UIC defines groups as well as individuals. The file protection mask for each file then says whether the owner, a member of the same group, or any other user can read, write or delete the file. The file protection can be set or changed by the owner of a file. More complex schemes are also possible, but the idea is the same. Access to each file can be controlled by the owner of the file.

4 Command Processors

Command processors are programs that accept user input command lines and either perform the commands or arrange for execution of other programs that perform the commands. A command processor may be a simple program loader (CP/M, MS-DOS), or a sophisticated batch processor (JCL, RT-11 BATCH, CCL), or a "higher level" programming language (DCL, the UNIX shells). A batch command processor reads a file of commands and executes them all sequentially and automatically, so that a whole sequence of programs (jobs) can be run without operator intervention. Some command languages provide conditional jumps and repeat loops also, so that they are in effect higher level languages whose primitives are entire programs and operating system commands. This is useful because it enables you to define yet higher level commands consisting of sequences or combinations of programs. You can also redirect input and output from each program to/from files when the job runs, or run the same program repeatedly on a series of different files.

The first operating systems (ca. 1950) provided only the simplest commands, and were oriented to simply load programs, compile them if necessary, and run them in sequence. This style of computing is called batch processing. By 1960 there were time-sharing operating systems, in which several jobs could run concurrently. Although we think of time-sharing in the sense of a user sitting at a terminal interacting with his/her program and the command processor through the keyboard and display, early systems simply provided batch processing in which multiple jobs were active at once. This nevertheless required the solution of problems of allocating and sharing resources, job synchronization and interjob communication. The resulting sophisticated command languages provided powerful (if somewhat obscure) techniques for doing all these tasks.

Interactive command processors have evolved with the capabilities of CRT displays. Most notable is the recognition that there is an alternative to typing a command at a keyboard, with characters appearing on a screen. Instead some computer systems[1] provide a pointing device (called a mouse) that you use to move an indicator on the screen. Also on the screen are icons (symbols) for the commands and data that you would otherwise type. Instead of typing a command you move the indicator to the command icon and press a button to do that command. More sophisticated systems come with window managers that provide several parallel activities each with its own region of the screen (a window). The window style user interface has become so popular that there is now a standard (the X window system) for both the window functions and the way in which a program interacts with the window manager software [12].

5 Memory Management

Memory managers are program fragments that allocate, deallocate and protect portions of the computer's short term memory. There are several reasons for including memory management in an operating system. The first is to protect the operating system code from corruption by a user program that may have errors in it. Another is to provide sharing of memory resources among several tasks or programs that are all running on a multi-tasking system. A third is to provide a transparent way for a very large program to run in a very small memory (virtual memory).

In order to protect the operating system code and data from corruption by a user program, it must be possible to protect areas of memory from access by user programs. These areas are accessible only through "system calls" that execute part of the operating system code. That code then checks the requested operation and data to determine if they are a valid request, and then carries out the request if valid. In order for an operating system to provide this, there must be some hardware mechanism for determining when an instruction attempts to read or write a location in the memory that is not supposed to be accessible.

Several kinds of memory management may be found in modern operating systems. These are described in the following sections.

5.1 Absolute Addressing

A computer's main memory is addressed by assigning a number to each memory location. In an absolute addressing scheme, the number is the physical address of the corresponding word or byte. When a program is copied into the computer memory in order to run, the addresses in the program refer to physical memory locations and the program will only work correctly if located at the right place in memory. A relocatable program file contains relative addresses, but when the program is copied (loaded) into memory, the addresses must be changed (bound) to the physical addresses corresponding to the area the program was actually loaded. So, when the program runs it still refers to actual physical addresses. Examples of operating systems that use this form of addressing include MS-DOS and RT-11.

[1] Apple's Macintosh is a popular example, but is actually a descendant of the Alto computer, developed at XEROX Palo Alto Research Center [11]

5.2 Partitions

The dividing up of memory into smaller regions that are assigned to different tasks is called "partitioning". The regions are called partitions. This kind of system works well for a computer whose central processor can address only a small basic memory address range (the older 16 bit minicomputers could only address 65,536 bytes of main memory, though the same minicomputers could accomodate as much as 4 megabytes of physical memory). The computer includes memory mapping registers that make a block of program memory addresses translate to a range of physical memory addresses. When each task gets a turn to run, the mapping registers are written with the information to map that task's collection of blocks of memory. The task still could not be bigger than 65,536 bytes but it could use any collection of regions of memory that added up to that amount. Thus each of many tasks could occupy a different collection of regions of memory. Since each task can map only its own regions, tasks cannot interfere with other tasks. Examples inlcude RSX11/M and OS/2.

5.3 Virtual Memory

A computer system that has a large address word can in principle address a correspondingly large physical address space. For example a 32 bit address word can handle up to 4 Gigabytes (four billion bytes) of memory. On the other hand, until recently, memory was fairly expensive and few computer systems had more than a few megabytes of memory capacity. The lack of large amounts of actual memory was a serious problem, since some important programs were written to operate on very large data sets, including medical images, astronomical data, satellite photos, etc. Also, as the capability of compilers to handle large program codes increased, the programs themselves grew, so that programs of 50,000 lines of code to 500,000 lines of code were no longer unusual.

An early solution to the problem of large programs was to segment the data and overlay the programs. The data are handled in smaller units if possible. The programs are divided into subroutines. The subroutines are put into groups and each group is assigned to an overlay segment. There can be many groups assigned to the same segment, but only if no subroutine in one group calls a subroutine in another group in the same segment. Then those groups can each use the same segment without interacting, and each group gets overwritten when a subroutine is called in another group. The difficulty is that the programmer must do the work of dividing the program into subroutines, organizing the subroutines so that they fit the scheme. Some obscure problems can arise from misassigning a subroutine to an overlay segment.

The concept of virtual memory eliminates the need for the programmer to do these jobs. The program is written as if there really were gigabytes of memory. The addresses in the program are "virtual addresses". The computer hardware has address translation registers that translate from virtual addresses to physical addresses, using a memory page table. The data or program code might be in a physical memory location, or it might be on a disk (in a special file called a page file, or in some other file). The disk is used as an "extended memory", but not directly. If the required data or code is on the disk, a page or pages of the disk data are read into an area of physical memory, and then used there. Pages of physical memory that have data that was not needed for a while may be written back to the disk, and that physical memory is freed up for other data. The important fact about this system is that the computer system and the operating system software take care of this automatically and the person who writes the program need not think about it. Thus you can run a program (for example) that requires a 40 megabyte address range on a computer with only 6 megabytes of memory. Of course you must have a large page file on your disk to hold the pages of data and programs that are temporarily not in use. Examples of virtual memory systems are the Digital Equipment VAX (VMS and Ultrix operating systems) and the Motorola 68020 microprocessor and various UNIX implementations for it.

5.4 Security

In any multitasking (or multi-user) system it is important to be able to prevent one task from interfering with the program code or data of another task, as well as protecting data in files. A way in which this is done for active tasks is that memory locations are mapped so that a task simply cannot address any location it does not own, unless the operating system gives it explicit permission. In order to do this the system must have a protected mode and an unprotected mode. In the unprotected mode (kernel mode) the system can access any resources. In the protected (user) mode the task can only access its own data. In order for tasks to share data, there needs to be a mechanism for each task to gain exclusive access to the data so that no inconsistency can arise, and the programs operate in the correct order. One such mechanism is called a semaphore [13,14]. A semaphore is a special program variable on which there are only two operations, WAIT and SIGNAL. The WAIT operation (Dijkstra called this "P") waits until the variable value is greater than zero and then subtracts 1 from it. The SIGNAL operation (Dijkstra called this "V") adds one to it.

Let us say for example that there is only one printer, and several programs need to print something. Only one program at a time should be able to use the printer, otherwise the outputs of the two programs would get garbled together. To solve this problem, a printer semaphore would be initialized to 1. If no program were using the printer, a program that needed to write to the printer would call the WAIT procedure. Since the semaphore is already greater than zero, the program decrements the semaphore, and uses the printer. When done, the program executes SIGNAL, thus restoring its previous value. Now suppose the printer is in use. When the second program executes WAIT, the semaphore is set to 0, so the program waits. When the first program is done, it executes SIGNAL, thus incrementing the semaphore, and now the second program can decrement the semaphore and proceed. Thus a program can be assured of exclusive access to a resource by operating on the semaphore controlling that resource.

In a larger context, it is necessary to provide authentication of input from a terminal or card reader or other input device that requests jobs to be run. This prevents unauthorized individuals from using the computer system altogether. This kind of security concerns granting access to the system by issuing usernames and passwords so that only those people possessing valid usernames and passwords can run programs or type commands. Here again the system depends on the availability of different modes of operation and protection of files from unauthorized access as mentioned in Section 3.5 above.

6 Task Schedulers - Concurrent Programming

A task scheduler is a collection of programs and program fragments that arrange for execution of multiple tasks, either in turn by time clock (time-sharing) or in response to events (e.g., device drivers) [13]. The task scheduler arranges to save the context of the current task, decides what task will execute next, and restores the context of that task. If no task is able to run, the processor may simply wait for some event to make a task ready to run or may execute a do-nothing task at the lowest possible priority (for consistency of design).

In a multiprogramming or multitasking environment, each task or program runs in a context called a process. The process contains information about what shared resources the process has allocated and has access to, as well as the program code and data it does not share with other processes. Except where a program explicitly uses some mechanism for synchronizing, you can not assume that two programs or even two processes running the same program will proceed at any particular rate with regard to each other. So you can't, for example, assume that program A will finish a computation and complete its output before program B attempts to read the output from A and use it as its input. Mistakes in assuming that input occurs in a certain sequence with specified timing can lead to serious consequences [15].

The classic illustration of the problems of concurrent programming is the "dining philosophers" problem [16]. In a monastery there are five monks who are philosophers. The philosophers would be happy to spend their time deep in thought except that it is occasionally necessary to eat. Each philosopher spends his time alternately eating and thinking, with no fixed cycle time or amount of time spent on each activity, so they are in no way synchronized with each other. The dining room contains a circular table around which there are five bowls, five chopsticks, and in the center of the table a large bowl that is continually replenished with rice (the mechanism for that is beyond the scope of this article). This scenario is a metaphor for a multiprocessing computer system, in which the philosophers are tasks, and the chopsticks are resources such as shared memory, disks, printers, etc.

To eat rice requires two chopsticks. A philosopher may use only the chopsticks that are on either side of his bowl, and only one philosopher may use a chopstick at any time. This scenario illustrates three of the central concepts and problems of concurrent programming: mutual exclusion, deadlock, and lockout.

The fact that only one philosopher at a time should pick up a chopstick means that we need some mechanism by which a process (a philosopher) requests a resource, but that the resource never gets allocated to more than one requestor at a time. Mutual exclusion is the means by which only one process is allowed access to a resource at a time. If there were no such mechanism the operating system's data would soon become inconsistent because two programs modify the same data in an uncontrolled way. The computer system would then not produce reliable and trustworthy results.

It is possible that five philosophers come into the dining room together, and each picks up a chopstick with his left hand, then waits until a right hand chopstick is available. If the philosophers are stubborn, they will all starve, since no right hand chopstick can ever become available. This illustrates the phenomenon of deadlock.

Deadlock is the situation where two or more processes are blocked from executing because each has allocated a resource needed by the other(s) or each is waiting for one of the others to complete a computation. The wait will never end because it is circular. In a multiprocessing system it is necessary to insure that this can never happen with regard to allocation of system resources, and it is important for the mechanism to be available to programmers so their programs can avoid the situation with regard to data shared by the programs.

Two philosophers may conspire to starve another philosopher. Let us suppose that in order to avoid deadlock, the philosophers agree that no one will pick up one chopstick unless he can pick up both. Then philosophers 1 and 3 eat alternately. This means that philosopher 2 never gets to eat, since at any time, at least one of the two chopsticks on either side of him is in use. Lockout is the possibility that two or more processes will conspire (accidentally or intentionally) to prevent another process from ever acquiring access to the resource(s) it needs to proceed.

The philosophers' problem can be solved by using semaphores. Put a semaphore at the door, whose initial value is 4. Then when a philosopher wishes to enter the dining room, he executes WAIT, and when he leaves he executes SIGNAL. Then no more than four philosophers will be seated at the table, and there can be no deadlock. Then the rule requiring that both chopsticks are picked up together is not necessary, and lockout is prevented as well.

Proper use of such synchronization mechanisms is important in applications programs as well as in operating system design, particularly in real time control systems. For further discussion of how mutual exclusion can be implemented and used, see [13].

7 Utilities

Utilities are programs and program fragments that perform often-needed generic tasks, so that a user (programmer) need not write his/her own (e.g., file copying, sorting, program language compilers, text editors). While technically a utility program is not part of the operating system, most operating systems come with a small or large set of them, and these utilities are often identified with (or in some people's minds, characterize) the operating system. An example is the C language [17] compiler, which is strongly identified with the UNIX operating system. Much of UNIX is written in C and the compiler and support routines for C in the UNIX operating system are very complete and efficient. Another example is the EDT text editor for the VAX/VMS operating system.

Utilities perform generally useful functions (usually on files) such as sorting records in a file, searching a file for a data item, creating, editing and deleting files, listing contents of directories, on-line help, defragmentation of disks, copying files from disk to tape and back, for backup or transport. Although the operating system vendor usually provides some number of utilities often these kinds of programs are also available from third party vendors.

8 Standalone Systems

For applications of computers in which the computer is used to control some equipment or in which the computer manages data acquisition, it is not appropriate or even possible sometimes to run programs in the context of an operating system. It may be necessary to have the entire program self-contained and able to run with no disk drives or terminals or command processors etc. On the other hand, the application may still require multi-tasking, and standard device inputs and outputs. It would be useful to have an operating system kernel and development tools that can be used to build such a stand alone system without having to write all the generic code oneself. Such development toolkits exist for most popular microprocessors as well as for some larger systems. You use the toolkit software on a host computer (a single user or time-sharing computer), write and compile your program, and then you use a system building utility to create the loadable binary version of the final product. It may be programmed into ROM (read only memory) or it may be loadable from tape or disk or even a computer network.

9 Portability

Portability is the ability to make a program written for one system run on another system. There are several levels of portability, including executable image portability, object code portability and source portability.

9.1 Executable Image Portability

Executable image portability means that a program written on one computer system and compiled and linked into a binary executable file will run unchanged on another system, without recompiling or relinking. In order for this to be possible, the two computers must have the same exact CPU type or equivalent CPU's in the sense that the two CPU's can execute the same exact binary instructions in the same way. For example, both computers might have a Motorola 68000 CPU. On the other hand, binary code that was created for a Motorola 68000 will not work on an Intel 80386. In addition, the very lowest level details of how a program transfers data to/from devices (keyboard, display, disk, etc.) must be identical, so even two systems that have both Motorola 68000 CPU chips might not have executable image compatibility.

Some manufacturers have made an effort to produce a range of computer systems all with identical architecture, so that all systems are compatible at the level of executable images. The architecture includes not only the CPU instruction set but the mechanism for memory addressing and other details as well. Two examples are the IBM System/370 and the DEC VAX, where there are computer system models that span several orders of magnitude in speed, storage and other characteristics.

9.2 Object Code Compatibility

Object code compatibility refers to a more intermediate level, in which the two computers must have identical or equivalent instruction sets, the details of input/output could be different. In this case, there must be a common interface to the I/O system.

9.3 Source Code Compatibility

Sometimes one can write a program that is at source level independent of any particular system and can ultimately run on systems that have even different CPU's. In order to achieve this, you usually must write your program to follow ANSI standards for language definitions (if they exist), and you must be careful not to take advantage of implementation details, like how many bits are in a word. The program must be recompiled and relinked on each machine in order for it to run.

9.4 Operating System Portability

Portability is often discussed in a more ambitious way. This concerns the ability to make an entire operating system or programming environment run on a different computer system. If the computer system hardware is configured identically to the original, or reasonably close, there is usually little problem. The computer's basic instruction set must be the same, the methods for input and output from/to peripheral devices must be the same or compatible, and the peripheral devices must be the same or similar. Sometimes it is possible to separate the operating system device driver components from the other components, and replace this part so that essentially the same operating system runs with different peripheral devices and means of connecting them.

If the computer instruction set is also different from the original, the operating system programs and program fragments must be recompiled (at least) and sometimes significant changes must be made to adapt processor dependent code to the new processor. Nevertheless, some operating systems have been originally written for one system and modified for many others, with a fair degree of success. The most notable example is the UNIX operating system, which was first designed for a Digital Equipment Corp. PDP-7 computer and has evolved to versions that run on everything from desk-top personal computers to the Cray supercomputers.

Although UNIX is ubiquitous, one should not assume that there are no problems. UNIX implementations differ often in fundamental ways, so that programs that run with one version may not run with another version [18]. Even UNIX implementations for the same central processor may differ in (for example) subroutine linkages, use of co-processors, differences in memory management implementations, etc.

9.5 Data File Portability

Sometimes it is more important to be able to move data between different operating systems, file systems or computers, than to have the same programs run on different systems. In fact it may be necessary for the output data from one program to be readable as input for another program, on the same system or on a different system. The easiest way to make this possible is to design programs (or use programs) that read and write data in the form of text files. Text files are as close to universal as can be imagined. They allow the programmer and user to concentrate on the content of the file, not on the detailed format. This is particularly true when using programs that read free field input, without requiring that data items line up in particular columns. The programs that produce the data could then arrange it to make it readable by humans as well. As noted in Section 3.1 these files can be printed and edited as well.

9.6 How to Achieve Program Portability

When writing your own program, you can take some steps to avoid pitfalls that make programs for one computer system incompatible with another system. It is also possible to write a computer program that is so intimately dependent on the configuration details of a computer (what drives the files are on, what the names of the files and directories are, etc.) that the program won't run correctly even on another system with the same identical central processor, bus, peripherals (disk drives, interfaces, graphic displays), operating system and language compilers. Here are some do's and don'ts that promote (but do not guarantee) program portability across different operating systems as well as different computers.

- Don't use operating system features.

- Don't use device dependent features.
- Don't use system-dependent language extensions.
- Don't use binary files with the data in some arcane format.
- Do use standard programming languages.
- Do use layered software design.
- Do encapsulate system dependent and device dependent calls.
- Do use text files and label items with keywords.

By "layered design" we mean that the program is written in logical layers, with each layer connecting to the next only by simple well defined means, so that the details of each layer don't matter to the other layers. For example, a subroutine may need some data to do a computation. It calls another subroutine to get the data. The subroutine that gets the data contains the details about the organization of data in files, but the calling routine need not be concerned with that. If the files are changed or the method of reading the files changes, the computation subroutine does not need to change.

10 How to choose an operating system

While it is impossible to provide a catalog or "buyer's guide", it is possible to summarize some of the considerations in selecting a computer and its operating system. While some computer systems are available with a choice of operating systems, others have only one operating system software package available. Therefore, it is wise to consider first what your software needs are. Here is a list of some of these items.

- Do you need a computer that is multi-user or single-user?
- Do you need multi-tasking capability? This can be a big asset if you routinely do jobs that can run in a background mode while you use a text editor or other interactive programs.
- Will the computer be used in a network environment? If so, you will need to be sure that the operating system and software are compatible in some way with the existing environment.
- Will you be using large databases? If so, it will be important to have a computer with virtual memory, and an operating system that supports it. Also important is that the computer and the operating system be able to support large capacity disk drives. For example, there are now disk drives with a capacity for hundreds of megabytes.
- For large programs, similar considerations apply.

Often you will see articles comparing the features, and "look and feel" offered by different operating systems. These articles discuss the command interpreters, the organization of the files and directories, the kinds of utility programs that are available with each system. These differences, while important to the user, are not fundamental differences but differences of style. All modern operating systems have the same basic components (kernel, device drivers, file system) and use the same principles. As an example, two popular operating systems for high performance workstations and time-sharing systems are the DEC VMS operating system and the UNIX family of operating systems. These two operating systems are identical from the point of view of the checklist above, but are very different in style. Discussion of the stylistic differences is beyond the scope of this article.

It may be of interest to the reader with more experience to learn about current research in operating system design. Directions that are likely to be significant in the future include distributed file systems [19], remote procedure calls [20], and how to utilize systems of multiple processors [21].

11 Acknowledgements

It is a pleasure to acknowledge advice and friendly critique from Dr. Jonathan Jacky and Dr. George Starkschall.

References

[1] Peter J. Denning and Robert L. Brown. Operating systems. *Scientific American*, 251(3):94–106, September 1984.

[2] George W. Sherouse. Computer languages. In *Computers in Medical Physics*, American Institute of Physics, Inc., 1989.

[3] George W. Sherouse. Communications and networks. In *Computers in Medical Physics*, American Institute of Physics, Inc., 1989.

[4] Dennis M. Jennings, Lawrence H. Landweber, Ira H. Fuchs, David J. Farber, and W. Richards Adrion. Computer networking for scientists. *Science*, 231:943–950, 1986.

[5] John E. Yates. A time-sharing system for the PDP-1 computer. In *DECUS Proceedings 1962*, 1963. Reprinted in DECUS Proceedings Spring 1986.

[6] Edsger W. Dijkstra. The structure of the THE multiprogramming system. *Communications of the ACM*, 11(5):341–346, 1968.

[7] Peter J. Denning. Are operating systems obsolete? *Communications of the ACM*, 25(4):225–227, 1982.

[8] Alan Shaw. Special section on operating systems. *Communications of the ACM*, 31(3):255–257, 1988. Guest Editorial.

[9] Frederick P. Brooks, Jr. *The Mythical Man-Month.* Addison-Wesley, 1975.

[10] Dennis M. Ritchie and Ken Thompson. The UNIX time-sharing system. *Communications of the ACM*, 17(7):365–375, 1974.

[11] C. P. Thacker, B. W. McCreight, R. F. Lampson, R. F. Sproull, and D. R. Boggs. Alto: a personal computer. In D. Siewiorek, C. G. Bell, and A. Newell, editors, *Computer Structures: Principles and Examples*, McGraw-Hill, New York, 1982.

[12] R. W. Scheifler and J. Gettys. The X window system. *ACM Transactions on Graphics*, 5(2):79–109, 1986.

[13] R. C. Holt. *Concurrent Euclid, The UNIX System, and Tunis.* Addison-Wesley, 1983.

[14] Edsger W. Dijkstra. Cooperating sequential processes. In F. Genuys, editor, *Programming Languages*, Academic Press, 1968.

[15] E. J. Joyce. Malfunction 54: unraveling deadly medical mystery of computerized accelerator gone awry. *American Medical News*, 1986. Oct. 3, page 1.

[16] Edsger W. Dijkstra. Heirarchical ordering of sequential processes. *Acta Informatica*, 1:115–138, 1971.

[17] Brian W. Kernighan and D. M. Ritchie. *The C Programming Language.* Prentice-Hall, 1978.

[18] Robert M. McClure. The portability of UNIX applications programs and other modern folk tales. In *Proceedings of COMPCON 1986*, pages 14–16, 1986.

[19] David K. Gifford, Roger M. Needham, and Michael D. Schroeder. The cedar file system. *Communications of the ACM*, 31:288–298, 1988.

[20] David Notkin, Andrew P. Black, Edward D. Lazowska, Henry M. Levy, Jan Sanislo, and John Zahorjan. Interconnecting heterogeneous computer systems. *Communications of the ACM*, 31:258–273, 1988.

[21] Channing H. Russell and Pamela J. Waterman. Variations on UNIX for parallel-processing computers. *Communications of the ACM*, 30:1048–1055, 1987.

Computer Languages

George W. Sherouse
Radiation Oncology, North Carolina Memorial Hospital,
University of North Carolina, Chapel Hill, NC 27599-7512

INTRODUCTION

Despite all the mystique that often surrounds computers, a computer is only a machine: a box full of electronic switches that some *person* has to flip in the right order to generate desired results. There is a fundamental cultural conflict between computers and people which makes the flipping of those switches in the right order a conceptually difficult task, explaining the need for specialists called programmers who excel in switch flipping. Computers ''understand'' only the electronic state of their component parts. It is left to humans to cast arbitrary problems into a set of electronic states in order to effectively use a computing machine for mechanization of problem-solving. The computer does not negotiate or compromise. The human must help himself by developing efficient automatic mechanisms for large-scale translation of algorithmic problem solving into machine instructions. We call those mechanisms *computer languages.*

The intent of this presentation is to demystify the world of computer languages, to suggest a framework for understanding why there is not likely to ever be a universal language of computing, to give some feeling for the strengths of currently available languages, and hopefully to provide some guidance for people in search of an appropriate language for their own work.

1. MACHINE ABSTRACTION

The major attribute of any computing language which distinguishes it from others is the extent, flavor, and specificity of *machine abstraction* which it places between the programmer and the machine. The least abstract languages require the programmer to have intimate knowledge of the machine on which the program will run and to cast his problem in terms which map the solution directly to the capabilities of that specific machine. As the extent of abstraction expands, the programmer is given more and more latitude to cast solutions in a way which maps naturally to the way in which the problem domain is conceptualized by experts in that problem domain, with decreasing regard for the architecture of the machine. It is important to remember that (at least in today's common programming environments) no matter what language is used or how deep the abstraction, the set of instructions which is ultimately executed by the machine is fundamentally the same. What changes is the latitude which the programmer is allowed in conceiving and encoding a solution to a problem, and the extent to which the programmer's abstract instructions must be processed to produce the final machine instructions. Typically, languages which provide very little abstraction of the machine are called *low-level* and more abstract languages are called *high-level.*

Sometimes increased machine abstraction in a language is accompanied by increased specificity of the language to a given problem domain. Language specificity greatly enhances usefulness in a well defined problem domain, but restricts productive use of the language to that domain. One can imagine, for instance, a language for financial computation. In that language `AMORTIZE` and `DEPRECIATE` might be language primitives which encapsulate all pertinent IRS regulations regarding corporate capital investment, leaving the programmer free to `DEPRECIATE` without any specific concern or even knowledge of those regulations. It would not make sense, however, to attempt to use that language for radiation dose calculations, even though `DEPRECIATE` might be redefinable to calculate photon beam attenuation. We simply do not conceive of beam attenuation as depreciation, hence that specific abstraction becomes an obstacle to use of the language for other than its narrow intended problem domain.

In the following sections we discuss how programming languages are implemented, and then give an overview of several specific languages and programming environments.

2. COMPILERS AND INTERPRETERS

To make a broad generalization, the transformation of people-readable text into numerical machine instructions can be done in one of two ways: by interpretation or by compilation.

When a program is compiled, a utility program called a compiler takes the text of the program as input and transforms those abstract instructions into a list of machine instructions, commonly known as *object code.* Then another utility, called a linker, combines the object code with other pieces of object code from various libraries (for instance, the routines that read and write data), makes one long list of instructions, makes sure that references to specific addresses in the code are adjusted to reflect the modified length of the code, and produces an executable program.

Compiled code is in some sense mechanically generated assembler language. When it is executed it will be the only thing the computer does and so will run at the computer's top speed.

The development cycle for compiled languages involves four repeated steps and uses at least three different utilities. First, one must compose and edit the text of the program. Second the programmer leaves the editor to run the compiler on the new text. Third, the resulting new object code is linked to produce a runnable program. Fourth, the program is executed and tested. Some computing environments are tailored to support this type of activity. For instance, the editor I normally use for program development allows me to compile, link, and run a program from within the editor. For other environments, such as that normally provided by a microcomputer, the use of compiled languages can be hampered by the unwieldy mechanics of the development cycle.

An alternative strategy is to use an interpreter. An interpreter is a utility which is best thought of as a simulation of an imaginary machine, and the commands of the interpreted language as the primitive instructions of that machine. The interpreter then takes program text as input and executes the program instructions just as a real computer would execute object code. The task of mapping the instruction set of the imaginary machine to sets of instructions on the physical machine is the role of the interpreter.

Many interpreted languages have the advantage that it is very quick and easy to modify the program and re-execute it since no recompiling or relinking is required. In fact most interpreters have built-in editing functions. Interpreted language implementations do, however, suffer from the major disadvantage that they tend to be much slower than equivalent compiled languages. That is equivalent to saying that the software simulation of a machine is not as efficient as a real machine – clearly a predictable result.

Some intermediate strategies exist. For instance, in the so-called p-code system, which probably reached its highest visibility with the Apple-II language system, source code is compiled not into the native instructions of the host computer, but into the native instructions of an imaginary p-machine. The p-machine instructions are then executed by a p-code interpreter. Execution of the p-code is much more efficient than direct execution of the raw source code would be, but not as efficient as execution of the machine's native instruction set.

Another interesting intermediate strategy is to use an interpreter for program development in order to exploit the rapid development cycle, and then to compile the verified program in order to exploit the speed of a compiled program. Compilers exist for a number of languages which are normally interpreted, including many microcomputer BASIC implementations. Some languages, however, embody a machine abstraction which is so far removed from the norm that they cannot (yet) be compiled.

3. ASSEMBLER LANGUAGE

The simplest, or least abstract, computer language that most people encounter is one which implements a mnemonic coding system for a machine's native instruction set. These languages are usually referred to as *assembler* languages and the mnemonic tokens they provide are called *opcodes* for *op*eration *codes*. There is a one-to-one correspondence between the primitive functions of the machine (move data, add, subtract, test data) and the opcodes. Opcodes are cryptic codes like `MOV` for "MOVe data", `BNE` for "Branch if datum is Not Equal to zero", and `SOB` for "Subtract One from a register and Branch to indicated address if the result is not equal to 0". The list of opcode instructions is then directly translated (by a utility called an assembler) into a list of numerical machine instructions which can be executed by the machine. An assembler program is a complete, detailed, step-by-step list of specific commands to the machine describing each desired change in the machine's state.

Assembler languages have severe limitations as means for achieving solutions to arbitrary problems. The language provides essentially no abstraction of the machine. That forces the programmer to pay very close attention to the state of the machine at all times, thus diverting mental energy which could otherwise be directed toward producing more innovative and robust solutions to problems. Secondly, the problem must be broken down to extremely small grains in order to code it. Even the simple addition of two numbers will normally require three machine instructions to execute, thus the process of writing an assembler language program can be very long, tedious, and error-prone. It is also not possible to take assembler programs written for one type of machine and use them on another machine (except, as we have seen, by writing an interpreter). As a rule, assembler language should only be used in situations where very efficient direct access to the hardware is required (e.g., for control of physical devices or for maximum execution speed).

4. GENERAL-PURPOSE AND SCIENTIFIC LANGUAGES

For a group of scientists and engineers, the most important class of languages are those commonly known as "general-purpose". We all learned at least one of them in college and can, if necessary, still make some degree of sense of programs written in these languages. They are general-purpose in the sense that they provide machine abstractions which are general enough to accommodate a wide range of problems, but still concrete enough to allow efficient translation of programs into machine instructions. That is not to say that they do not put their own peculiar "spin" on problem solving, but rather that the abstraction is for the most part mechanism-specific rather than problem-specific.

A significant advantage of general-purpose languages over assembler language is that to a large extent they separate the program from the hardware. The programmer need not know the state of the hardware at any given time, the specific addresses of his data, or even necessarily what type of machine the program is intended for, although many of these languages do have mechanisms by which the programmer can access particular hardware locations if need be. Programs in these higher-level languages are thus much more likely to be *portable* from one type of hardware to another compared with assembler programs. Further, these languages allow the specification of seemingly simple instructions which in fact generate quite lengthy assembler language programs.

The languages discussed in the following sections have much more in common with one another than they have differences, at least partly because they represent a variety of attempts to provide the same facilities but by different mechanisms. Each language supports the same basic data types: boolean, character, integer, floating point, double precision floating point, and arrays of those data. They provide the same basic set of operators (boolean tests, addition, subtraction, division, multiplication) for scalars, and have no explicit support for functions on arrays. They use a notation borrowed from algebra, e.g., `A=(B+C)/D`. They are all (as are assembler languages) *procedural*, that is,the flow of program control must be explicitly specified by the programmer.

The significant differences arise mostly from the historical and philosophical contexts in which the languages arose. Over time, the prevalent thinking with regard to what mechanisms are important in a computer language has evolved. At each plateau in that evolution one or more major new languages has emerged from an existence proof.

I offer one caveat to the following discussions. In the interest of conveying the flavor of each of these languages, I have described many of these languages in a somewhat stereotyped way. My intent in this discussion is to convey the characteristics of the languages at their most distinctive, not as they have been modified by assimilation of ideas from other languages. Many of these languages have changed substantially over time, particularly as vendor-specific changes have been introduced into commercial language products, so some of the limitations cited are no longer as real as they may once have been.

4.1. FORTRAN

Since its introduction in 1957, FORTRAN (for mathematical FORmula TRANslation system) in a variety of incarnations has been the main programming language of the scientific and engineering communities. It was originally developed by IBM as an attempt to provide efficient mechanisms for coding scientific problems. Since its introduction it has been revamped a number of times. FORTRAN-IV was and probably still is the most widely used dialect. The most recent incarnation, FORTRAN-77, represents an attempt to modernize the language with some concepts discussed in the following sections, but the language remains true to and somewhat hamstrung by its long heritage.

FORTRAN provided a number of important features that are now easily taken for granted. The most important by far is the ability to describe mathematical manipulations in the natural language of those manipulations, namely quasi-algebraic notation. This abstraction is seen from the syntax of the assignment operators (e.g., `X=Y+Z`), array references (the Ith element of `X` is `X(I)`), and subroutine calls (`X=SIN(THETA)`), to the less visible but very important optimization of most compilers for generation of efficient numeric code.

At the same time, FORTRAN has a variety of shortcomings and misfeatures that decreases its appeal compared to more recently-introduced languages. The most serious of these are:

(1) Implicit typing – Most computing hardware embodies a fundamental distinction between integers and real (floating point) numbers. The two classes of numeric representation are treated very differently by the hardware. From a mathematical point of view, however, the integers are merely a proper subset of the reals. Most languages deal with this problem by providing the programmer the ability to explicitly store a number as an integer or as a real, and by providing automatic type translation when data of the two types appear in the same expression. In FORTRAN, any combination of 6 letters and digits (beginning with a letter) is a valid *predefined* variable whose type is established implicitly by the language based on the first letter of the variable name. Names starting with letters `I` through `N` are integer, others are real. These conventions

for data typing, while they are consistent with the basic abstraction of the language, have a variety of unfortunate side effects. The most visible is an awkward tendency for programmers to use names like `KOUNT` or `ISUM` in order to have somewhat mnemonic variable names for integers. Others are more insidious. Since any combination of letters and digits can be a variable name, typographical errors can yield legal but incorrect code in a way which is very difficult to detect. For instance, a mistyped attempt to increment a counter such as `KOUNT=KOUMT+1` would compile and execute without problem, except that `KOUNT` would always be wrong. Even worse, the mistyping of a punctuation mark can make huge differences in the meaning of a code fragment, again because of the automatic generation of variables. Consider that `DO100I=1,10` specifies that a series of instructions should be executed ten times, while `DO100I=1.10` will assign the value 1.1 to the real variable `DO100I` and execute the aforementioned set of instructions exactly once. These kinds of errors can result in what are known politely as *subtle* or *obscure* bugs – that is, errors in a program which yield incorrect answers for obscure reasons and that can require absurdly long times to discover and repair.

(2) Simple-minded flow control – The mechanisms for specifying the flow of program control provided by FORTRAN mirror the mechanisms which exist in conventional (von Neumann) hardware. The program is just a long list of instructions which are normally executed sequentially. If at some point it is necessary to make a decision and execute some series of program instructions *conditionally* based on that decision, then it is necessary to transfer control to a different location in the sequential list of instructions. In assembler languages, this process is called a conditional branch. In FORTRAN it is a `GO TO`. The use of conditional (and unconditional) branching in a program, perhaps to a location several pages away in a large traditional FORTRAN program, can make the flow of control nearly impossible to follow. The result can be what is known in the trade as "spaghetti code" in honor of the way threads of control intertwine and then sort of lie there in a sticky heap. This pitfall of FORTRAN is the result of a failure to sufficiently abstract hardware flow control.

(3) Lack of facilities for text manipulation – While many FORTRAN implementations have extensions to deal with text, the language was never intended for text processing. As a result, FORTRAN is an extraordinarily bad choice as a language for interactive screen manipulation, database applications, or other text-intensive applications.

4.2. BASIC

One of the most widely-known higher-level languages, and one which enjoyed a glorious renaissance with the birth of the PC, is the Beginner's All-purpose Symbolic Interactive Coding system, or BASIC. It was introduced in 1964 as a teaching language and owes much of its character to FORTRAN, on which it was clearly modeled. Its name is descriptive of its intended use. It is a fairly simple language to learn, and thrives in environments which are in one way or another too small for more advanced languages. Since it is usually interpreted rather than compiled, BASIC programming and debugging can be very easy and quick for simple programs.

Since it is so small and easy to learn, BASIC is a common choice for many kinds of microprocessor-controlled devices. Unfortunately, BASIC shares many of FORTRAN's worst problems as listed above. Most implementations do have very strong text-manipulation features, though. Implementations for microcomputers also tend to include extensions for doing graphics, device control, and other machine-specific tasks. Those extensions are almost without exception specific to the vendor and so impede portability of BASIC programs.

4.3. Pascal

Pascal (named for the mathematician Blaise Pascal) was originally developed in 1970 by Nicklaus Wirth as a teaching language, emphasizing *structured programming*, a concept which was quite new at that time but is now a common consideration in language design. The emphasis of structured programming is on producing solutions which can be proven to be correct and software which can be easily verified. The most visible effect of this emphasis is the use of block-structured flow control. Pascal was the first widely-used language to provide mechanisms for support of structured programming, though these mechanisms first appeared in ALGOL.

As previously noted, the very limited facilities for control of program flow which are provided by BASIC and FORTRAN can yield programs which are essentially unreadable and certainly unverifiable. A central concept of structured programming is that program execution proceeds linearly from beginning to end, even though the resulting machine instructions surely will not. Branching, as with the FORTRAN `GO TO`, is considered extremely bad practice, though it is grudgingly supported by the language. In place of branching, one uses a series of program clauses or *blocks* which are conditionally executed. Thus `GO TO` was supplanted by constructs such as `IF-THEN-ELSE`, `WHILE-DO`, `REPEAT-UNTIL`, and so forth.

Pascal also was the first widely used language to provid mechanisms for construction of new data types by combination of the primitive data types and/or by restriction of the domains of those primitive types. These mechanisms are provided in Pascal through the `RECORD` and `TYPE` primitives respectively. Use of these so-called *structured* or *user-defined* data types allows much more effective and efficient manipulation of sets of related data than is possible with more primitive data types. For instance, a datatype for inventory control can be imagined which contains a character member for a description of an item, an integer member giving its part number and a floating point member giving its retail price. If these data are dealt with as instances of this structured data type, then the bookkeeping overhead in passing and manipulating these data are greatly reduced. Significantly, the use of structured types also encourages more modular software design.

A related feature of Pascal is something called *strong typing*. Pascal compilers are very strict about performing operations only on variables for which those operations are defined. Further it is required that every variable be explicitly declared as being of some type before it is used. For instance, an attempt to assign a floating point value to an integer variable will be flagged as an error, whereas in less strongly-typed languages the value would be silently truncated. Pascal requires explicit type conversion (also known as *coercion* or *casting* in other languages) for assignment of one type of data to a variable of another type. There are even different operators for floating point division and integer division. This can sometimes be a nuisance, especially when the two "different" types are really equivalent derived types of the same primitive type, but is also a powerful mechanism for assuring code quality. It makes it just that much harder for accidents and typos to pass unnoticed.

Another important feature of Pascal is its support for sets and boolean operations on sets. One can specify a variable which contains a set of objects of any data type. Objects of that type can then be added to or removed from the set, and operations such as union and intersection can be performed between the set and other sets of the same type.

The importance of some of these concepts is such that Pascal is now almost universally accepted as the language of choice for teaching computing concepts. That is not surprising since that was Wirth's intention in the design of the language. Unfortunately, Wirth's philosophical purity on that point as manifest in his decision to not include some essential features such as file I/O, has interfered with adoption of Pascal for applications beyond education. Many very good, fully-capable Pascal compilers are commercially available today but all are necessarily non-standard extensions of the language defined by Jensen and Wirth.

4.4. Modula-2

Modula-2 is Wirth's successor to Pascal. It is also to some extent a teaching language, though it is more capable of handling real-world problems. The language grew out of Wirth's interest in programming of concurrent processes. Its major additional features with respect to Pascal are:

(1) The concept of a *module* which is a generalization of the Pascal procedure. Modules have separate definition and implementation parts, a mechanism similar to the data hiding facility of object-oriented languages discussed below.

(2) The concept of the *process* which is central to the support of concurrent processes.

(3) Mechanisms for direct manipulation of hardware, something not allowed by Pascal's strict machine abstraction.

4.5. C

The C programming language was developed by Kernighan and Ritchie at Bell Laboratories in the hope that it could be used by colleagues in their project to produce a machine-independent operating system (what became UNIX). It is, in fact, based very closely on the instruction set of the PDP-11 computer series, the computers on which UNIX was developed. It clearly started life as a sort of structured assembler language. C is a very small language in terms of the number of primitive operations it provides, but by design is a very versatile and expressive language. C is perhaps unique in the variety of ways in which the same operation can be expressed. This is both its glory and its downfall: good programmers can write programs which are very easy to read and maintain, and careless hackers can write totally unfathomable code. While this is true in most languages, C provides copious opportunity for writing code which is either self-documenting or self-obfuscating.

C provides the usual set of mathematical operators. In addition it has an additional set of unique operators. For instance `++` and `--` are the increment and decrement operators (e.g. `index++`, `--counter`). If a variable is preceded by one of them, its value is incremented or decremented before it is used. If the operator follows the variable, the current value is used and then incremented or decremented. Similarly, any of the standard operators (`+`, `-`, `*`, etc.) can be combined with `=` to succinctly express modification of a variable. For instance "`index += 3`" is equivalent to "`index = index + 3`". The same applies to the rich set of bitwise operators such as bitwise or (|), bitwise and (`&`), shift left (<<), shift right (>>), etc.

C is sometimes called a *middle-level* language because it can be used to directly manipulate hardware or just as easily in applications where Pascal would be appropriate. In fact, to many C is Pascal without the straitjacket. To others it is Pascal without the safety net. Both points of view are valid. C is a weakly-typed, structured language with capabilities for user-defined data typing similar to Pascal's.

One of the most important features of the C language is its capacity for redefinition. Essentially any character string can be redefined to give it any meaning. This facility is particularly useful for writing programs which are to be ported from one environment to another. For instance, a C graphics program can be written which uses only stereotyped graphics calls like `DRAW(x,y)` and `MOVE(x,y)` which are then defined through the macro facility to be

---

UNIX is a trademark of Bell Laboratories.

PDP-11 is a trademark of Digital Equipment Corporation

appropriate to a given environment. Note that this is different from using calls to generic subroutines. Macros expand in place and so can use any variables which are defined in that context.

4.6. C++

Just as structured programming, both as a concept and as a buzzword, was a dominant issue in language development in the 1970s, *object-oriented programming* (OOP) is the topic of the late 1980s. Behind the mystique, OOP is really just a logical extension of structured programming, and one step further in the support of deep machine abstraction. The central feature of OOP is a shift from thinking of programming in terms of programs which act on passive data, to environments in which objects interact. An object is a structured data type which includes not only data but definitions of the operations which can be performed on those data. Such a definition is called a *class* and objects of a class are called *instances* of that class. A class of object can be defined in such a way as to keep all, some, or none of its procedures (called *methods*) and data inaccessible to (*hidden* from) other parts of the program. Thus, the *interface* to the object, that is the way other objects interact with it, is all that need be published. The methods and data can be kept secure. This approach has far-reaching ramifications in terms of software modularity and hence provability. Consider, for instance, the difference in the problem abstraction between the approach of writing a program to sort some data and that of telling the data to go sort itself. The latter case describes the situation of a programmer in an OOP environment. The sort algorithm and other details are irrelevant to the programmer who uses the class, and so can be transparently changed at any time by the programmer responsible for maintenance of the class. As long as the published interface is not changed, the two programmers will not interfere with one another.

Some language implementations are now becoming available which provide explicit mechanisms for OOP. An important example is the successor to C, inventively called C++. C++ is very nearly a proper superset of C but provides a number of new facilities. Most important is the ability to define, instantiate, and manipulate objects. A close second in importance is the introduction of strong typing. These improvements are accomplished without sacrificing C's flexibility and power, particularly its provisions for directly manipulating hardware. AT&T has announced plans to rewrite UNIX in C++.

5. SPECIAL-PURPOSE LANGUAGES

As suggested in the introduction, not all problems are best addressed using general-purpose languages. In some domains the use of a language with an appropriate abstraction of the machine can make a significant difference in programming efficiency and reliability. Some important examples are briefly discussed in this section.

5.1. Business languages

The languages of the previous section are not appropriate for business applications. A different set of abstractions is required for languages which must be optimized to store and sort huge volumes of data (inventories, accounts, etc.) and to produce reasonable textual summaries of those data (reports).

While a number of languages exist for business, the longstanding dominant player is COBOL (for COmmon Business Oriented Language). Developed in 1960, it is now probably the most widely-used language in the world. It is a very verbose language, with listings resembling pages of a newspaper. This was done intentionally in the name of improved readability for non-programmers.

Another important language for business is MUMPS, a language developed in the late 1960s at the Massachusetts General Hospital. The intent of MUMPS was to provide an interactive query language which was sufficiently English-like that physicians could interrogate databases in an intuitive way. The effort was not particularly successful but the language has survived, mostly because of its simplicity. There is essentially no concept of type (data are normally treated as text) and very few commands. MUMPS is an interpreted procedural language, often compared to BASIC.

5.2. Declarative languages

I have said that C++ supports much deeper abstraction than, say, FORTRAN. It is reasonable to wonder then what the most abstract language would look like. Any answer I gave to that question would be provocative so I will resist temptation. But it is worthwhile to discuss a class of languages which, in contrast to the procedural languages we have already discussed, provide only (or mostly) facts and queries regarding the implications of those facts, leaving the formulation of mechanism for deriving those implications up to the language system. This represents a nearly total disengagement of the mechanisms for stating the problem from the underlying mechanisms of the machine. As a class these languages are referred to as *declarative* (or descriptive or nonprocedural). They are most often associated with work in Artificial Intelligence (AI). Because of their deep abstraction of the machine they are usually implemented as interpreters.

The oldest, most common, and to some extent least pure of the declarative languages is LISP (for LISt Processing language). LISP is an untyped language with a very simple syntax. A LISP program is merely a set of nested lists. The first element of each list is normally presumed to be a function which is to be applied to the remaining elements of the list. Data type is then imposed on the items in the list by the function. LISP excels at symbolic manipulation and so is a common choice for implementation of expert systems.

A more recent and purely declarative language is PROLOG (for PROgramming in LOGic). A PROLOG program consists of a series of assertions which together define a universe. For example, the statement "`Parent(x,y) ← Father(x,y)`" says that if `x` is the father of `y` then `x` is the parent of `y`. A statement with nothing on the left side of the arrow is called a *goal*, and the system upon seeing a goal will produce a list of all players which meet the specified criteria. Note that no mechanism (procedures) need be specified for this to happen, only the facts and the goals.

Currently declarative languages tend to be large and slow. This is partly because the mechanisms they use, such as exhaustive search, are inherently consumptive of resources. Fortunately, many of these needs can be efficiently addressed through more efficient hardware support. In is for this reason that one can now find "LISP machines" commercially available. It is also worth noting that the mechanisms required for PROLOG are well suited to implementation on machines capable of processing multiple threads of control simultaneously (i.e., parallel computers).

5.3. Ada

Ada (named for Augusta Ada, Countess of Lovelace, daughter of Lord Byron, and reputedly the world's first programmer) is a language developed in response to a specification formulated by a committee on behalf of the United States Department of Defense. It is in some sense a general-purpose language with the same concerns as those discussed in that section above. I have listed it as special-purpose, though, because in practice it is used almost exclusively by defense contractors as a condition of their contracts. To meet the extensive criteria set by the design committee, the language is very large and as a result very difficult to learn and difficult to read. It has been said that no matter what language you like to use, you will find it in Ada.

That is no accident; one of the reasons for formulating Ada in the first place was to stop the proliferation of incompatible languages in defense systems. Unfortunately, this implies that the one superlanguage must include constructs which support a wide variety of different machine abstractions for different problem domains. So, it is inevitable that effective abstractions from a wide variety of domain-specific languages would have to be included. Critics maintain that this functionality overload serves to obscure rather than clarify.

5.4. Other special-purpose languages

Three other languages deserve a brief mention for their unique market niches. Forth is a language originally developed by Charles H. Moore for control of the telescopes at Kitt Peak Observatory. It has since gained wider acceptance as a good language for robotics and other automated control systems. It is an interpreted stack-based language which uses Reverse Polish Notation (RPN). It is an entirely fair analogy to say that Forth is to BASIC as Hewlett Packard calculators are to Texas Instruments calculators.

LOGO is a language which is interesting for its incorporation of the display into the language as a primitive. It has become very popular as a language for teaching elementary school children basic concepts of geometry and computing. The *turtle*, a graphic icon which can be directed to move around and draw things on the screen is a central feature of LOGO. LOGO also served as one of the inspirations for the work at Xerox Palo Alto Research Center (PARC) which spawned many of the ideas Apple incorporated into the now ubiquitous Macintosh (windows, mouse, etc.) and also spawned the influential Smalltalk language.

At the other end of the spectrum, APL (for A Programming Language) is a language which provides a very deep abstraction of the machine as a matrix arithmetic engine. APL provides a rich set of functions for constructing and manipulating arrays of numbers. It is remarkable, among other reasons, for the fact that all commands are represented by greek letters or other symbols. Thus, a terminal capable of displaying the APL character set is required in order to use the language.

APL and Forth (and tersely-written C) are sometimes half-jokingly called ''write-only'' languages. This is a reference to the problems which can arise in maintenance of code which is written in a language whose syntax departs significantly from English and/or algebraic notation. For serious users the expressiveness of these machine abstractions can outweigh the risk.

6. COMPILERS FOR NEW ARCHITECTURES

There is currently a great deal of activity in the development of computer systems with multiple processing threads. Coincident with that development has come a new specialty involved with methods for detecting and exploiting parallelism in programs. Several compilers are now commercially available from vendors of parallel architecture hardware which claim to optimize FORTRAN-77, C, and other languages with minor or no modification to existing code. The success of these compilers can vary widely as a function of the structure of the program. A more reliable approach is to recode the program in a language which explicitly expresses parallelism to the compiler. Clearly the abstraction of parallel architectures by appropriate constructs in high-level languages is still in its infancy.

7. CHOOSING A LANGUAGE FOR PROGRAM DEVELOPMENT

There is no ''right'' language for most computing applications that medical physicists are likely to be involved with. In choosing a language, therefore, one must use one's own judgement and experience. Unfortunately, issues which have nothing to do with the appropriateness of a given language in a given situation can often drive the ultimate decision. For instance, a sufficiently robust implementation of a desired language may not exist for a particular computer system. Or, worse yet, administrative decisions regarding development environments may limit one's choices. But in the absence of this type of influence, the choice is best made based on the intended use of the software and on personal productivity and style.

A number of points are worth repeating:

(1) If you program only rarely and only small tasks, a small interactive interpreted language like BASIC or Forth is a good choice. Think of your PC as an extended calculator.

(2) If your intent is to program seriously but infrequently, a general-purpose language with strong type-checking and support of structured control is a good choice. Pascal is a standout in this regard, though Modula-2 and Ada have their proponents.

(3) If you are dealing exclusively with compute-intensive tasks of a scientific nature and have significant self-discipline, or alternatively if for reasons of job security or downright contrariness you *want* to use a language which is the rough equivalent of a bull in a china shop, FORTRAN (preferably FORTRAN-77), can be an acceptable alternative. This is particularly true in environments where one hopes to employ optimizing compilers for special-purpose, parallel architecture hardware since FORTRAN is often the first language commercially supported on new hardware.

(4) If you have a special problem for which a special language has been developed you ought to seriously consider using it. In particular the use of languages like Forth for robotics, LISP for symbolic processing, APL for matrix arithmetic, or PROLOG for logic programming can greatly simplify program development.

(5) If your intent is to program seriously over a wide range of applications, from direct machine manipulation (operating systems, process control), to general-purpose programming, to scientific computing, and/or if you are concerned with generating software which is easily ported to other operating environments, the C programming language and its object-oriented successor C++ are the languages of choice.

For those still in doubt, I have provided a helpful everyperson's guide to computer languages in an appendix.

SUMMARY

Computers would not be nearly so valuable to us today as they clearly are if it were not possible to express solutions to problems in terms of an abstract machine. This abstraction is made possible by the use of computer languages. Computer languages provide the ability to describe solutions using a stylized subset of the natural language of the problem domain. The translation of those instructions to machine instructions can then be automated.

One major thrust of future language research and development is clear. Languages are providing deeper abstraction, and in turn are becoming more succinct, and in turn more readable. This trend reflects increasing recognition of the fundamental importance of software quality assurance. To the extent that computer languages support abstract problem solving without regard to the specific mechanisms to be used in those solutions, they become more verifiable. This is an issue that should concern us all. As more and more computers are used in the control of our environment, the possibility of their containing unverified software poses a greater and greater threat to our well-being. We have seen dramatic examples of this in the computerized control of radiotherapy machines. Consider for a moment the potential risk from the plethora of devices and systems which directly affect our safety and security (automatic teller machines, space-based directed energy weapons, microwave ovens, et al.) and are ultimately only as reliable as the software that controls them.

SELECTED REFERENCES

Most well-established languages are formally defined by documents from standards organizations such as ANSI. Many newer languages, however, are defined by publications from the creators of the languages. These reports often serve as de facto standards until such time as the language can be formally standardized. So it is not unusual to read references to, for instance, "K&R C" referring to the C language as defined by Kernighan and Ritchie. The following references are important definitions of their respective languages and so are essential tools for anyone who wishes to use the languages.

References

1. Jensen, Kathleen and Wirth, Nicklaus, *Pascal: user manual and report,* Springer-Verlag (1975). ISBN 0-387-90144-2
2. Kernighan, Brian W. and Ritchie, Dennis M., *The C Programming Language,* Prentice-Hall (1978). ISBN 0-13-110163-3
3. Stroustrup, Bjarne, *The C++ Programming Language,* Addison-Wesley (1986). ISBN 0-201-12078-X
4. Wirth, Nicklaus, *Programming in Modula-2,* Springer-Verlag (1982). ISBN 0-387-12206-0

ACKNOWLEDGEMENTS

The author wishes to acknowledge the contributions of the following people to this manuscript: Matthew Barkley, Russ Tuck, and Don Ragan for helpful descriptions of specific languages; James Coggins for screening out bald-faced lies and advising on conceptual purity; Ed Chaney and Liz Sherouse for proofreading and editorial advice.

APPENDIX I: A SIMPLE PROGRAM IN SIX LANGUAGES

The following six pages list implementations of a very simple program in six of the languages to which I have access. The example was chosen to show how several basic ideas are handled by the various languages including comment lines, variable declaration, input and output of numbers, addition, multiplication, assignment, subprogram declaration and use, and decision-based flow control. This example does not illustrate any of the strengths of any of the languages and in that sense does them all an injustice.

```
100 rem
110 rem  This is BASIC.
120 rem
130 rem   This program:
140 rem   1. reads two integers from the keyboard
150 rem   2. for each integer prints whether the number is
160 rem      less than, equal to, or greater than 0.
170 rem   3. calculates the sum of the two integers and prints it
180 rem   4. calculates the product and prints it
190 rem
200 input I1
210 input I2
220 let I = I1
230 gosub 310
240 let I = I2
250 gosub 310
260 let S = I1 + I2
270 let P = I1 * I2
280 print 'The sum is ', S, '.'
290 print 'The product is ', P, '.'
300 stop
310 if I < 0 then print I, ' is less than zero.'
320 if I = 0 then print I, ' is equal to zero.'
330 if I > 0 then print I, ' is greater than zero.'
340 return
```

```
      PROGRAM EXAMP
C
C  This is FORTRAN.
C
C   This program:
C   1. reads two integers from the keyboard
C   2. for each integer prints whether the number is
C       less than, equal to, or greater than 0.
C   3. calculates the sum of the two integers and prints it
C   4. calculates the product and prints it
C
      READ (5, 1000) INT1
      READ (5, 1000) INT2

      CALL PRTZ(INT1)
      CALL PRTZ(INT2)

      ISUM = INT1 + INT2
      IPROD = INT1 * INT2

      WRITE (6, 2000) ISUM
      WRITE (6, 3000) IPROD

      STOP
 1000 FORMAT (I10)
 2000 FORMAT (" The sum is ", I10, ".")
 3000 FORMAT (" The product is ", I10, ".")
      END

      SUBROUTINE PRTZ(I)
      WRITE (6, 4000) I
      IF (I) 100, 200, 300
 100  WRITE (6, 5000)
      RETURN
 200  WRITE (6, 6000)
      RETURN
 300  WRITE (6, 7000)
      RETURN
 4000 FORMAT ("+", I10, " is ")
 5000 FORMAT ("less than zero.")
 6000 FORMAT ("equal to zero.")
 7000 FORMAT ("greater than zero.")
      END
```

```
program examp(input, output);

{
  This is Pascal.

  This program:
  1. reads two integers from the keyboard
  2. for each integer prints whether the number is
     less than, equal to, or greater than 0.
  3. calculates the sum of the two integers and prints it
  4. calculates the product and prints it
}

var
    Integer1, Integer2, Sum, Product: integer;

procedure PrintRelationToZero (Integer: integer);
begin
    write (Integer, ' is ');

    if Integer < 0 then
     writeln ('less than zero.')
    else
    begin
     if Integer = 0 then
         writeln ('equal to zero.')
     else
         writeln ('greater than zero.')
    end
end;

begin
    read (Integer1);
    read (Integer2);

    PrintRelationToZero (Integer1);
    PrintRelationToZero (Integer2);

    Sum := Integer1 + Integer2;
    Product := Integer1 * Integer2;

    writeln ('The sum is ', Sum, '.');
    writeln ('The product is ', Product, '.');

end.
```

```
#include <stdio.h>

main ()
/*
  This is the C programming language.

  This program:
  1. reads two integers from the keyboard
  2. for each integer prints whether the number is
     less than, equal to, or greater than 0.
  3. calculates the sum of the two integers and prints it
  4. calculates the product and prints it
*/
{
    int          integer_1, integer_2, sum, product;
    void   print_relation_to_zero ();

    scanf ("%d", &integer_1);
    scanf ("%d", &integer_2);

    print_relation_to_zero (integer_1);
    print_relation_to_zero (integer_2);

    sum = integer_1 + integer_2;
    product = integer_1 * integer_2;

    printf ("The sum is %d.\n", sum);
    printf ("The product is %d.\n", product);

    exit (0);
}

void print_relation_to_zero (integer)
int integer;
{
    printf ("%d is ", integer);

    if (integer < 0)
     printf ("less than zero.\n");
    else
     if (integer == 0)
         printf ("equal to zero.\n");
     else
         printf ("greater than zero.\n");
    return;
}
```

```
#include <stream.h>

main ()
/*
  This is the C++ programming language.

  This program:
  1. reads two integers from the keyboard
  2. for each integer prints whether the number is
      less than, equal to, or greater than 0.
  3. calculates the sum of the two integers and prints it
  4. calculates the product and prints it
*/
{
    void   print_relation_to_zero (int);

    int          integer_1, integer_2;
    cin >> integer_1;
    cin >> integer_2;

    print_relation_to_zero (integer_1);
    print_relation_to_zero (integer_2);

    int sum = integer_1 + integer_2;
    int product = integer_1 * integer_2;

    cout << "The sum is " << sum << ".\n";
    cout << "The product is " << product << ".\n";

    exit (0);
}

void print_relation_to_zero (int integer)
{
    cout << integer << " is ";

    if (integer < 0)
     cout << "less than zero.\n";
    else
     if (integer == 0)
         cout << "equal to zero.\n";
     else
         cout << "greater than zero.\n";
    return;
}
```

```
(prog ()
      ;;  This is LISP
      ;;
      ;;  This program:
      ;;  1. reads two integers from the keyboard
      ;;  2. for each integer prints whether the number is
      ;;   less than, equal to, or greater than 0.
      ;;  3. calculates the sum of the two integers and prints it
      ;;  4. calculates the product and prints it

      (defun print-relation-to-zero (arg)
      (princ (list arg " is "))

      (cond
        [(> arg 0) (princ "greater than zero.")]
        [(= arg 0) (princ "equal to zero.")]
        [(< arg 0) (princ "less than zero.")])
      (terpr))


      (setq integer-1 (read))
      (setq integer-2 (read))

      (print-relation-to-zero integer-1)
      (print-relation-to-zero integer-2)

      (setq sum (plus integer-1 integer-2))
      (setq product (times integer-1 integer-2))

      (princ (list "The sum is " sum "."))
      (terpr)
      (princ (list "The product is " product "."))
      (terpr))
```

APPENDIX II: SELECTING A PROGRAMMING LANGUAGE MADE EASY

Daniel Solomon & David Rosenblueth
Department of Computer Science, University of Waterloo
Waterloo, Ontario, Canada N2L 3G1
(condensed by G. Sherouse)

With such a large selection of programming languages it can be difficult to choose one for a particular project. Reading the manuals to evaluate the languages is a time consuming process. On the other hand, most people already have a fairly good idea of how various automobiles compare. So in order to assist those trying to choose a language, we have prepared a chart that matches programming languages with comparable automobiles.

Assembler	A Formula I race car. Very fast, but difficult to drive and expensive to maintain.
FORTRAN II	A Model T Ford. Once it was king of the road.
FORTRAN IV	A Model A Ford.
FORTRAN 77	A six-cylinder Ford Fairlane with standard transmission and no seat belts.
COBOL	A delivery van. It's bulky and ugly, but it does the work.
BASIC	A second-hand Rambler with a rebuilt engine and patched upholstery. Your dad bought it for you to learn to drive. You'll ditch the car as soon as you can afford a new one.
C	A black Firebird, the all-macho car. Comes with optional seat belts (lint) and optional fuzz buster (escape to assembler).
Pascal	A Volkswagon Beetle. It's small but sturdy. Was once popular with intellectuals.
Modula II	A Volkswagen Rabbit with a trailer hitch.
LISP	An electric car. It's simple but slow. Seat belts are not available.
PROLOG/LUCID	Prototype concept-cars.
FORTH	A go-cart.
LOGO	A kiddie's replica of a Rolls Royce. Comes with a real engine and a working horn.
APL	A double-decker bus. Its takes rows and columns of passengers to the same place all at the same time. But, it drives only in reverse gear, and is instrumented in Greek.
Ada	An army-green Mercedes-Benz staff car. Power steering, power brakes and automatic transmission are all standard. No other colors or options are available. If it's good enough for the generals, it's good enough for you. Manufacturing delays due to difficulties reading the design specification are starting to clear up.

SOFTWARE AND COMPUTER QUALITY ASSURANCE

Philip H. Heintz, Ph.D.
N.T. Enloe Memorial Hospital
1500 The Esplanade
Chico, CA 95926

ABSTRACT

The importance of quality control in the computer industry is only being realized today. This paper addresses the problems and concepts associated with quality control in the computer field. Special emphasis is placed on the responsibilities of the program developer. Focus is concentrated on the troubled areas within software quality control and concepts are discussed to cope with these problems. The end user's responsibility is outlined.

"Quality Assurance" or "Quality Control" are terms used a great deal today, especially in the medical field. Government agencies such as the Food and Drug Administration and the Center for Devices and Radiological Health along with the Joint Commission on Accreditation of Healthcare Organizations, the American Association of Physicists in Medicine and even the public are clamoring for quality products, peer review and quality control[1&2]. In fact, much of what a medical physicist does today is quality assurance. When a physicist does acceptance testing of a new piece of equipment, routine equipment calibrations (daily, weekly, monthly and even annually), and radiation therapy chart checking he or she is performing quality assurance. Thus, the medical physicist is very familiar with the usefulness, necessity and boredom of quality assurance.

Probably the most important reason for quality assurance is <u>safety</u> followed by <u>correctness</u> and <u>reliability</u>, but cost is often the overriding factor. Whenever quality assurance is performed, it must address these areas. In most products quality assurance is directly affected or dictated by economics. Figure 1 shows that economics is a tradeoff between the cost of quality assurance and the cost of defects. The minimum in the total cost of a product is the optimum economic return of that product. This may or may not be the most reliable or even the most acceptable end product. As medical physicists, we have an analogous problem when designing radiation shielding. We must travel the fine line between the cost of the facility and the safety of the personnel.

What is quality assurance or quality control as it applies to the computer industry? To answer this question, let us divide the subject into two areas: hardware and software. Each of these areas can then be subdivided into the responsibility of the manufacturer and that of the end user as shown in Figure 2. Although this paper deals primarily with software quality assurance it is very difficult to discuss software quality assurance without briefly discussing hardware because they are heavily entwined. Since hardware is the easiest to discuss, let us address hardware first, then devote most of the time to software.

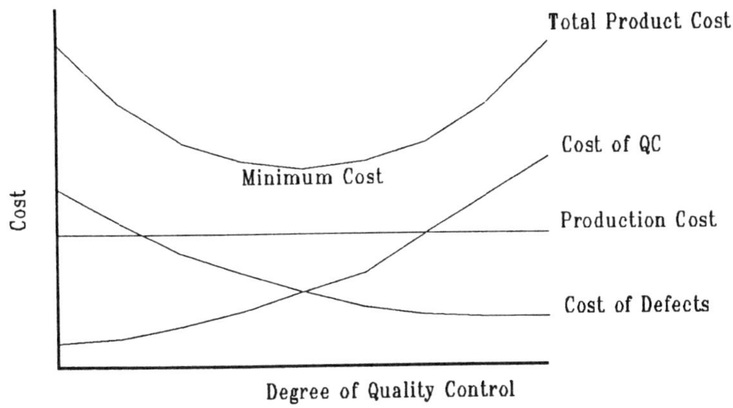

Effect of Quality Control & Defects
On Cost of Production

Figure 1

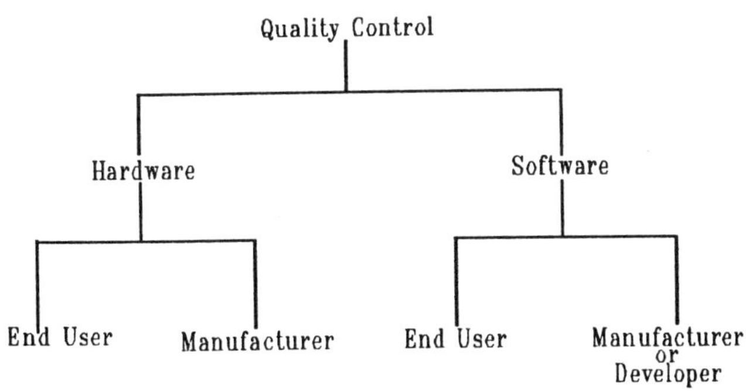

Quality Control Responsibilities

Figure 2

Quality Assurance for Computer Hardware

Quality assurance for computer hardware is fairly easy, as it involves the design, manufacture, and use of a physical product. Strict and well-established procedures apply to the manufacture of most physical products. However, computer hardware technology is changing. In order to enhance the hardware performance the hardware has become more sophisticated and more complex. Microcode or software burned into hardware is being used more frequently. What this means is that hardware quality control is becoming less like that needed for physical products and more like that needed for software.

Statistical quality control is one of the major technologies used for establishing a quality control program. Dr. W. Edwards Deming[4] introduced statistical quality control during World War II for the procurement of military products. His approach has been broadened and applied to almost every form of manufacturing process today. These procedures have been molded to develop acceptance and performance tests of products by the end user. The Japanese consider Deming the father of statistical quality control, as they are certainly masters in this technology. Only of late are the U.S. manufacturers starting to fully implement Deming's ideas. Most hardware manufacturers follow Deming's techniques during manufacturing to measure and evaluate their quality control.

One good indication of the quality control in a hardware product is in its warranty. Generally, the longer and more comprehensive the warranty, the better the quality of the product. For example, in the automobile industry, five years or fifty thousand miles, parts and labor free, is not uncommon. The price of extending the warranty is a further measure of the quality of the product.

As the computer hardware market has become more competitive and quality has improved, warranties have been extended. Recall when IBM introduced the PC-AT it had a full year warranty. This was a significant change from the original 90-day warranty offered on the PC. Other manufacturers followed IBM. Now, some computer manufacturers have two to five year warranties on their hardware. Even hard drives with so many moving parts are warranted for over one year.

To perform quality assurance on computer hardware, the end user must have certain tools. The manufacturer usually provides a library of diagnostic programs to test the hardware. These tools should be supplemented by standard test cases designed to test the specific needs of the end user's hardware. Such programs are available from magazines like PC Magazine, Byte Magazine, DEC Professional and others. With these tools, the end user can evaluate the product during acceptance testing to make sure that the hardware meets the manufacturer's specifications and reliability. These performance results can be supplemented with the results shown in these same magazines which test a variety of products for performance and reliability. The same tools should be run on a regular basis to evaluate the ongoing integrity of the hardware. Generally, when hardware fails it is very apparent (hard disk crash, power supply failure, etc.). With the help of the quality control tools, many times these failures can be anticipated and prevented by routine quality assurance testing. Yet, as the hardware or microcode becomes more sophisticated the probability of a less apparent failure occurring increases. Therefore, it is more important than ever to perform routine quality control tests on all aspects of the hardware including the CPU and floating point hardware.

Quality Control for Software

Now to the more complex subject of software "quality control". As I mentioned under hardware, the warranty is a good measure of the quality of the software. Let me paraphrase a common software warranty which might read:

> This program is sold as is, without warranty of any kind, either expressed or implied, including but not limited to, warranties of merchantilability and fitness for your purpose. The entire risk as to the quality and performance of the program is with you. If the program is found to be defective you and not the developer or an authorized dealer assume the entire cost of all necessary servicing repair or correction. This warranty gives you specific legal rights and you may also have other rights which vary from state to state.

This is purely a facetious warranty exploiting the worst traits of a warranty. Surprisingly enough for many software products it is close to reality. Read your warranties even from the prominent software companies. When applied to high risk products such as medical devices many warranties reduce to the one above primarily because of the high visibility of a medical device and potential for a legal suit. Fortunately, most reputable software companies make every effort to correct errors and omissions promptly.

Why are the warranties being offered by the manufacturing industries so different than those offered by the software industries? The answer is quite simple. Statistical quality control is widely used in manufacturing industries but very seldom used in the software industry. Today, much of the software is written by small groups and amateurs. As such, none of the good design practices and established quality control procedures are being followed. Once this software is commercially introduced, it is often unreliable and very hard to maintain. It is often undocumented and not well organized.

R.H. Thayer[9] has pointed out twenty problem areas that must be addressed before software quality control can be achieved. Table 1 lists and explains these items.

Awareness of the aforementioned problems has stirred sincere interest in improving quality control and assurance, especially in the medical field where software failures have the potential to cause severe injuries. Recently, there have been several books published which address good programming practices and software quality control problem[1,2,3,4,7,8 & 10]. A new field called quality programming is emerging. This field incorporates quality control for the development of high-quality and cost-effective software. As this field grows and the procedures are accepted, software users will be able to demand a meaningful software warranty. As in many fields, the competitive marketplace will force the software industry to accept these demands or the product will not survive.

Most medical physicists are both end users and developers. As developers, we write software to solve special problems at our own institutions. Even though the software is not a commercial product, it may be used by others in our own institution or exchanged with physicists at other institutions. Sometimes the software is developed into a commercial product. Each one of us, as such, needs to be aware of the steps that must be taken to develop good software quality control. Therefore, let me address the problem in developing good quality software before we address the end user's responsibilities.

1. Software Developers Responsibility

Planning is the first step in writing even a simple program. An estimate of the time involved should be made. Development time is divided into three areas: analysis and design, 40%, implementation, 20%, integration and testing, 40%. The relationship of each area is shown in Figure 3. Notice this is a closed loop process. Continuous reexamination and redesign is necessary.

Table 1

Software Engineering Quality Control Problems[9]

Planning

1. **Requirements:** Requirement specifications are frequently incomplete, ambiguous, inconsistent, and/or unmeasurable.
2. **Success:** Success criteria for a software development are frequently inappropriate, resulting in "poor-quality" delivered software which is not maintainable, unreliable, difficult to use, relatively undocumented, and so on.
3. **Project:** Planning for software engineering projects is generally poor.
4. **Cost:** The ability to estimate accurately the resources required to accomplish a software development is poor.
5. **Schedule:** The ability to estimate accurately the delivery time on a software development is poor.
6. **Design:** Decision rules for use in selecting the correct software design techniques, equipment, and aids to be used in designing software are not available.
7. **Test:** Decision rules for use in selecting the correct procedures, strategies, and tools to be used in testing software developed are not available.
8. **Maintainability:** Procedures, techniques, and strategies for designing maintainable software are not available.
9. **Warranty:** Methods to guarantee or warranty that the delivered software will "work" for the user are not available.
10. **Control:** Procedures, methods, and techniques for designing a project control system that will enable project managers to successfully control their project are not readily available.

Organizing

11. **Type:** Decision rules for selecting the proper organizational structure, for example, project, matrix, function, are not available.
12. **Accountability:** The accountability structure in many software engineering projects is poor, leaving some question as to who is responsible for various project functions.

Staffing

13. **Project Manager:** Procedures and techniques for the selection of project managers are poor.
14. **Techniques:** Decision rules for use in selecting the correct management techniques for software engineering project management are not available.

Controlling

15. **Visibility:** Procedures, techniques, strategies, and aids that will provide visibility of progress (not just resources used) to the project manager are not available.
16. **Reliability:** There is no practical way to show the delivered software meets a given reliability criteria. Measurements or indexes of reliability are not available and there is no way to predict software failure.
17. **Maintainability:** There is no practical way to show that a given program is more maintainable than another. Measurements or indexes of maintainability that can be used as an element of software design are not available.
18. **Goodness:** There is no practical way to show that one program is better than another. Measurements or indexes of "goodness" of code that can be used as an element of software design are not available.
19. **Programmers:** Standards and techniques for measuring the quality of performance and the quantity of production expected from programmers and data processing analysts are not available.
20. **Tracing:** Techniques and aids that provide an acceptable means of tracing software development from initial requirements to completed code are not generally available.

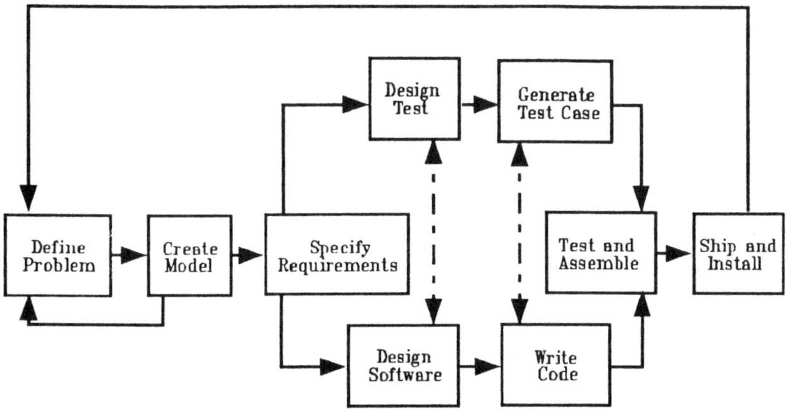

Programming Process

Figure 3

It is important that a program developer understand and implement a complete design plan including some form of quality assurance before embarking on the text of the project. Many developers (large and small) jump right into implementation, ignoring analysis, design, testing and integration. They often let the user debug the program. The final product certainly suffers from these deficiencies. Figure 4 shows a relationship between product quality characteristics and the software life cycle. Notice that it too is a closed loop process. Also notice that with every step, a quality characteristic is associated. The software quality characteristics in Box A can be defined precisely as those of any manufacturing product. Some of the dimensions that can be used to define a software product's unit characteristics are: correctness, accuracy, useability, communications. Features desirable in the system quality characteristics, or Box B include: reliability, understanding, efficiency, structuredness, consistency, robustness, testability, human engineering, modifiability, maintainability and portability.

Many of these system quality characteristics may not be measurable and conformance to the user's requirements may require a subjective evaluation. However, the conformance of the systems software product to the user's needs is definable. The quality of the end product can be examined. Each module can and should be classified as either acceptable or defective. The defective parts must be repaired.

In order to achieve quality programming, one must implement a complete software life cycle incorporating the principles of software quality assurance. One cannot just sit down and write a quick program and expect to have a quality product used by others. It must be well thought-out using a model like that shown in Figure 4. Such a process involves the following areas: modeling, requirement specification, concurrent software design and test design, concurrent implementation of software design and test design, test and integration, software acceptance. Following delivery, the cycle is closed and the product is re-evaluated and updated in the continuing process. It is extremely important to look at each stage in this closed loop process, especially the definition of the problem, modeling and writing the required specification.

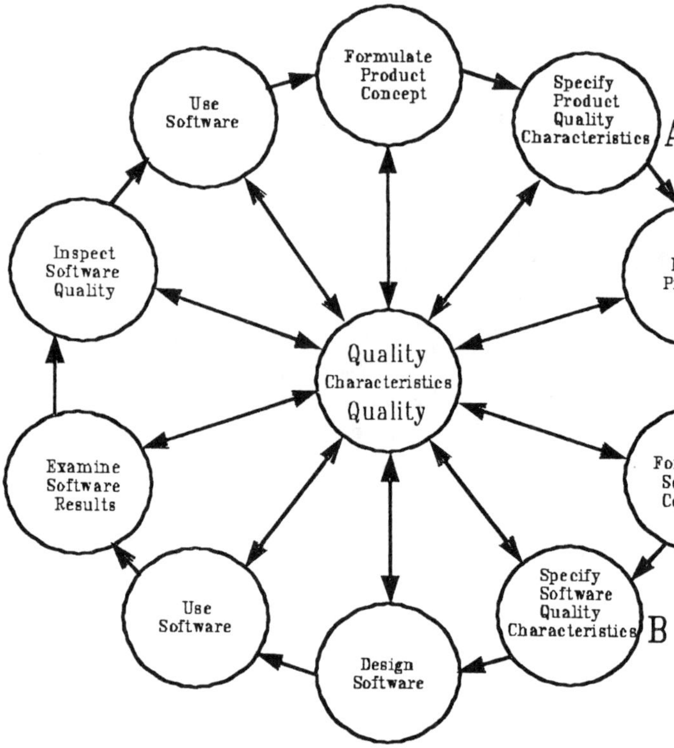

Software Life Cycle

Figure 4

The level to which these principles apply depends first on the safety risk of the software and secondly, the size of the project. The FDA (Food and Drug Administration) recently implemented review of software controlled medical devices, including linear accelerator and treatment planning computers[5]. Their levels of concern are based on safety and are divided into major, moderate and minor risk products. Linear accelerators and treatment planning computers fall into the category of major concern. Under the new guidelines, before a device can be approved for marketing the manufacturers must submit written adherence to the basic design plan described above along with a safety analysis. The completeness of the report depends upon the level of concern. Their primary concern is safety.

The ANSI/IEEE standard 730 entitled "Standard for Software Quality Assurance Plans"[1&3] follows this same approach with more emphasis on the design approach and size of the project. Adherence and documentation of the design process depends upon the man-hours involved in development. Size can be broken into three groups: one person project, less than ten person-years and greater than ten person-years. Clearly the larger the project the more complex the design plan and quality assurance program.

2. Software Design Specifications

The first and probably the most important step in implementing the design of any software system is to establish the specifications. The more complex the project, the more detailed the specifications should be. For example, in designing a radiation therapy treatment planning system the design must include a description of the calculational model and the accuracy of the implementation of the model along with input and output information. From these specifications a software quality control program can be established. Often in radiation therapy the only tests done for quality assurance are to verify that the final answer is correct. If the answer is not correct then the program is said to be faulty. Yet, the program may very well be calculating according to the specifications, it is just that the model or data are not correct under the conditions tested. In actual practice less than 20% of the code in a radiation therapy planning system is for calculation. Most of the code is input, output and screen display. Detailed specification and quality control of this part of the program is just as important as the calculational section.

A key part of software quality control is testing. Exhaustive testing is impractical if not impossible. Testing a few simple cases often implies that all cases will behave in a like manner. This is an unsafe assumption. Error rates vary from less than one error per thousand lines of code to one error per 300 lines of code.[8] A treatment planning system may have as many as 50-100,000 lines of code. That means there could be 50 - 3000 errors in a treatment planning system. It is for this reason that designing tests to find all of the errors is nearly impossible. Keep in mind that releases to fix previous errors often introduce additional new errors.

3. Software Design

Some simple rules can be followed to design software using good quality control techniques.

1. Establish detailed written specifications for the software. If this is a custom program for an end user be sure the specifications are approved by the user.
2. Expand the specification with more detail and break it up into modules.
3. Draw a flow chart showing program flow and data flow within the program. Also show how each module fits into the program.
4. Perform a safety analysis of the software. Evaluate possible errors including input, output and computational. Establish limits for these errors and incorporate these into the design specification.
5. Establish a time table for development and assign responsibilities for the project, including testing and final approval.
6. Design test cases to evaluate the design.
7. Each module should be able to be tested independently allowing all of the parts to be tested before the final product is assembled.
8. Document all of the steps listed above.

The basic design is now complete and there should be five written documents. A salient feature of each module should be put into a check sheet for evaluation after the module is completed.

Programming begins according to the time table and assignments specified. As each module is completed it is tested against the specification.

Upon completion the final product is evaluated against the original design specification. It is further evaluated using the safety evaluation.

Once satisfied that the product is complete it is released for Alpha testing by an independent party. When the testing is complete and the program modified based on the report then it is released for Beta testing. Again the program is modified based on the results of the Beta test. Following completion of the tests and final corrections made then the product is ready to be released.

Do not forget that all software products must be accompanied by written documentation, i.e., an operator's manual. At the same time as the coding is being done an operator's manual can be written. You might think it strange to write it in parallel with the coding but if the specifications are complete then much of the manual can be written at this point. The details can be added once the program is complete.

Final release of a software program should require approval from one person who checks to be sure it is complete and tested. It must have full documentation and approval from all agencies, i.e., FDA for medical devices.

This procedure sounds cumbersome for small inhouse projects. If the basic steps are followed then the inhouse projects turn out to be quality products, easy to support and use. Therefore, I recommend following these basic steps even for personal programs.

My approach to the problem is to start with complete design specifications then divide the design into independent modules which can be tested, assemble the modules and test the final product. As the product grows new modules can be designed and added to the system after they are tested. Test results and specifications should be available to the end user of the software.

End-User Responsibilities

The end user has a definite responsibility in the process for software quality assurance. Unlike hardware, which wears out, software should not change without hardware failure or user input. Software is like the text in a book: It should never change. Therefore, the end user's software quality assurance program must be different from that of hardware. The bulk of the end user's quality assurance program is involved in defining the specifications of the software and evaluating the performance based on the design criteria. In other words, acceptance testing. I emphasize testing again. Good testing cases and procedures must be established. Following acceptance testing is education. One must be sure the users know how to use the system. New users must be trained on an ongoing basis. A routine quality assurance schedule of testing hardware and software must be established to be sure no software code changes have occurred. These tests should test for safety, reliability and accuracy. Before a "new release" is put into operation, complete acceptance testing must be performed as if it were a new product. Specific attention should be made to evaluating the "new features" and the elimination of those unwanted features (bugs). Remember: Every software program has at least one bug.

Details in establishing a quality control program depend upon the origin of the software. Ready-made or off-the-shelf software requires a slightly different program from that of custom made software.

For off-the-shelf software the end user should obtain detailed product specifications from the manufacturer and design procedures to test the accuracy and applicability of the product against their needs and the manufacturer's specifications. Part of the acceptance testing must evaluate the path to report non-compliance with the specifications.

Custom software must be tested against the specifications agreed upon at the outset of the development contract. Procedures to test against the design specifications must be developed. A path must be established to correct the area of non-compliance.

In both cases the end user "should determine that the software product was developed by persons practicing systematic development methods in which the product is documented in sufficient detail that it can be audited by yourself, by regulatory agencies or by other quality assurance staff independent of the developers."[11] Medical products receiving pre-market approval from the FDA since June 1988 have gone through such an audit procedure.

In summary, quality assurance in the computer field is separated into hardware and software, but in the final analysis, quality assurance in the complete system is the most important. It is the responsibility of both the computer manufacturer and the end user to implement a satisfactory quality assurance program. Because the quality assurance programs for most manufacturers are not well-developed today, the end user has a much higher responsibility to see to it that the computer system operates safely, accurately, and efficiently. Probably the one task that the end user can do to assure him or herself of the best software product is to develop a very good description (specification) of the tasks that software is supposed to perform and on which hardware it is to operate. Convey this to the software developer, then evaluate the software once you receive it. All too often, the end user leaves the task of writing the specifications to the software developer resulting in an unhappy customer and an unusable system.

References

[1] Branstad, Martha and Patricia B. Powell, "Software Engineering Project Standards", *IEEE Transactions on Software Engineering*, Vol. SE-10, No. 1, January 1984.

[2] Bryan, William and Stanley Siegel, "Making Software Visible, Operational, and Maintainable in a Small Project Environment", *IEEE Transactions on Software Engineering*, Vol SE-10, No 1, January 1984.

[3] Buckley, Fletcher, "A Standard for Software Quality Assurance Plans", *IEEE STD730 Computer*, August 1979.

[4] Cho, Chin-Kuei, *An Introduction to Software Quality Control*, Wiley, New York, c 1980.

[5] FDA, "Reviewer Guidance for Computer-Controlled Medical Devices", a Draft, August 1, 1988.

[6] Holden, Constance, "Regulating Software for Medical Devices", *Science*, Vol. 234, October 3, 1986.

[7] Jacky, J.P. and I.J. Kalet, "Assuring the Quality of Critical Software", I.A.D. Bruinvis, P.H. van der Giesson, H.J. van Kleffens, F.W. Wittkamper,eds. Proceeding of the 9th International Conference on the Use of computers in Radiation Therapy. Amsterdam, North-Holland, 1987, pp.49-52.

[8] Petschenik, N.H., "Practical Priorities in Software Testing", *IEEE Software*, September 1985, pp. 18-23.

[9] Thayer, R.H., A Pyster and R.C. Wood, "The Challenge of Software Engineering Project Management," *IEEE Computer,* Vol. 13, No. 8, August 1980.

[10] Lamb, David Alex, Software Engineering: Planning for Change, Prentice Hall, 1988.

[11] Jacky, J.P., private communication, September 1989.

Communications and Networks

George W. Sherouse
Radiation Oncology, North Carolina Memorial Hospital,
University of North Carolina, Chapel Hill, NC 27599-7512
sherouse@godot.radonc.unc.edu

INTRODUCTION

There was a time not too long ago when the most expensive part of a department's computing arsenal was the central computer (CPU) itself. The computer was configured with terminals to support many users, thus encouraging the maximum use of an expensive shared resource. In the last five years things have changed dramatically. Now, thanks to the rapid development of the microcomputer industry, the most expensive resources in the computing arsenal are the peripherals (such as laser printers and large disk drives) and the *software and data* which the machines manipulate. In response, the emphasis in resource sharing has shifted toward convenient sharing of data and peripherals among a group of single-user computers. When a collection of machines and peripherals are interconnected so that they can transparently and automatically exchange data, we say they form a **network**.

In this presentation I will discuss a variety of mechanisms, conventions, and applications of data communication among machines, beginning with the very simplest machine-to-machine transfers, progressing through the variety of options available for local resource sharing, to a discussion of some very large national and international networks. No effort is made to review, specify, or recommend specific products. There is also an unavoidable bias toward systems for minicomputers and the UNIX operating system, for which I beg the reader's indulgence.

1. POINT-TO-POINT SERIAL COMMUNICATIONS

The most basic method for communicating data from one machine to another is through the use of a serial connection. Nearly all computers and peripheral devices come equipped with serial ports which provide data communications in reasonable compliance with the RS-232C standard. Serial connections between physically proximal machines can be made quickly and with a high degree of confidence using only three wires. One can even connect machines across considerable distances using telephone lines and modems.

Physical connection, however, is necessary but not sufficient to affect reliable data transfer. Some software is also required. At a minimum, one needs to instruct one computer to send data out and the other to catch it. Nominally, those functions are provided by commonly available utilities like COPY, PRINT, cat, etc. However, merely sending data and trusting that they will all arrive intact is overly optimistic. There is, for instance, a good possibility that the sending machine will be able to send faster than the receiving machine can receive. An additional mechanism, called flow control, must be provided to let the receiver inform the sender when a

---

Due to the nature of this topic, a great many trademarked items are discussed. To the best of my knowledge the following is an exhaustive list of the registered trademarks mentioned in this manuscript, with my apologies for any omissions or errors: DEC, VAX, PDP, DECnet – Digital Equipment Corporation; Ethernet – Xerox Corporation [ethernet (lower case e) is a generic name for Ethernet-like technology]; Kermit – Henson Productions; Apple, Appletalk, Macintosh – Apple Computer; NFS – Sun Microsystems; UNIX – AT&T Bell Laboratories.

pause is required.

Ideally, a dedicated serial line with flow control, a utility on one end to send data and a utility on the other end to catch the data would be sufficient for reliable data transfer. This, in fact, happens all the time; it is how computers drive printers and other peripherals.

For the purposes of data transfer between computers, it is desirable to have some additional mechanisms for ensuring that no errors are introduced in the data by the transmission process. This is a particular problem when using modems across noisy phone lines, but errors can be seen even in dedicated wires. A number of protocols exist for providing error-free transmission across serial connections. Two notable examples are Kermit and XMODEM, protocols for which abundant public domain software implementations exist. I will use Kermit as my example in the rest of this discussion.

It is important to emphasize the difference between a protocol and a particular software implementation of a protocol. A protocol is a set of conventions specifying a logical sequence of events which make communication possible. Those conventions can cover a wide variety of issues, but their importance is to set policy on how data are presented and intact transmission verified. Each individual programmer who implements the protocol chooses the way in which the resulting program will use the resources of the target machine to comply with the protocol.

Kermit[1,2] (named for no apparent reason after the famous frog) was developed at Columbia University to fill a need for reliable data transfer between mainframes and microcomputers. The basic principles employed by Kermit appear in some form in most protocols which provide error-free data exchange across error-prone physical connections. In the Kermit protocol, data are transmitted in packets which consist of framing characters and blocks of data. The framing characters include a start-of-block marker, a byte count, a packet serial number, a packet type field to indicate whether the packet contains data or a command, and a checksum[†]. Together these data are sufficient to detect missing or redundant packets (serial number) and damaged packets (byte count and checksum).

All data are transmitted as printable ASCII characters. Non-printable characters and binary data are represented by two- or three-character sequences of printable characters. For example, the non-printable ASCII bell (control-G) is represented by the two-character sequence #G.

All packets must be acknowledged. If the receiving machine detects an error, it asks for a retransmission of the missing or damaged packet. Otherwise, it acknowledges intact receipt and the sender proceeds to the next packet. If the receiving machine does not respond in a reasonable time the sender assumes the packet was lost and retransmits.

A variety of control messages can be passed, from simple instructions informing the remote machine of the name of a file, to such services as requests for remote directory listings. Control messages also provide the minimum functionality required for the remote machine to be controlled from the local one. The state in which the remote machine receives instructions exclusively from the local machine is called **server** mode, a concept we will see again in other contexts. Once the remote machine is placed into server mode, the local machine can request transfers of files in either direction in much the same way that files would normally be copied on a single machine.

Another interesting feature of the Kermit protocol which has counterparts in other protocols is an autoconfiguration dialogue which takes place between the two machines when they first connect. The Kermit implementations running on the two machines negotiate a variety of parameter settings such as maximum packet length, maximum time to wait for an acknowledgement,

---

†The checksum is the lower 7 bits of the arithmetic sum of the ASCII codes of all the characters in the packet.

and the codes to be used to represent 8-bit binary data and non-printable characters. Thus the two machines automatically configure one another to assure proper interpretation of packets.

Public domain implementations of the Kermit protocol are available for an astounding variety of machines and operating systems. Many implementations are of similar or superior quality to corresponding commercial products and include such embellishments as autodialing of modems from a stored directory and support of stored command files for the automation of oft-executed command sequences. The complete distribution (now several 2400 foot 9 track tapes) can be obtained from Columbia University. Implementations for specific machines can usually be found through users' groups or simply by asking around.

2. LOCAL-AREA NETWORKS

The term "local-area network", or LAN, refers to a system for interconnecting pieces of hardware which are located in close geographic proximity to one another, typically in the same department or building. In principle, a LAN is a mechanism for establishing and breaking point-to-point connections (either logical or physical) as needed to effect communication between arbitrary pairs of machines (nodes) in the network. The following sections will discuss a number of the issues which must be addressed in the implementation of a LAN. We begin with a review of a model of the general networking problem and then discuss some specific approaches to various layers of the model. For an in-depth discussion of LAN technology see the excellent textbook by Stallings[4].

2.1. The OSI Reference Model

By the late 1970s a number of vendors had produced *ad hoc* solutions to the problem of connecting large numbers of computers and peripherals into a LAN. This fragmentation of the LAN marketplace, and the potential for disastrous failure of various proprietary systems to interconnect in the future, prompted the International Organization for Standardization (ISO) to form a subcommittee in 1977 charged with the task of proposing a standard framework which could facilitate reconciliation of various LAN technologies while accommodating existing de facto standards. That effort is called Open Systems Interconnection or OSI. The strategy of the OSI group is to divide the process of transferring data from one machine to another into a series of subproblems or **layers**. As a first step, OSI developed the seven-layer Reference Model for LANs, summarized in Table I, as a basis for discussion and comparison of different technologies.

Logically, software implementing a given layer communicates directly with its peer on the remote machine using a layer-specific protocol. To actually send a message to its peer, the software which implements that layer forms the message in accordance with the peer-to-peer protocol and then passes it down to the software implementing the next lower layer on its local machine. After passing through all intervening layers, the now multiply-encapsulated message reaches the physical layer and is transmitted to the remote machine where the inverse operations occur.

An analogy can be made to a person sending a letter, though it should be taken more metaphorically than literally. The process begins when a high-layer entity (a person) has a message to send to a peer. The message is encoded using an accepted protocol (written English). The encoded message is segmented to fit the constraints of the physical medium (the writer may need to use more than one page of stationery). The segmented message is encapsulated with its destination address (placed in an envelope) and then presented to the physical transportation layer for conveyance (and short-term storage) by standard means. When the letter arrives at its destination, it is decapsulated and decoded to reveal the intended message to the high-layer entity.

Table I. The OSI Reference Model		
Layer	Name	Service
7	Application	General user access to network services
6	Presentation	Syntactic interpretation of interprocess messages on behalf of Application layer including encryption/decryption, protocol conversion, packing/unpacking of data
5	Session	Control structure for interprocess communication including sequencing, prioritizing, and buffering of messages
4	Transport	Reliable, transparent transfer of interprocess messages including error detection and flow control, perhaps segmentation of messages into smaller chunks if required by lower levels.
3	Network	Translation between logical addresses and physical routing of data chunks through network
2	Data Link	Reliable transfer of data chunks across physical link including formation of transmittable data packets, error detection, and flow control
1	Physical	Transfer of unstructured bit stream across physical medium; deals with mechanical, electrical, and functional characteristics of the physical medium

In general, the subtleties of the Reference Model concern only those people who are directly involved in development of network protocols, standards, and software. It is, however, a very useful framework for discussion of existing and proposed network systems and services. The following, slightly more detailed description of the layers is offered in that light.

Physical layer

Specifications at the physical layer describe the physical connection of one machine to another and the mechanisms by which bit streams are passed across the connection. This can include such things as the type of wire to be used, the types of connectors to be used with that wire, the scheme by which data are encoded and transferred, transfer rates, etc. The well-known RS-232C specification for serial communications is an example of a physical layer standard.

Data link layer

The data link layer transforms the unstructured bit stream provided in the physical layer into a reliable *data* stream. This involves the encapsulation of data into packets which contain extra fields to enable error detection, establishment and maintenance of the physical connection, and handshaking and flow control between the two physical entities to assure that packets are sent only when they are expected and that they are received without error. The communication protocol of Kermit can be viewed as a data link mechanism.

Network

This layer translates requests for logical interprocess connections from higher levels into requests for specific physical connections to the data link level. A telephone directory serves part of the function of a network layer entity. In cases where the physical connection is a dedicated line, the network layer has very little to do other than to initiate connections. In situations where the physical connection may involve a number of connections across a number of different networks (an **internet**) the network layer must also provide internet routing services, specifying which intermediate machines must be used. In fact, internet management is now specifically included in the OSI model as a sublayer of the network layer. Note that there is a natural division between hardware-specific concerns in layers 1-3 and more abstract process-specific concerns in layers 4-7. For that reason, the latter are generally referred to as the higher levels.

Transport
: The transport layer assures reliable, error-free communication of data across a logical network connection. This may require, for instance, that the messages generated at higher levels be broken up into smaller chunks that can be accommodated by the protocols at the lower levels. Additional flow control and error-correction services are provided at this level to ensure that interprocess messages are communicated in the proper order without loss, duplication, or error. The amount of work required at the transport layer is strongly dependent on the robustness of the network layer. If the network layer can provide reliable links, then a very small transport layer may be adequate. If there is significant potential for corruption of messages, the transport layer must be correspondingly more robust.

Session
: At the session layer logical connections (called sessions) between peer processes on different machines are managed. This layer determines the proper order and priority of interprocess messages.

Presentation
: It is often necessary to establish application-specific protocols for the messages passed between peer processes. This might include encryption and decryption of sensitive information, packing and unpacking of large data sets (such as medical images) to reduce transmission overhead, or a variety of other schemes for encoding of messages. In addition it might be necessary to resolve differences in data format between client machines, or to emulate some particular device or pseudodevice. The presentation layer is responsible for these translations.

Application
: The application layer provides users with commands and/or programs which can be used for accessing network services. Utilities which initiate cross-network file transfers or provide terminal session support are good examples of application layer entities.

2.2. Project 802

The establishment of standard protocols for the various layers of the Reference Model is an ongoing task. Perhaps of most immediate interest to end-users in the U.S. are the series of standards developed by the Local-Network Committee of the Computer Society of the Institute of Electrical and Electronics Engineers (IEEE). Established in 1980, Project 802 had the unenviable task of developing standards for the LAN industry while not disenfranchising vendors with significant research investments or market share. The result was adoption of a series of standards corresponding to the lower layers of the OSI Reference Model. Three of those standards correspond to the three dominant commercial technologies of the time: Ethernet (backed by Xerox, Intel, DEC, and numerous other mini- and microcomputer vendors), token bus (endorsed by General Motors), and token ring (IBM). Standards for the higher layers are still being hammered out by the Project 802 group and by OSI.

2.3. Physical aspects

This section describes in some detail the physical aspects of the major network technologies in widespread use today. The subsequent major section covers the predominant protocols used by those systems to arbitrate the use of the physical media.

2.3.1. Media

One important choice to be made when implementing a LAN is the physical medium which will be used to connect nodes. There are currently four types of cable in common use for LANs: twisted pair, 50Ω coaxial, 75Ω coaxial, and fiber optic. Each type imposes different constraints on the mechanisms and scale of the network. Table II gives a very rough order-of-magnitude assessment of the cost/performance characteristics of various media. The cited costs reflect only the addition of a node to an existing network, not the cost of the backbone media or electronics of the network. Note that while connection costs vary widely, the cost/performance ratio is roughly constant across the board.

Twisted pair is essentially the same wire used for conventional telephone systems. It consists of two small-gauge insulated wires twisted together to provide improved environmental noise rejection. This medium is readily available, relatively inexpensive, and easy to install. It is not unusual to find extra twisted pair cable preinstalled in a building to provide redundant cable for telephones. Even if new cabling must be laid, the space required is minimal and no special tools are required for termination and connection.

Twisted pair is most appropriate for small local networks (a few tens of nodes in close proximity) and where comparatively slow transmission rates are acceptable. Data rates of 1 Mbps (million bits per second) are typical for department-sized LANs, though 10 Mbps can be achieved for short runs (up to 100 meters). These criteria have made twisted pair technology most visible in office environments.

Coaxial cable (coax for short) is the most common medium for larger LANs including hundreds or even thousands of nodes over distances as great as a few kilometers. Two types of coax are in common use: 50Ω for baseband systems and standard CATV 75Ω cable for broadband systems. Coax can support data rates of up to 50 Mbps and so represents a better solution than twisted pair for LANs with heavy use (many nodes and/or large data transfers) or where rapid transfer is important (e.g. process control).

Fiber optic technology is a relatively recent development and is only beginning to become a reasonable option for intrabuilding interconnection. That is partly because of technical problems related to tapping and splitting the cable. On the other hand, the very high data rate capabilities of fiber have already made it an accepted medium for high-speed point-to-point connection, for example between two geographically separate subLANs. With current technology data rates in excess of 2 Gbps can be achieved. Standards for 100 Mbps local fiber networks are under development.

2.3.2. Topology

Another important design choice for a network is the topology, or physical interconnection scheme to be used. There are three schemes in common use: star, ring, and bus.

Table II. Order-of-magnitude Cost/Performance of LAN Media

Medium	Maximum Speed (Mbps)	Maximum Range (kilometers)	cost/node (dollars)	cost/Mbps (dollars)
twisted pair	1	1	< 100	< 100
	10	.03	100	10
coax	50	10	500	10
fiber optic	1000	100	25,000	25

In a **star** topology, each node is connected to a central switching station which arbitrates connections. This is the topology of telephone systems: each telephone is a node which can initiate a dedicated connection to any other node (phone) by telling the central switch the address of the remote node (dialing the number). Just as with telephones, a number of point-to-point connections can exist simultaneously. For LANs, star topology is usually implemented using twisted pair cable, though some of the current limitations of fiber optic interconnection can be circumvented using a star topology since no tapping or splitting is required.

A **ring** LAN forms a closed loop consisting of point-to-point connections between pairs of nodes. Each node is connected to exactly two neighbors, like a circle of people holding hands. The concept of a direct physical connection does not exist in a ring as it does in a star. Messages are passed around the circle from node to node, and so conceivably there can be as many messages circulating at once as there are nodes. Each node in the network is responsible for maintaining continuity between its neighbors. If a node fails, the entire network is brought down until the faulty node can be either physically or logically bypassed.

Since connections in a ring LAN are point-to-point (no taps or branches required) nearly any transmission medium can be used. Twisted pair is a common choice, but coax or fiber can be used if higher data rates or large distances between nodes are required.

The simplest topology in many respects is the **bus**. One main cable is laid around the entire area to be served by the LAN and nodes simply attach to (tap) that ''backbone'' cable where needed. Thus, unlike the star or ring, a bus topology implies that all nodes share the one main cable and so only one pair of nodes can communicate at a time. The bus is the special case of a **tree** in which there are no branches. An advantage of the bus topology is that operation of the network does not depend on any particular node. Nodes can be added or removed with no disruption of service to the other nodes.

2.3.3. Mechanics of data transfer

An issue which is closely related both to the topology of a network and to the type of medium used is that of electrical signal propagation. In this section we examine some alternative strategies for putting meaningful electricity on the wire.

2.3.3.1. Baseband coax One of the first bus architecture LANs, and still the dominant player in many LAN markets is Ethernet. One of the definitive characteristics of the Ethernet technology (and the closely related 802.3 standard) is the use of 50Ω coaxial cable to implement a single digital channel for communication. Since a baseband wire supports only one channel, nodes must take turns in using the wire. This is analogous to cars taking turns to use a narrow one-way bridge on a country road.

50Ω coax was chosen rather than the more common 75Ω CATV coax in order to reduce reflection of digital signals from the taps. The cable is typically tapped using so-called ''vampire taps'' which clamp onto the coax and penetrate the layers of insulation to make electrical contact. The tap includes a **transceiver** which contains the electronics for transmitting and receiving signals on the backbone, and is connected to the node device (computer, peripheral, etc.) by a tranceiver cable.

Digital signals propagate bidirectionally from their origination point on the backbone coax and are absorbed by terminators at the ends. For the standard 10 Mbps data rate the recommended maximum backbone length is 500 meters and the maximum number of taps is 100. The taps are spaced at multiples of 2.5 meters to prevent constructive interference of reflections from adjacent taps.

2.3.3.2. Broadband coax If a baseband system is a narrow country bridge, broadband is a superhighway with traffic peacefully coexisting in separate parallel "lanes". In contrast to the single digital channel of baseband, broadband systems employ analog encoding and have a number of logical channels discriminated by frequency in the same way that various TV channels are discriminated in cable TV systems. It should come as no surprise, then, that 75Ω CATV coax is the cable typically used.

Reduced attenuation of the analog signals in the cable as compared with the digital baseband signals allows broadband networks to be much longer without the need for repeaters, on the order of a few tens of kilometers. The propagation of signals in a broadband system is necessarily unidirectional because of constraints imposed by amplifier design issues, so to achieve the desired connectivity between arbitrary pairs of nodes it is necessary to use a central node, called a **headend**, to redistribute signals. The headend must be one end of a bus topology or the root of a tree topology so that it has a "clear shot" to and from every node. Nodes transmit all packets to the headend on an *inbound* channel. The headend then retransmits the packet on a different *outbound* channel. All nodes transmit on one channel and receive on another. Note that the headend is not a switch as the central node in a star topology would be, but serves a function more analogous to that of a superhighway interchange. Since traffic can only travel in one direction, packets have to get off at the interchange and turn around (change frequency) in order to go in the other direction toward their intended destinations.

In addition to the ready availability of cable and accessories, broadband systems have the potential advantage that channels which are not used for data can be used for the things normally associated with CATV such as video, radio, etc. It is also worth noting that a baseband network can be implemented using a single channel of a broadband system, leaving the rest of the cable's channels free for other uses.

2.3.3.3. Ring networks I have already made the analogy of ring architectures to a circle of people passing notes. Let us examine more closely the mechanisms required for message circulation.

From the ring's point of view there are two logical functions to be performed, insertion of messages and removal of messages. For the ring to function properly, mechanisms must be provided to ensure 1) that messages are only inserted if there is sufficient idle capacity to accommodate them, 2) that the messages get removed after they have reached their destinations or clearly have no hope of reaching their destinations (due to a bad or damaged address or failure of the destination node, for instance) so that they do not clog the ring by circulating forever, and 3) that the circle will be unbroken (by and by) despite the failure of participating nodes or the entry of new nodes.

From each node's point of view, there are four functions to be performed: 1) initiate messages for other nodes, 2) recognize and read messages intended for the node, 3) remove messages for which the node has responsibility, and 4) keep track of connectivity (that is, maintain a functional link to a downstream neighbor).

Both sets of demands can be fulfilled by nodes which have two states. In listen mode, the node waits for a message to come in, examines the message to see if it needs to do something special with the message, and either does that something (like removing or copying the message) or just passes it on to the next node in the ring. In transmit mode, the node hands its packet off to its neighbor. While transmitting, it listens for incoming bits and either removes them (if it recognizes them as the beginning bits of the message it is still sending) or buffers them. When transmission of the packet is complete, any buffered bits are either read or sent on, depending on their intended destination.

2.3.3.4. Repeaters, bridges, and gateways Just as there is a real need to connect computers and other devices on a LAN, there is also very often a need to connect LANs into extended LANs or Wide-Area Networks (WANs). There are three dominant strategies for achieving interLAN connectivity, and they are often confused.

A **repeater** is the simplest device for connecting two networks, either to extend service in a single administrative domain beyond the normal distance limits of a LAN, or to interconnect LANs in different administrative domains. In a situation where two instances of the same network technology (say two ethernets) are to be connected, a device called a repeater which bidirectionally amplifies and passes on (repeats) traffic can be inserted between the two networks. To each subLAN, the repeater is just another node. The effect is a logically contiguous network; all nodes on both subnetworks appear to be on the same LAN, and all traffic originating in each subLAN appears on both. Repeaters typically come in pairs, one attached to each LAN, which are physically connected by coax or fiber and so can be very far apart compared with the limited size of a LAN.

A slightly more sophisticated device, a **bridge**, serves the same purpose as a repeater but with the added feature that it contains enough intelligence to *isolate* local traffic to the originating LAN. Techniques vary, but the basic bridge strategy involves listening to traffic on both sides of the bridge and taking note of the originating addresses encoded in the packets on the subLANs. In this way a table is built of node addresses for each side of the bridge. Using that table, the bridge is able to prevent local traffic from leaking from one subLAN to the other by only passing packets across the bridge if the origination and destination nodes are not on the same side of the bridge.

A third strategy is required if the two networks to be interconnected are not of the same type, by virtue of differences in physical specifications or in communications protocols. In that case an active agent, normally but not necessarily a general-purpose computer, is connected to both networks and software is provided for transforming packets of information from one technology into the corresponding packets of the other technology. The computer in that situation is said to be an internetwork **gateway**.

2.4. Medium access protocols

In a previous section we discussed the physical mechanisms which are provided by several LAN technologies. In this section we will cover the protocols used in LANs with shared media to assure that every node gets a chance to use the medium and that every message is received intact by its intended destination node. The underlying principle of these protocols is that if nodes are going to share a single wire, a means must be provided by which it can be guaranteed that only one node uses the wire at any given time.

2.4.1. CSMA/CD

The mechanism used by Ethernet and IEEE 802.3 is called Carrier Sense Multiple Access with Collision Detection or CSMA/CD. The protocol is actually much simpler than its name. When a node wishes to send a packet it first listens to the cable to see if there is any activity. As soon as the wire is idle, the node transmits immediately. While the node transmits, it listens to make sure that its packet is not being garbled by some other node transmitting at the same time. If there is a collision, the node immediately stops its transmission in midpacket and transmits a jamming signal so that all nodes realize there has been a collision. The node then waits a random time and attempts to retransmit. Since each node chooses its random time independently, there is a good chance that the same collision will not recur and that each node will be able to transmit successfully on its next attempt. As an added precaution against deadlock, the random wait time is made longer with each unsuccessful attempt to transmit the same packet.

2.4.2. Token bus

An alternative approach to shared access is the use of token passing. A variety of schemes exist but their common thread is that, in contrast to the collision detection of CSMA, token passing schemes *avoid* collisions by circulating a **token** which grants nodes the right to use the wire for some slice of time. Token-passing in a bus architecture, discussed in this section, is the basis of the IEEE 802.4 standard. The token ring is described in the next section. The mechanics of token passing are too complex to describe in detail here, but I will attempt to illustrate some major features.

In a token bus, the nodes are organized into a *logical* ring, generally based on sequencing of unique addresses assigned to each node. The physical order of the nodes on the bus is irrelevant. Under normal circumstances, a bus-grant token circulates continuously from node to node. The node which holds the token has control of the bus until it either voluntarily relinquishes it or its allotted time expires. It then passes the token on to its successor. While a node controls the bus it can send and receive as many messages as it needs to.

Many things have to be monitored and managed in this scheme, for instance: 1) when the network is first powered up none of the nodes know who their predecessors or successors are and so have no idea from whom they should expect packets or to whom they should pass them, 2) if a node fails (e.g. gets turned off) it can no longer pass tokens and so the network is jammed, 3) if a new node wishes to join the network it has to break in to the circle somehow, 4) if the token gets lost the network will jam, and 5) if more than one token should appear on the network, chaos will ensue.

Of course, these issues have all been addressed by the standard. The general strategy is that if any node detects that something has gone wrong it initiates actions to either rediscover the identities of its neighbors or to reinitialize the whole ring depending on the nature of the fault. Since the maintenance of the ring is decentralized it can become fairly complex. That is, in fact, one of the most often cited disadvantages of token-passing techniques. A supposed advantage is that medium access in token-passing systems are (mostly) deterministic so it is assured that each node will get its fair share of access at regular intervals. Another potential advantage is that medium access can be prioritized, for instance by assigning longer access time windows to more important nodes. These attributes makes token-passing attractive for process control and other time-critical applications where events need to be serviced promptly.

2.4.3. Token ring

The token ring is the access protocol chosen by Project 802 for ring networks (IEEE 802.5). In most respects it is very similar to token bus. I will briefly discuss the differences.

First, the mechanism used in token bus, wherein a node can send out messages as long as it holds the token, will not work in a ring. Everything released onto the ring will circulate indefinitely if not removed by some active element. Accordingly, the token-passing protocol is modified as follows. A single "free" token circulates. If a node needs to transmit, it waits until it sees the token, changes it to a "busy" token, and places the packet to be transmitted on the ring right behind the busy token. The other nodes do not see a free token and so will only listen. The node for which the packet is intended will copy the packet as it goes by and mark it as copied. The transmitting node will place a free token on the ring at the end of the transmitted packet and will remove the busy token and the message packet when they come back around.

Token ring systems have all the same problems of maintaining connectivity and of maintaining the token as discussed above for token bus. A slightly more democratic process is used to monitor the ring, though. The details of the mechanism are complex, but the essence is that one node is elected lord of the ring and has responsibility for sensing token faults and initiating recovery procedures. Of course, as in any good democracy, the other nodes are responsible for making sure that the token monitor is doing its job. If not, a successor is promptly elected.

2.5. Higher layer protocols

It has been said that the great thing about standards is that there are so many to choose from. Nowhere is that more apparent than in the higher layer protocols used by various LAN systems. A number of systems exist which implement fully functional networks. In that sense, various *ad hoc* standards exist which, by definition, implement the functionality of all seven layers of the OSI Reference Model. Unfortunately, most existing systems do not divide the tasks in the same way as did the OSI group. Some even violate the strict layering of the OSI model by supporting direct communication between peer layers of different processes or between nonconsecutive layers in a single process without passing through all intermediate layers.

It is not unusual to see discussions of specific network systems accompanied by figures illustrating the correspondence or lack thereof between the system in question and the OSI Reference. At one level, this is in keeping with the intent of the Reference Model to serve as a basis for comparison. However, it also points out how far we still are from real standards for the higher layers of the model. Notable players in today's LAN market include DEC (DECnet), IBM, Apple (Appletalk) and perhaps surprisingly the U.S. Department of Defense (TCP/IP). Details of the various proprietary protocols and system designs are available from the vendors in question. TCP/IP, though, bears a bit more discussion because of its contribution to the evolving OSI standard for the network and transport layers of the Reference Model.

When the Department of Defense speaks, contractors listen. In the case of higher layer protocols, DoD has spoken through the Defense Advanced Research Projects Agency. The DARPA Internet, discussed further in the section on major national networks below, uses protocols for reliable transmission of data across logical point-to-point connections and for information exchange using connectionless services. Connectionless services were not initially included in the OSI model but can be quite useful in applications where information is to be made available to anyone who cares to listen or where information is sought without regard to which network entity provides it. Examples of this type of service from UNIX are utilities which broadcast information regarding current system status to other machines and collect the same statistics from those machines. Implementation of this service as connectionless implies that the local host need not know in advance which machines to listen to, it just listens to anybody who speaks up.

The internet sublayer protocol used by the Internet, responsible for message routing, is called Internet Protocol (IP). The transport layer protocol which supports reliable point-to-point connections is called Transmission Control Protocol (TCP) and the transport layer protocol which supports unreliable connectionless communications (datagrams) is called Uniform Datagram Protocol (UDP). Both TCP and UDP assign a unique address to a particular service of a particular host machine.

The combination of IP and TCP, commonly called TCP/IP, has become widely accepted by virtue of its ready availability to the academic community in BSD UNIX, its use in the DARPA Internet, and the DoD's requirement that TCP/IP be supported on equipment it purchases. One can now purchase software to support TCP/IP for most mini and microcomputer environments.

It is worth re-emphasizing that many vendor-specific and other higher layer protocols are nominally compatible with the same lower layer protocols. It is conceivable to have a number of workstations communicating via TCP/IP, some Macintosh computers using Appletalk, and some VAX computers speaking DECnet all on the same ethernet. The various homogeneous groups should not interfere with one another. They won't be able to talk to one another across protocol boundaries, either.

2.6. ACR-NEMA and PACS

No discussion of networks for a medical physics audience would be complete without some mention of the ACR-NEMA standard for Picture Archiving and Communication Systems (PACS). PACS refers to systems for storing, fetching, and transmitting images, usually in the context of a hospital Radiology department. These systems are examples of LANs, but with the significant special constraint that they be capable of very high data rates so that very large image sets can be transmitted in very short times. Typical nodes in the PACS LAN are data sinks such as viewing workstations and digital storage devices, and digital image sources such as CT scanners, MR scanners, digital radiography units, and film digitizers. In a fully-capable PACS network it would be possible to transfer images to and from any of a variety of scanners, archives, and viewing consoles both in the Radiology department and in other departments such as Radiation Oncology. Perhaps the most attractive promise of a networked Radiology department is the elimination of lost or ''borrowed'' images.

Clearly, the problem of interconnecting these diverse devices can be greatly simplified if standards are set for the presentation format of the images. It was in this spirit that the American College of Radiology (ACR) and the National Electrical Manufacturers Association (NEMA) set out to define standards for data exchange. The resulting ACR-NEMA Standard defines a six-layered model for image transfer corresponding to all seven layers of the OSI Reference Model. The standard specifies image formats, formats for administrative data to accompany images, the electrical characteristics of the network interface, and various communications protocols. It is important to note, however, that the ACR-NEMA standard does not specify, or even require, a network technology for data transmission. The standard only goes as far as the cable and plug coming out of the machine. The mechanism for getting data from one plug to another is a local decision, and will normally rely on one of the LAN technologies already discussed.

So far manufacturer compliance with the standard in product offerings is slow. This is partly because the standard, while relatively complete, is still undergoing validation. It also appears to be a function of the current lack of demand for ACR-NEMA from the marketplace. As specifiers and consumers of imaging devices, medical physicists should be aware that reasonable interconnectivity of those devices, in compliance with industry-wide standards, will only come to the marketplace if it is made an issue in purchasing decisions.

2.7. Services

Issues of system design and standards aside, what ultimately counts is the set of services which a network provides, and the related question of how easy it is to create new services. The most popular services are file transfer, remote login, disk sharing, printer sharing, electronic mail, and electronic news.

In some sense the most basic network services are file transfer and remote login. File transfer is usually implemented as an extension of the normal file copy command, with a syntax change to allow the name of the source and/or destination machine to be specified. For instance in BSD UNIX ''cp file1 file2'' is normally used to copy file1 to file2. The network copy version is ''rcp godot:file1 file2'' (the r is for remote) to specify that file1 on remote machine godot be copied to file2 on the local machine. Similarly, remote login is often implemented as an extension to the normal login procedure, e.g. ''rlogin godot'' instead of ''login''. The effect of remote login is to (logically at least) attach the user's terminal to a terminal port on the remote machine.

Disk sharing and printer sharing are more transparent to the user. In most implementations, the machine to which the disk or printer is attached is called the server. Users are granted access to the peripheral exactly as though it were physically attached exclusively to each machine. The details of remote access are invisibly handled by the operating system and/or dedicated background server processes. One widely-used system for remote disk usage is the Network File

System (NFS) from Sun Microsystems which is now available from a number of sources for a variety of machines and operating systems.

The ability to share resources across networks offers a clear advantage in optimizing the use of those resources. It also has the significant effect of reducing the overall maintenance burden of the system. For instance, it is much easier to maintain one shared copy of software or data than to maintain copies on each of a set of machines. Reducing the overhead involved in maintaining copies on all machines in turn reduces the burden of routine system backups. It is important to realize that maintaining software and data for a LAN is at least as big a job as, if not bigger than, maintaining a single large machine.

Electronic mail and electronic news are often touted as virtues of LANs. The transparent delivery of mail and news between machines, just like peripheral sharing, involves the use of hidden system software to do the actual work. In both cases, textual messages can be composed at the computer and then sent either to a specific person's private mailbox (mail) or to a public area where anyone can read them (news). These mechanisms are very useful in situations where the analogous physical functions, namely a written note or memo, are appropriate. They are limited, however, by the inclination of the intended reader to regularly take the time to read and respond to mail or news. It is a mistake to justify investment in a LAN with the fantasy that the department business manager who is always off at meetings somewhere and never returns his calls will be suddenly more accessible.

In the BSD UNIX operating system (and analogously in other environments) it is relatively easy for a programmer to implement new network services. The usual model for a network service is that of two (or more) processes connected by a dedicated (logical) communication channel. The process which requests the service is called the **client**, the provider of the service is the **server**, and the connection is a pair of **sockets**. The programmer writes his server to open and listen to a socket at the address assigned to that service. The server can either be started manually by a user or automatically by the system. When a client process wishes to use the service, it opens a socket and attempts to connect to the server's socket. Once the connection is made, the client and server can exchange any required information, such as details of the request or results of the service, across the socket pair and then close the connection. Once the service has been delivered, the client goes away satiated and the server goes back to listening for requests.

An important feature of this scheme is that the client and server need not be on the same machine. Thus services such as dose calculation for a treatment planning system can be implemented as network services and then run on any machine in the network. This strategy, which is used by the /usr/planning Radiotherapy Treatment Design System (RTDS) developed by our group at UNC, allows a number of relatively slower treatment planning stations to share a much faster machine for time-intensive calculations or to perform the calculations locally, at the user's discretion. The same strategy is used by the X Window System from MIT to provide network-transparent display management for workstations; each display has a server associated with it and any client which wishes to create or modify a window does so by communicating with the server for that display.

3. MAJOR NATIONAL AND INTERNATIONAL NETWORKS

A surprisingly large number of national and international networks exist for corporate, academic, military, and other uses, and most of them are interconnected through gateway nodes. An even more surprising number of people, several members of the faculty of this summer school included, use one or more of these networks routinely in their day-to-day work. I will confine my descriptions to the three networks to which medical physicists are most likely to have access, but refer the interested reader to a very thorough recent survey article by Quarterman and Hoskins[3].

3.1. ARPA Internet

One of the oldest internets, the most pervasive network in the UNIX community, and one of the most influential in terms of fundamental global network research, is the ARPA Internet (the upper case I is significant). The Internet began as the ARPANET in 1969 to fill certain common needs of DARPA contractors. It has since spun off a strictly military subgroup called MILNET and has been augmented by the commercial network service CSNET. It is estimated that the Internet now serves a few thousand computers and tens of thousands of users, both inside and outside the military-industrial complex. The major services supported are mail, file transfer, and remote login. It is possible for me to sit down at my workstation in North Carolina, login to a machine in Massachusetts, interactively obtain directory listings, and select and transfer perhaps tens of megabytes of publicly available software back to my machine. There is no apparent difference to me between remote login to machines in the Computer Science Department across campus and login to machines at Berkeley across the country, except for a certain sluggishness of response. Electronic mail normally travels across the country in under 5 minutes.

The Internet proper consists of the ARPANET and MILNET backbone sites which are connected via 56 kbps dedicated lines. These sites may in turn serve as gateways for LANs or extended LANs, representing a wide variety of technologies such as ethernet, token ring, and others, thus granting Internet access to large numbers of dependent hosts. The details of host management for the dependent sites are the responsibility of the backbone site. There is a central administrative organization with which each host is registered by a unique name and network address.

A hierarchical naming convention, including domain and subdomain specification for each host, has recently been implemented. The name consists of two to four fields, more to less specific from left to right. The top-level domains specify the type of organization to which the host belongs: COM for commercial, EDU for educational, GOV for government, MIL for DoD, and ORG for others. The next two fields may contain subdomain specifications. For instance, the University of North Carolina at Chapel Hill is unc.edu, and the Radiation Oncology Department is radonc.unc.edu. The central computer in Radiation Oncology is godot.radonc.unc.edu and my Internet mail address is sherouse@godot.radonc.unc.edu. Each host is assigned a unique (numerical) Internet address. Host and domain numbers are also hierarchical. Radonc.unc.edu is 128.109.146, godot is 128.109.146.1. The translation between host names and host addresses was table-driven in the past but, in part due to the unwieldy size of the table, many hosts now use a distributed name service to dynamically resolve addresses.

DARPA has recently announced that for most purposes the present ARPANET is to be taken down in the near future. The Internet will continue, partly through funding from other government agencies such as NSF.

3.2. BITNET

In contrast to the publicly-funded Internet, BITNET is a cooperative network of mostly university sites in more than 20 countries. BITNET began in 1981 between City University of New York (CUNY) and Yale University. It now serves approximately 1500 sites worldwide. The cost of BITNET membership is the lease on a dedicated phone line to an existing BITNET site and a willingness to serve as host to at least one future site. A site must also buy or emulate IBM's communications protocol, Remote Spooling and Communications Subsystem (RSCS), though BITNET is apparently considering migrating to TCP/IP. The use of the network is otherwise unrestricted. BITNET has a central administration which looks after host registration and policy decisions. The Canadian and European arms of BITNET are, respectively, NETNORTH and EARN.

The main services provided by BITNET are mail (using the same protocol as Internet), file transfer, and semi-interactive teleconferencing. Host addressing is flat, that is host names have only one field and so all host names must be unique across the whole network, though Internet-style subdomain support is apparently planned for the near future. For example, Trevor Cradduck's BITNET address is TREVORC@UWOVAX.BITNET. Communication between hosts is via 9600 bps leased lines. The topology, interestingly, is a tree (that is, while a site may serve many dependent sites, there are no loops in the connectivity). Mail and file transfers are reasonably reliable and take tens of minutes to a few hours.

3.3. UUCP/USENET

The UUCP network and its Frankenstein offspring USENET are a minor miracle of the computing age. The UNIX-to-UNIX CoPy (UUCP) program was originally implemented as a mechanism for automatic file transfer and remote command execution across dial-up serial links between UNIX machines (much like Kermit in many respects). This ability to move files combined with the ability to remotely execute commands that specify what is to be done with those files is all that is required for electronic mail transfer across phone lines. Furthermore, with knowledge of the connectivity of machines it is possible to save on long distance (locally at least) by requesting that the remote machine forward the mail to another host. Thus, by the proliferation of phone connections between machines, the UUCP network has become arguably the largest fully functional anarchy in the world. The UUCP network has no central authority and is maintained exclusively by community effort. It is estimated that there are close to 10,000 hosts representing something like 250,000 users in the UUCP network, though these estimates are necessarily very unreliable.

Historically the host naming of UUCP has been flat (Internet-style addressing is expected soon) and addresses are specified as a series of intermediate sites, called a path. For instance, I usually give my UUCP address as ...!decvax!mcnc!godot!sherouse indicating that godot is connected to mcnc, which is connected to decvax, which is a site everybody knows how to get to. Clearly, this type of specification places a heavy burden on the sites that everybody knows how to get to (the so-called backbone sites). For that reason, the volunteer UUCP mapping project has developed software and issues maps which allow one to locally divine "optimal" paths to other UUCP sites.

The only services of UUCP are mail, file transfer, and remote job execution. It would probably not be nearly so popular if not for the USENET news system which is based on UUCP.

USENET began in 1980 as a connection between two machines, one at the University of North Carolina and the other at Duke University (15 miles away in Durham). It now consists of several thousand hosts, and uses a number of transmission protocols in addition to the original UUCP. Like the UUCP network, USENET has no central administration. Policy decisions are made partly by public debate in the appropriate newsgroup, and partly by fiat by a group of volunteer USENET managers known widely as the net.gods (pronounced net-dot-gods, a play on the USENET convention for naming newsgroups). In fact, any system manager in the USENET community can issue commands to create and delete newsgroups, but the rain of electronic hate mail that results from unilateral (even accidental) actions of that type is a strong disincentive. The software has been rewritten a number of times by volunteers and is freely distributed using the USENET.

USENET news is divided into a hierarchy of **newsgroups** containing open public discussions on an plethora of topics. The top level divisions are: comp for computing-related topics, news for discussion of USENET, talk for verbose discussions of a mostly socio-political nature, rec for discussion of various forms of recreation, sci for science, soc for somewhat more scholarly discussion of social issues, and misc for everything else. This division is intended to allow sites to choose to only subscribe to "technical" news in order to limit their overhead (see volume statistics below). Some notable newsgroups are comp.unix.wizards where you can ask questions of a technical nature regarding details of the UNIX operating system, and rec.nude for the latest in nudism news. All told, there are roughly 300 newsgroups.

All news is stored locally on each participating machine; there is no central repository and no guarantee of delivery. Machines connect to one another periodically (varying from on-demand to once nightly) and exchange any news items which they have come into possession of since the previous connection (a sort of electronic gossip). Most sites share news with more than one other site. In this respect, USENET functions very much like BITNET, though the topology of USENET is fairly complex, by no means a simple tree. Traffic analysis suggests that 90% of news items reach 90% of USENET hosts within 3 days, though delivery order is nondeterministic. It is not uncommon, for instance, to see replies to questions before seeing the corresponding questions due to topological anomalies.

The major costs of USENET membership are the disk space to store news and the long-distance phone charges between geographically distant nodes. Typical news traffic for a full subscription is on the order of *several megabytes a day*. Since news is normally expired after two weeks, the steady-state storage burden for a full subscription is roughly *40-50 megabytes per machine*. Still, end-node participation in USENET easily pays for itself every time a difficult local problem is resolved overnight by a consultation with "the net", a very common occurrence among USENET users. The backbone sites, however, bear the bulk of the expense of transcontinental communication links and so tend to take a dim view of the more frivolous uses of USENET resources. For that reason drastic changes are often threatened.

CONCLUSION

The days of the isolated computer are clearly behind us. The current wisdom regarding deployment of departmental computing resources holds that a network of reasonably capable workstations augmented by special-purpose server machines for disk storage, common peripheral devices, high-speed computation, high-performance graphics, and other special functions is a cost-effective and usually preferable alternative to the conventional model of one large shared computer. The workstation LAN model also naturally accommodates incremental expansion and upgrade much more easily than a single large system. The standards, technology, and products required to make this model a reality are readily available in the marketplace.

At the same time, the barriers to cost-effective interconnection of machines on a national and international basis are falling away rapidly. A number of medical physicists already make use of large scale computer networks†. The potential for closely-coupled collaborative work across wide geographic separation, which has been successfully exploited in a number of other disciplines, is only beginning to be tapped in our specialty.

ACKNOWLEDGEMENTS

I am profoundly grateful to the following friends and associates who reviewed this manuscript for form and/or substance: Phil Stancil, Joel Tepper, Liz Sherouse, Ed Chaney, Bob Thompson, Brad Hemminger, and James Coggins.

†Trevor Cradduck maintains an electronic mailing list for medical physicists involved in Nuclear Medicine and I maintain a similar list for Medical Physics (mostly therapy). Any electronic mail sent to a mailing list is automatically redistributed to all subscribers. For subscription information contact TREVORC@UWOVAX.BITNET regarding the nucmed list or medphys-request@godot.radonc.unc.edu for the medphys list. To send a message to all subscribers use NUCMED@UWOVAX.BITNET or medphys@godot.radonc.unc.edu, respectively.

References

1. Frank da Cruz and Bill Catchings, ''Kermit: A file-transfer protocol for universities; Part 1: Design considerations and specifications,'' *Byte,* p. 255 (June, 1984).
2. Frank da Cruz and Bill Catchings, ''Kermit: A file-transfer protocol for universities; Part 2: States and transitions, heuristic rules, and examples,'' *Byte,* p. 143 (July, 1984).
3. John S. Quarterman and Josiah C. Hoskins, ''Notable Computer Networks,'' *Communicatons of the ACM* **29**(10) pp. 932-71 (October, 1986).
4. William Stallings, *Data and computer communications,* Macmillan (1985). ISBN 0-02-415440-7

Reader's Guide to Cryptic Acronyms	
ACR	American College of Radiology
ASCII	American Standard Code for Information Interchange
AT&T	American Telephone and Telegraph
BITNET	Because It's Time NETwork
BSD	Berkeley Software Distribution
CATV	Community Antenna TeleVision (Cable TV)
CPU	Central Processing Unit
CSMA/CD	Carrier Sense Multiple Access with Collision Detection
DARPA	Defense Advanced Research Projects Agency
DEC	Digital Equipment Corporation
DoD	United States Department of Defense
IBM	International Business Machines
IEEE	Institute of Electrical and Electronics Engineers
IP	Internet Protocol
ISO	International Organization for Standardization
LAN	Local-Area Network (a tautology of Local Network)
Mbps	Million bits per second
MIT	Massachusetts Institute of Technology
modem	modulator/demodulator
NEMA	National Electrical Manufacturers Association
NFS	Network File System
NSF	National Science Foundation
OSI	Open Systems Interconnection
PACS	Picture Archival and Communication System
RF	Radio-Frequency
RSCS	Remote Spooling and Communications Subsystem
RTDS	Radiotherapy Treatment Design System
TCP	Transmission Control Protocol
UDP	Uniform Datagram Protocol
UUCP	UNIX-to-UNIX CoPy
WAN	Wide-Area Network

THE ACR-NEMA DIGITAL IMAGING AND COMMUNICATIONS STANDARDS

Steven C. Horii, M.D.
Georgetown University Hospital
Washington, D.C. 20007

ABSTRACT

The rapid growth of digital imaging techniques coupled with the pressure to develop PACS led the American College of Radiology (ACR) and the National Electrical Manufacturers Association (NEMA) to develop a means by which different manufacturers could provide image and associated data in a common format over a common interface. In addition to development of this "parent" standard, the committee formed by ACR-NEMA has done (and continues to do) additional work on related standards including documents for data compression and magnetic tape. This paper will describe the history of the ACR-NEMA effort and provide an overview of the existing standards and work in progress.

INTRODUCTION

A combination of advances in computer hardware and software, expansion of digital techniques in medical imaging, and interest in improving the delivery efficiency of medical imaging services led naturally to electronic picture archiving and communications systems (PACS). Enthusiasm for PACS was very pronounced initially [1] and though the interest in the systems has not waned, the hope for widespread PACS within a few years of the initial meeting has not been met [2]. Though perhaps the most important reason for the development proceeding more slowly than first envisioned is cost, there is little doubt that the inability to interconnect imaging equipment from different vendors has also been a factor.

This problem, the lack of hardware and software standards for image interchange, prompted the formation of the ACR-NEMA Digital Imaging and Communications Standards Committee. The original goal of the Committee was to produce a standard which would allow imaging equipment from different manufacturers to exchange image and associated data.

HISTORY

At the first PACS meeting, held in 1982, the problem of acquiring image and image descriptive data from different manufacturers' equipment was already apparent. Most groups doing PACS related work were using digitized video (which has the disadvantage of limited dynamic range) or magnetic tape (which is comparatively slow). The latter method also often meant "breaking" the manufacturer's "code", which was really understanding the tape format and pixel packing method. The desire to acquire digital data in a direct manner was strong, but vendors were not unreasonably cautious about revealing how their software worked. The potential users of PACS, the radiology community, asked of the American College of Radiology whether a solution could be found. The engineering contingent within the manufacturer camp also began to realize that there was little to be gained by keeping image formats proprietary. The result was the formation of the ACR-NEMA Digital Imaging and Communications Standards Committee in early 1983.

The original charter of the Committee was to develop an interface so that image and related information could be exchanged between a piece of medical imaging equipment and whatever was on the other side of the interface [3]. In defining this as the charter, the Committee restricted its range of activity. The resulting standard would not be a full PACS standard, nor a database standard, nor a network standard. However, the ability to bring digital image information out of a piece of equipment over standard hardware using a standard format was felt to be an important enough goal.

The Committee empowered three Working Groups to accomplish the task defined. The Working Groups were assigned to hardware and protocols (WG I), data format (WG II), and system performance specifications (WG III). The full Committee assured coordination of the groups and the participation of members in more than one group ensured that communication between groups would be simple.

The Working Groups began their tasks by examining existing standards. This effort continued until quite late in development (the MAP/TOP Standard was considered even after the draft ACR-NEMA document had been written) since it was the intent to use other standards if possible. Other standards examined included IEEE 488, Ethernet, Intelligent Peripheral Interface (IPI), SCSI, DEC DR11W, DEC Unibus, and Multibus for hardware and AAPM Magnetic Tape, LANDSAT, NATO magnetic tape, and the Japanese JAMIT standard for data formats. While some ideas were drawn from these, none in particular was satisfactory for all the requirements defined. In addition, because of antitrust regulations in the United States, NEMA must operate under certain constraints. This prohibited the use of certain proprietary interfaces, or interfaces which would unfairly benefit a single manufacturer.

The result of this work was that a new standard had to be defined and the Working Groups set out to get this task done as rapidly as possible. The Working Groups went through numbers of proposals by circulating them to each other and the engineering staffs at the parent companies. Within two years a standard had evolved which was felt to meet all the requirements set forth by the Committee. Following Committee approval, the document was approved by vote of the member companies and ACR representatives. It was distributed at the Radiological Society of North America meeting in 1985 and published as a NEMA Standard (ACR-NEMA 300-1985) in the same year .[4]

ACR-NEMA 300-1985: THE PARENT STANDARD

This Standard, titled "Digital Imaging and Communications" covers the data format, protocol, and hardware interface necessary for transmission of data between medical imaging equipment. The interface plane is defined to be between the two devices as shown in Figure 1.

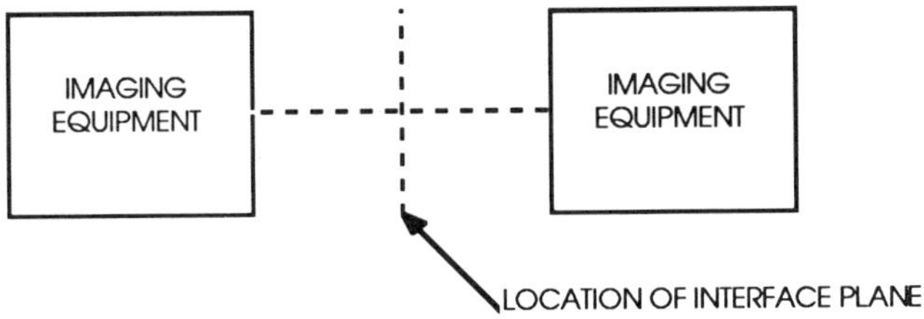

FIGURE 1: LOCATION OF THE ACR-NEMA INTERFACE PLANE

Though not intended as a network standard, it was certainly anticipated that for PACS use, connection to a network of some form would be essential. In this case, the device on one side of the interface would be a network interface unit (NIU). The network in use would determine the nature of the NIU. It was also recognized that such a unit would have to have significant local processing power, probably on the order of a microcomputer. However, this was felt to be reasonable, given the magnitude of the interfacing problem. A diagram of how the ACR-NEMA interface might be used in a network environment is shown in Figure 2.

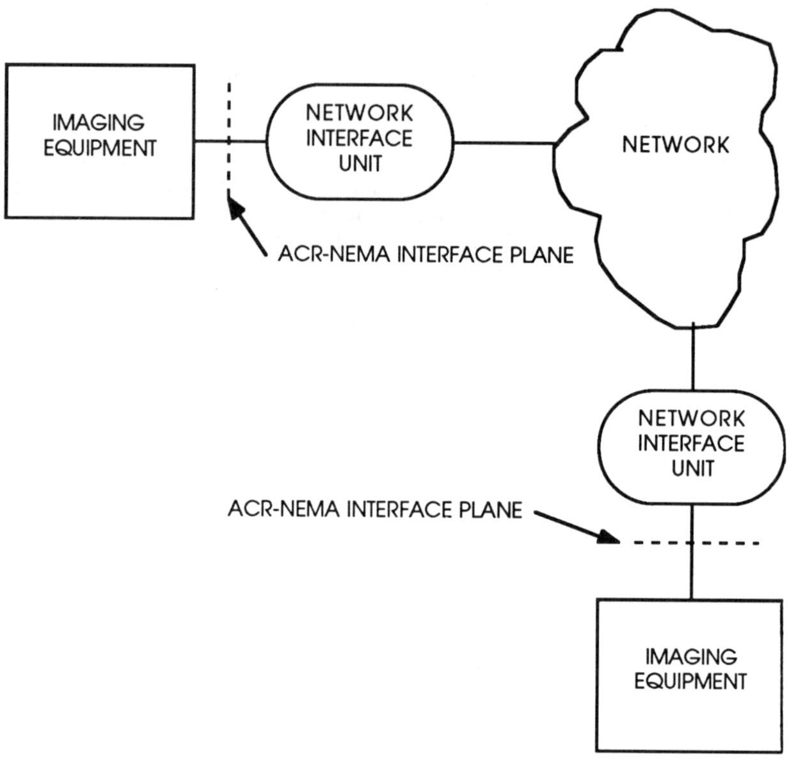

FIGURE 2: LOCATION OF ACR-NEMA INTERFACE PLANE WITH A NETWORK

In designing the interface, the Working Groups used as a model the International Standards Organization Open Systems Interconnect (ISO-OSI) Reference Model.[5] In this model, the functions of a network interconnect are split into well-defined layers. Each layer performs a specific set of operations, providing services to the layer above and below. On each side of an interface, the same layers are performing the same functions, thus to the programmer, the layers can be thought of as communicating directly with their equivalents. This "peer-to-peer" communication actually occurs in a vertical direction in the layers, and true connection is only at the physical layer.

The ACR-NEMA Standard builds upon the ISO-OSI Reference Model in that specific layer functions were defined. However, the ACR-NEMA Standard is not ISO compatible. An attempt was made to adhere to the layer functionality of the ISO-OSI model, but is not complete. The ACR-NEMA document refers to named layers and divides functions accordingly. The correspondence between the ISO-OSI layers and ACR-NEMA layers is shown in Figure 3.

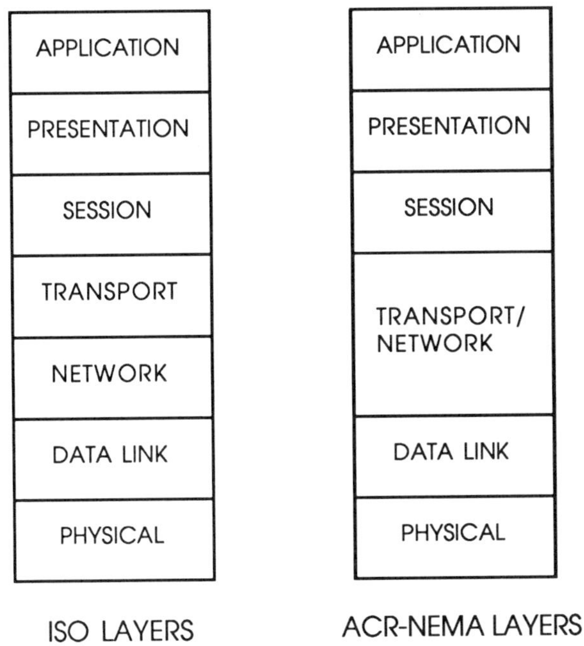

FIGURE 3: SIMILARITY OF ISO AND ACR-NEMA LAYERS

The discussion that follows will start at the topmost (or Application) layer and proceed downward to the physical layer.

When an examination is performed upon a patient, certain data about the patient is accumulated along with information on when and how the images were acquired, and along with the images themselves, forms a set of data amenable to display and interpretation (a data set in ACR-NEMA terminology). In the Standard, this information is associated with commands, and the whole is called a message. The commands indicate certain fundamental operations on the data: SEND, GET, FIND, and MOVE. SEND is the basic command, allowing an imaging device to output its data. GET allows a requesting device to obtain a specific set of data. FIND is a primitive search tool in that it will locate all sets of data which match the information requested. MOVE allows one device to request a second to send data to a third. In the Standard, message assembly and formatting takes place in the Application and Presentation layers.

The message format itself consists of data elements collected into groups. Each data element (for example, patient name) consists of two numeric fields forming a key, a length field, and the element value. The key (which is fixed length) consists of the group number and element number (in hexadecimal). The length field is a 32 bit integer which gives the length of the value field in bytes. The value field is therefore of variable length, the only restriction being that lengths must be an even number of bytes (padded if necessary). Figure 4 shows an example of the element structure.

Group	Element	Length	Value	Meaning
0008	0010	000C 0000	4341 2D52 454E 414D 3120 302E	"ACR-NEMA 1.0"

FIGURE 4: ACR-NEMA DATA ELEMENT STRUCTURE

Groups were defined in order to collect elements in a logical fashion. This was not entirely successful as it seemed some elements could then be located in two places. Since elements are not duplicated (except with specific elements in group 0000 hex) this resulted in imperfect groupings. The result, however, is that the group-element keys are unique. The structure also provides for expansion in that some "room" was left between assigned group and element numbers to allow for additions. To provide for user-defined or manufacturer specific data, a structure of shadow groups was defined. This also provided a mechanism by which new groups or elements could be used until considered for incorporation into revisions of the Standard. The defined groups of the Standard are given in Table 1.

Table 1
ACR-NEMA Standard Groups

Group Number	Type of Information	Description of Group Contents
0000	Command	Information which provides for end-to-end communication capability between two pieces of equipment
0008	Identifying	Information which provides unique image identification parameters to the user
0010	Patient	Information related to the patient
0018	Acquisition	Information related to the image acquisition device and imaging procedure
0020	Relationship	Information relating the image to the location within the patient and to other associated images
0028	Image presentation	Information related to the manner in which the image can be presented or displayed in a consistent and reproducible way
4000	Text	ASCII text
6000-601E	Overlay	Information related to overlay pixel data associated with an image
7FE0	Pixel data	Image pixel data

Though it appears that the Standard contains a great many elements which need to be sent with each image, in fact, some elements are specific for certain types of images. Also, the elements were assigned types based on how important it was thought they were for interpretation. For example, pixel format information is essential whereas equipment manufacturer name is probably not. The former is considered a mandatory data item (a type 1) and the latter is optional (type 3). There are also type 2 elements which are considered mandatory, but which conceivably may not be known at the time of the examination. Unlike type 1 elements, which must be sent and must contain valid data, type 2 elements may be sent with a zero length, indicating that the value is not known. Defaults exist for many elements and fall into type 1D and 2D categories. The difference between these is that the default is always assumed if a type 1D element is not sent. It is not possible to send a zero length (indicating unknown) for a type 1D element whereas this is possible for a 2D (in which case the default is not assumed).

Once the message is assembled, it is sent down to the Session layer. The Standard has provided "hooks" in this layer for a multiple virtual channel protocol. Connection control words and message source and destination fields are defined in this layer for incorporation in layers below.

The Transport/Network layer takes the message with the information from the Session layer, and fragments the message into packets. Packets consist of a block of the message (0 to 2048 bytes) plus a packet descriptor word and block sequence number. The packet descriptor word contains the information on channel number as well as packet type (channel control) and packet service class (i.e., acknowledge each packet or not). Packet structure is shown in Figure 5.

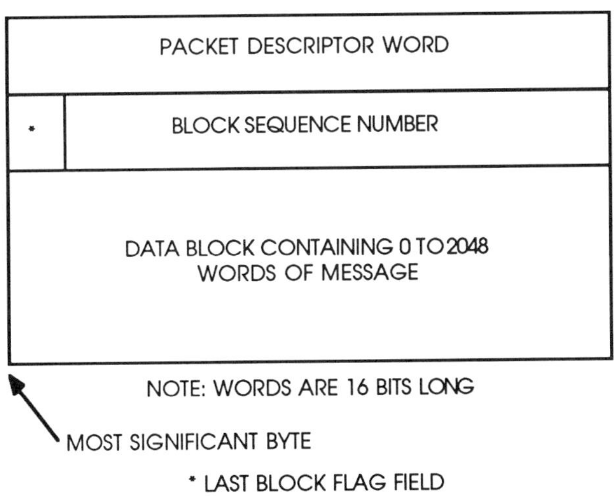

FIGURE 5: ACR-NEMA PACKET STRUCTURE

The now packetized data proceeds down to the Data link layer. Each packet is enclosed by a frame descriptor word and frame check sequence to form a frame. The frame descriptor word consists of a frame type (e.g., data and status) and a parameter field (length or status depending on frame type). The frame check sequence is a checksum. This allows for error detection, and with the addition of word parity information, enables single-bit error correction (though not required by the Standard). Frame structure is illustrated in Figure 6.

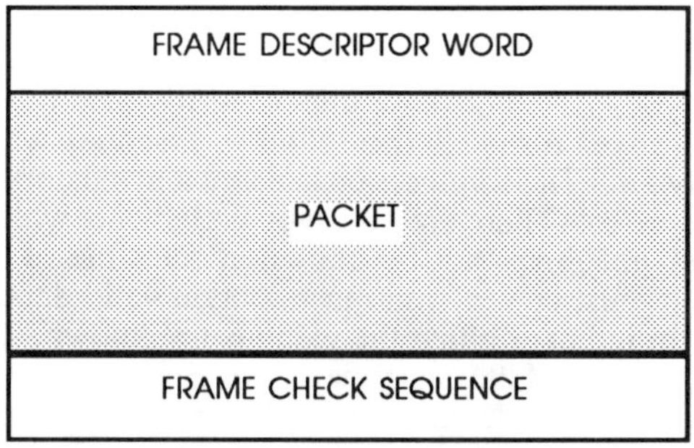

NOTE: ALL WORDS ARE 16 BITS LONG

FIGURE 6: ACR-NEMA FRAME

Frames are sent to the Physical layer which handles transmission across the interface. The physical interface is a 16 bit asynchronous parallel interface with control signals. The drivers for the signal lines are EIA RS485 standard, driving differential pairs for high noise immunity. The connector is a widely available 50-pin micro-type and must be female on the equipment. Since the interface is designed to operate in a master-to-master environment, the control lines cross over in the cable. With the specified drivers and a cable length of 15 meters, speeds of 8 megabytes per second can be achieved. Since the interface is asynchronous, however, much slower rates (limited only by the 1 second idle timeout) are accommodated. Figure 7 and 8 show the layout of the physical lines and the interface connector, respectively.

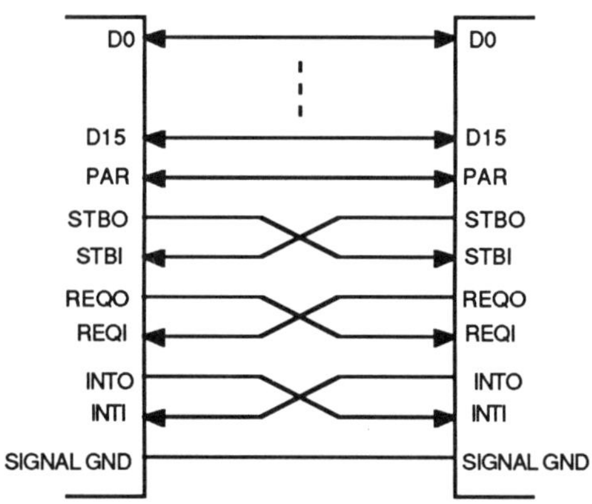

FIGURE 7: ACR-NEMA PHYSICAL INTERFACE SIGNAL LINES

FIGURE 8: ACR-NEMA PHYSICAL 50-PIN CONNECTOR

On the other side of the interface, the frames would proceed back up through the layers, eventually resulting in a reassembled message. Figure 9 shows overall flow through the layers.

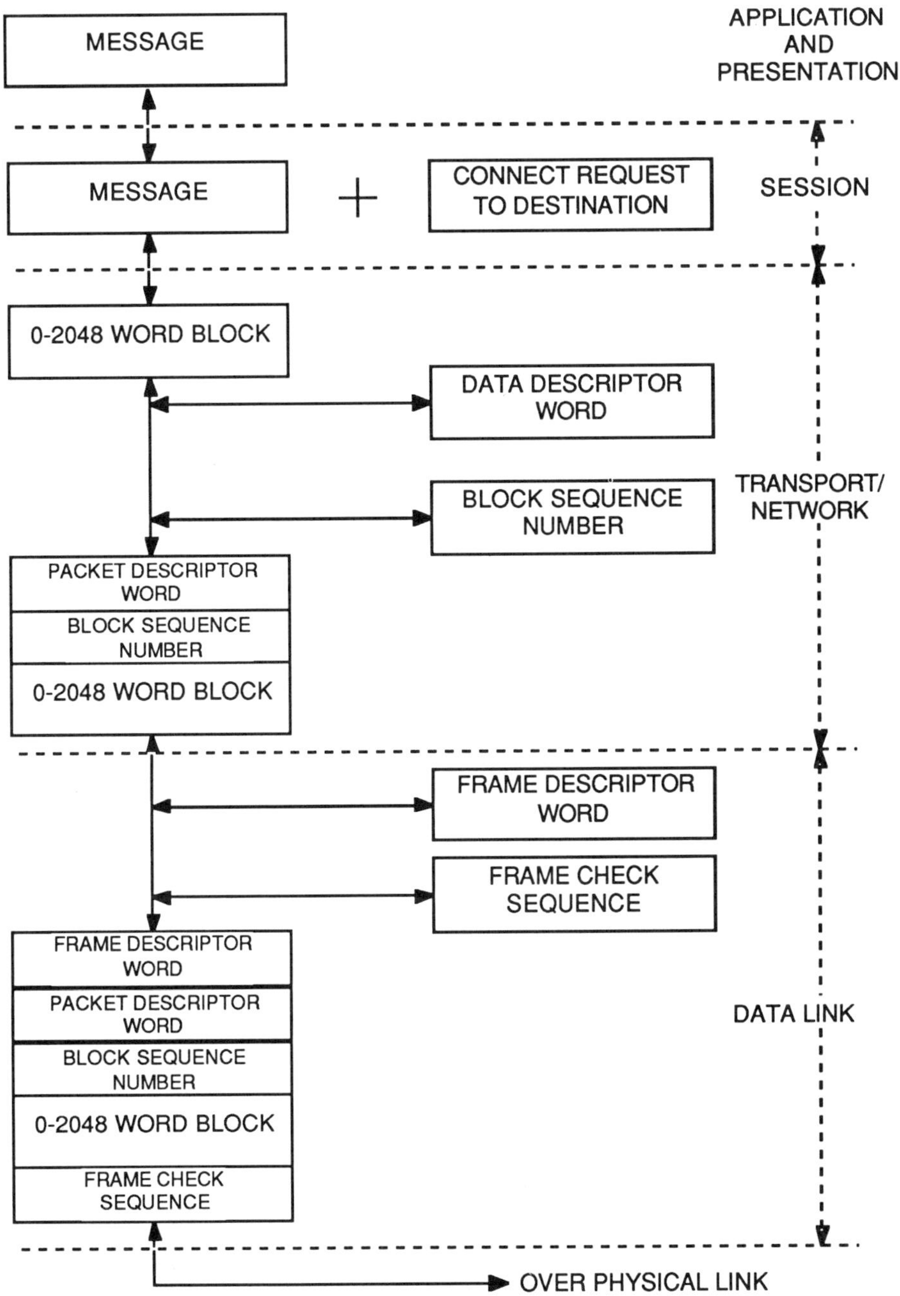

FIGURE 9: DATA FLOW THROUGH THE ACR-NEMA INTERFACE

While the Standard at first seems complex, in fact, a minimum for compatibility has been defined. This states that of the commands, only SEND need be implemented. Also, only a single channel need be supported. The full information on minimum requirements has been published by NEMA.[6]

THE OTHER STANDARDS: DATA COMPRESSION

Following publication of ACR-NEMA 300-1985, it was recognized that additional work was necessary as additional requirements were made known to the Committee. New Working Groups were formed to examine these new issues. Initially, Working Group IV was assigned data compression, Working Group V exchange media, and Working Group VI began to address validation.

The vast volumes of data generated by medical imaging departments put a strain on archival media, even optical disk. The communications systems which must handle the volume are also burdened. To help alleviate this problem, investigators have been examining data compression. This paper will not address the benefits of lossless versus lossy data compression as this was also not the task of WG IV. Rather, the Working Group sought to allow for data compression and decompression in a manner consistent with the existing Standard.

It was recognized that data compression consists of two steps. In the first the data is operated on so that it can be represented in a different way. For example, differences between pixels might be computed so that the data may be represented by these differences rather than the full pixel values. The second step consists of coding the converted data so that less space is needed for its storage. In reviewing compression methods applied to monochrome medical images, three popular conversion methods and three coding techniques were identified.

Supported conversion techniques are: differential pulse code modulation (DPCM), discrete cosine transform, and the pyramid (and its relative, the S) transform. DPCM is a differencing method which represents original data by a set of differences. The discrete cosine transform is a real valued transform for which fast algorithms exist. The pyramid and S transform are also differencing methods but compute sets of differences (in horizontal, vertical, and diagonal directions) from pixel data. The pyramid and S transforms have an advantage over others in that viewable images, reduced in resolution, appear at each step.

Once converted, the coding methods take advantage of redundancy in the data to do the compression. The compression standard supports run length, Huffman, and Lempel-Ziv methods. Depending on how the data are quantized, the compression may be bit-preserving (lossless) or non bit-preserving (lossy). An example is in quantizing the cosine transformed data. If the quantization extends out until transform values vanish, there will be no high-frequency information lost. Truncating the transform data before this point will cut off such high-frequency information.

Another popular compression method which is lossless but does not fit well into the above list of methods is perimeter encoding. In this technique, which is used for nonrectangular images, the zero values outside the actual image area are not saved. A map describing the number of pixels per line is sent instead. This method is also supported by this standard.

The design of this standard is such that methods may be combined if such combination will yield more compression. The compression standard consists of new groups and elements which allow a device to describe how compression was done. For proprietary algorithms, the standard provides a structure for sending these data about the compression used. The structure of the compression standard was designed to be flexible and allow for future growth.

As of the time of this writing, the final draft document for the compression standard has been circulated for a vote by members of the ACR-NEMA Committee.

EXCHANGE MEDIA

Working Group V of the ACR-NEMA Committee was charged with evaluating exchange media for possible standardization. When the Working Group was formed, the most popular exchange medium was one-half inch magnetic tape. Though there was considerable interest in optical disk, there was no industry standard for them. The interest in archival as well as exchange media came from the manufacturers who preferred to write one set of software for generating exchange or archive volumes.

Since there was no consensus on optical media standards at the time WG V began working, the group decided to consider magnetic tape first. Since the charter is to consider exchange media in general, as industry moves towards standards for optical media (for example, the 5.25-inch medium) the Working Group may continue to evolve standards.

As the full ACR-NEMA Committee had done, existing standards were first reviewed. The AAPM Magnetic Tape Format [7] was a very good possibility, but did not address all the issues required for exchange. The manufacturers thought that too much modification of existing operating system drivers would have to be done. LANDSAT, NATO,and Japanese JAMIT were among the other standards reviewed. None was thought to be as good as the AAPM standard for medical use. The Japanese standard, in particular, had fixed-length fields. If possible, the desire was simply to write out an ACR-NEMA message less the command group (this message subset is called a data set) onto the exchange medium. While manufacturers had widely differing formats for data, a common element proved to be use of the ANSI standards for magnetic tape. These covered both the physical media standardization and common labels and file structures. Using the ANSI standards, it was possible to fit the ACR-NEMA data set structure into the ANSI standard file structure with a "wrapper" consisting of ANSI labels. Figure 10 shows the overall structure of an ACR-NEMA compatible tape.

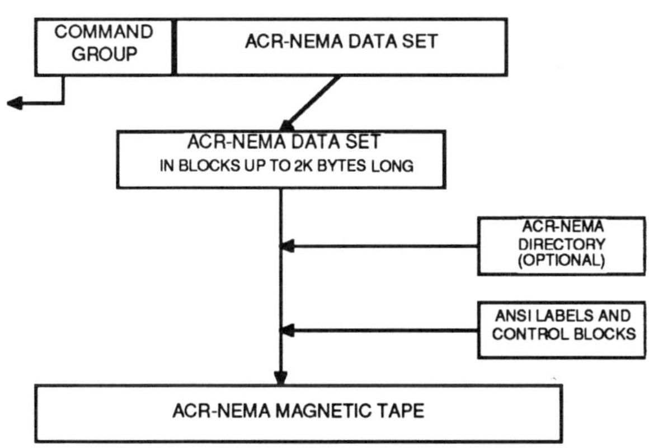

FIGURE 10: ACR-NEMA MAGNETIC TAPE STRUCTURE

One problem which had to be solved was the issue of how to build a directory structure and allow for updates of the directory as images are added to a tape. The directory structure developed is flexible in that it allows the user to specify the elements from the ACR-NEMA data set being used as directory entries in the first directory segment. The second segment of the directory consists of these elements for each data set on the tape. Additional allowance is made for non-standard directories, or using the "header" portion of the ACR-NEMA message as directory entries. At the simplest level, no directory is required.

The problem of tape updates is handled by allowing either dummy "filler" files or tape gaps to be written after the directory and certain labels. These allow for expansion as entries are added to the directory or labels. Since the directory type and update space may differ with different users, levels of implementation were defined. This allows users to specify which directory and expansion allowance method they are using. At the two simplest levels, the tape is not updated. The directory (if present) and data sets are written once. The two more complex levels have directories and either the filler file or tape gap method of allowing for expansion.

As of this writing, the draft of the magnetic tape standard is undergoing editing prior to letter ballot of the ACR-NEMA Committee members.

VALIDATION

Working Group VI was formed to examine issues related to validation of the "parent" standard. This includes both validation of the Standard itself, and assisting those implementing or testing hardware and software for the Standard. The Working Group is the clearinghouse for questions about the Standard and for suggestions regarding new groups or elements. This group is also examining issues related to revision of the Standard, and defines how revisions will be compatible with prior versions. The basic guideline has been that no element would be changed or replaced unless found to contain errors. Errata sheets for the Standard and a definition of the minimum requirements were two efforts of the Working Group.

The first revision of the Standard (version 2.0) has been letter balloted, approved, and published (ACR-NEMA 300-1988). This revision includes new elements for definition of an identification hierarchy, elements added for ultrasound, and commands and elements added to support laser film writers and other hardcopy devices. All changes are compatible with version 1.0 of the Standard.

This Working Group has been the busiest, and its work continues. As discussed in a later section, the revision for version 3.0 is presently being studied.

MULTIDIMENSIONAL DATA

With the increasing interest in images which are of multiple spatial dimensions, or which include parametric or time-sequenced data, Working Group VII was formed to examine how such data could be accommodated in ACR-NEMA 300-1985. The Standard specifies that images are rectangular with the number of dimensions allowed to be "2".

This Working Group has some working proposals for member use, but has suffered from difficulty assembling all members for meetings. Some of the newer work of WG VI may supersede the activity of this group. The Working Group is nonetheless seeking suggestions from interested users or manufacturers.

HIS/RIS/PACS INTERFACES

The newest of the Working Groups is defining a standard for interface of hospital information systems (HIS), radiology information systems (RIS), and PACS. Unlike the parent Standard, the product of WG VIII will be an application level standard. It will most likely build around the parent Standard so that ACR-NEMA messages (or parts) may cross from HIS to RIS to PACS and back.

The necessity of this effort is based on the requirement of having certain RIS or HIS based data in the PACS (patient demographic information, for example), and getting PACS data (such as time of examination) back to the RIS or HIS. Allowing direct exchange reduces the problem of operator errors in transferring information between systems. Currently, much patient information has to be re-entered in the PACS. Errors lead to database inconsistency which can result in an inability to access data or duplication of files. A similar situation pertains to PACS data which must be copied into the HIS or RIS.

Progress has been rapid for WG VIII, and a variety of proposals are before members for discussion.

FUTURE DIRECTIONS

The original ACR-NEMA Standard has undergone one revision, and has spawned two new standards with a third on the way. All of these have retained the structure of the parent Standard, and are compatible with it in various ways. However, as user requirements mature in PACS environments, even more changes to the Standard are anticipated.

The most ambitious effort at present is based on work done on a standard called the Standard Product Interconnect (SPI). Developed by Philips and Siemens in Europe, this standard is a superset of the ACR-NEMA Standard. Where the ACR-NEMA Standard did not address operation over networks, SPI does include this aspect of PACS. SPI also addresses the issue of using the ACR-NEMA data structure separate from the hardware part of the interface. In effect, this provides elements and commands necessary to send ACR-NEMA messages over other interfaces and networks. Working Group VI and other Working Groups with appropriate expertise are currently examining the SPI documents to determine if they can be used as the basis for the next version of the ACR-NEMA Standard. This does not obviate using the full ACR-NEMA Standard since SPI is a superset, but it allows "logical compatibility" for use over alternate hardware.

This also relates to a frequently asked question about availability of ACR-NEMA implementations. As of this writing, implementations for VME bus to ACR-NEMA, IBM PC/AT bus to ACR-NEMA, and Digital Equipment Corporation Q-bus to ACR-NEMA exist, though some are admittedly prototypes. The Japanese manufacturers have indicated that they are working on both full ACR-NEMA implementations as well as a minimum requirement version they call MIPS. Sanctioning the use of the ACR-NEMA data structure over non-ACR-NEMA interfaces creates a problem in that it has the potential to result in a multitude of hardware-incompatible devices. However, many of the ACR-NEMA Committee members see this change as a necessity as it will allow the ACR-NEMA data structure to be more widely used. The hardware interface will still be available, and for applications which require its speed, is still highly desirable.

The ACR-NEMA Committee welcomes comments on all of its efforts. In particular, WG VI is interested in hearing from those who are building ACR-NEMA interfaces. A large part of validation is based on discovering what problems befall those who are building harware or writing software. The various Working Groups can be contacted through:

National Electrical Manufacturers Association
2101 L Street, NW Suite 300
Washington, DC 20037
Attention: Staff Executive, MedPACS Section
(202) 457-1965

ACKNOWLEDGEMENT

The author wishes to acknowledge the members of the ACR, the manufacturer members of NEMA, nonmember contributors, and NEMA staff for the work on the Standards. The ACR members included: James Lenr, M.D., Gwilym Lodwick, M.D., John Loop, M.D., John Moore, Ph.D., Alan Rowberg, M.D., Bob Thompson, Yen Wang, M.D., and Jason Zielonka, M.D. Manufacturer representatives and nonmember participants included: Edgar Alzner, Ph.D., David Best, Ph.D., Hartwig Blume, Ph.D.,Jack Bowie, Samuel J. Dwyer, III, Ph.D., Allan Edwin, Paul Fromme, David Glassman, Ph.D.,Shankar Hegde, Ph.D., David Hill, Ph.D., John Hoffman, Will Hunt, Bob Lawrence, Robert McCarthy, Royal Morse, Owen Nelson, Ph.D., Herman Oosterwijk, Marcia Paisner, John Perry, Alan Rothlauf,Andrew Sherman, Ph.D., Chris Spilker, and Bill Wessel. NEMA staff members Cindy Fuscoe and Laura Lee Murphy made much of the work possible. The author wishes also to thank all those not named above who contributed their time and energy to the development of the ACR-NEMA Standards.

REFERENCES

1. SPIE Proceedings, Vol 318, parts I and II, Picture Archiving and Communication Systems (PACS) for Medical Applications, January, 1982.

2. SPIE Proceedings, Vol 914. part B, Medical Imaging II: Image Data Management and Display, February, 1988.

3. ACR-NEMA Dig. Imaging and Comm.Stds. Ctee., Minutes of Feb. 7 meeting, p 6, February, 1984.

4. ACR-NEMA Dig. Imaging and Comm. Stds. Ctee., ACR-NEMA 300-1985: Digital Imaging and Communications, November, 1985.

5. International Organization for Standardization, Document ISO 7498-1983, Information Processing Systems - Open Systems Interconnection - Basic Reference Model, 1983.

6. ACR-NEMA Dig. Imaging and Comm. Stds. Ctee., Application Guide: Minimum Requirements for Compatibility with the ACR-NEMA Digital Imaging and Communications Standard, 1987.

7. Maguire Jr., G.Q., Baxter, B.S., Hitchner, L.E., An AAPM standard magnetic tape format for digital image exchange, SPIE Proceedings, Vol. 318, PACS for Medical Applications, pp 284-293, 1982.

PERSONAL COMPUTERS IN MEDICAL PHYSICS

Philip H. Heintz, Ph.D.
N.T. Enloe Memorial Hospital
1500 The Esplanade
Chico, CA 95926

ABSTRACT

A brief overview of the personal computer (PC) as it relates to the medical field is presented. The architecture, speeds, uses, and operating systems of the PC are discussed in basic terms. Some comparisons of performance between computers are presented. For those interested in purchasing a PC for home or office suggestions are given on how to go about making decisions. This paper concludes with predictions for the future of the PC.

The advent of the personal computer has changed the way the world views electronic data processing. With so much of the world involved in computers, the technology is changing too rapidly for any one person to keep up with it all. My expertise is with the IBM PC/AT™ line of computers, but I will give an overview of where the whole PC technology is today and where it is headed.

There are three main personal computer lines based on the central processing chip. Table 1 lists these lines and some popular computer models.

Processor	Computer System
Motorola 6502	Apple II, Commodore 64
Intel 8088	IBM PC and compatibles
Motorola 68000	Apple Macintosh (Mac)

Central Processing Chips and Computers

Table 1

Comparing Computer Systems

Within each line of computers are a whole series of computers from slow, inexpensive models to very fast, expensive number crunchers. Table 2 shows comparable computer products by word size (8 bit, 16 bit, 32 bit). Clearly, the larger the word size the faster the processing speed within a computer line. This table also shows the clock speed of the processor. You cannot compare speed between computer lines by clock speed alone as will be shown later. Clock speeds are increasing dramatically on an almost continuous basis, so these values should be viewed in that light. Each line and/or manufacturer has its own interface protocol or bus structure as shown in the table. You cannot use interface cards from the Apple line of computers with the IBM PC line of computers and vice versa.

Each of the various computer lines has its own operating system (the program that you as a user use to communicate with the computer to run your application). Table 3 lists some available operating systems. For an Apple II system DOS 3.3 and ProDos are the most prominent operating systems. For the Apple Mac it is Mac DOS and for the IBM PC it is PC or MS DOS.

Processor	**Computer Manufacturer**	**Bus Structure**	**Clock Speed mHz**
<u>8 Bit Technology</u>			
6502	Apple II, IIe	Apple	1
	Commodore 64	None	
65C02	Apple IIGS	Apple	3
8088, 8086	IBM PC & compatibles	Extended Local Bus	4.77-10
<u>16 Bit Technology</u>			
68000	Apple Mac	None	7.72
	Atari-ST	None	8
80286 (80186)	IBM PC-AT & compatibles	Extended Local Bus	6-12
	PS/2 Model 50-60	Micro channel (16 bit CPU Bus)	10
<u>32 Bit Technology</u>			
68020	Apple Mac II	NuBus System Bus	16
80386	PC-386 machines	Extended Local Bus Proprietary Bus	16-25
	PS/2 Model 70-80	Micro channel (32 bit CPU Bus)	16

Computer Architecture

Table 2

Computer	Operating System	
Apple II	DOS 3.3, ProDOS, Pascal	
Apple Mac	Mac OS	
	AUX (UNIX)	multi-user/multi-task
IBM PC & compatibles	MS DOS (2.1-3.3)	single user
	OS/2	single user/multi-task
	Xenix (UNIX)	multi-user/multi-task

PC Operating Systems

Table 3

Task	Program
1. Word Processing	WordPerfect, MS Word, Appleworks
2. Spreadsheet	Lotus 1-2-3, Quattro, Surpass, VP-Planner Plus
3. Desktop Publishing	PageMaker, Ventura, First Publisher
4. Graphics	Micrografx, MS Paintbrush, Harvard Graphics, VP Graphics
5. Image Processing -video	Media Cybernetic Software
6. CAD	Auto CAD
7. Scientific Application Programs	Math CAD
8. Data Acquisition, Radiation Therapy Beam Data Acquisition System Pulse Height Analysis	
9. Music	Mostly Atari Software
10. Education	Mostly Apple II Software, IBM impact growing
11. Bulletin Boards	Compuserve, Kansas City Bulletin Board

Most Common Uses of the Personal Computer

Table 4

Computer Applications

Within each line of computers there are many applications. History tells us that Apple II's are used primarily for educational training in the elementary and secondary schools. The Apple Macintosh computers are easy to use, very friendly and are used extensively for desktop publishing, graphic design and word processing. The IBM PC is the most common and probably the least friendly of the three. It is popular because it is backed by IBM and is the least expensive and fastest hardware of the three. One reason it is so fast is that the operating system is simple. Its heritage dates back to mini computer operating systems. Because of this simplicity it is easy to develop software on the IBM PC system. As a result, the IBM PC class of computers are used for just about everything including word processing, spreadsheets, scientific applications, computer-aided-design, data acquisition and much more. In order to make the IBM PC more friendly several user interface programs have been developed. Menu programs such as Automenu from Magee Enterprises or Menu Maker by ETS Software can be used. If you like the MAC interface style windowing, Microsoft has Windows for the IBM PC computer.

The most common PC applications and popular programs in each category are listed in Table 4. There are thousands of programs available today, too many to start to mention in each category.

Computer/CPU	Speed	Floating Point Unit	Execution Time(sec)	Language	Remarks
IBM PC 8088	4.77mHz	None	342	C	Turbo C
		None	260	F77	Microsoft F774.0
		8087	18.8	F77	Microsoft F774.0
			13.2	C	Turbo C
			51.7	Pascal	Turbo Pascal 3.0
IBM AT 80286	8mHz	None	76	C	Turbo C
		80287	8.5	C	Turbo C
			9.5	F77	Microsoft F77 4.0
Zenith 80386	16mHz	80387	2.3	C	Turbo C
			1.3	F77	Microsoft F77 4.0
Compaq 80386	25mHz	80387	<1.0	C	Turbo C
PDP11/73		FPJ11-AA Chip	1.04	F77	DEC Fortran V5.0A
			8.3	F4	DEC Fortran V2.8 with EIS/FPU
		FPU microcode	4.1	F77	DEC Fortran V5.0A
PDP11/44		FPU-A	2.92	F77	DEC Fortran V5.0A
PDP11/34A		FPU-A	3.44	F77	with Cache memory
VAX 8700		None	0.32	F77	50 users
VAX 8200		Yes	0.86	F77	1 user
VAX 750		Yes	2.78	F77	1 user
MicroVAX II		Yes		F77	1 user
Apple Mac Plus		None		C	MAC C
Apple Mac II					

Simple Floating Point Performance[1]

Table 5

It is often important to compare computer performance. The sales literature often uses hardware processing speed as a way to make this comparison. Words like mips (million instructions per second) are used. In actual practice it is very hard to compare different computer lines. Not only does the hardware determine performance but the operating system and the support software, i.e., programming language used, also affect the performance. If you really want to compare performance you should find identical applications on the test systems and compare the time it takes to complete identical tasks. Tables 5 and 6 illustrate two approaches. Table 5 compares the execution of the program listed in Appendix A on numerous computer systems. This test was designed to compare the operational speed of floating point arithmetic on these computers using several different languages. From these results you can see that not only does the hardware affect performance but so does the language, even the version of the language. It is interesting to note that the new Compaq 386/25 computer with a floating point processor exceeds all of the other computers except the DEC VAX 8700.

Table 6 highlights the results presented in BYTE Magazine[2], comparing the performance of different brands of personal computers. Again, the Compaq 386 exceeds the performance of most other computers except in disk reading. What we see from these comparisons is that performance is hard to evaluate even within one computer line. In fact, the results may be meaningless if not interpreted correctly.

Computer/CPU	**Speed**	**Matrix Test (sec)**	**Sort Test (sec)**	**Floating Point Math (sec)**	**Disk I/O 1 Meg Read (sec)**
IBM PC 8088	4.77mHz	66.51	330.0	70.80	16.90
IBM AT 80286	8mHz	11.69	84.0	46.60	8.20
Compaq 80386	20mHz	3.06	27.0	7.01	2.88
Apple Mac Plus		79.30	179.7		7.50
Apple Mac SE		67.10	154.2		8.10
Apple Mac II (68020)		10.20	44.2	175.30	1.80

Performance Tests[3]

Table 6

Screen Resolution

Screen resolution is very important when selecting a computer system. You must decide if graphics or color graphics are going to be used. If so what level of resolution is important. Until recently the MAC had very good monochrome graphics, but no color. On the other hand only recently has the IBM PC line had good graphics at all. Now the IBM PC line has several types of graphics adapters as shown in Table 7.

Monochrome display is fine for word processing, simple graphics and spreadsheets. If you want to do image processing, complex graphics or sit back and relax with a challenging game then a color monitor is almost essential. I would never recommend an IBM-CGA because the text characters are poorly displayed on the screen and the graphics resolution is too coarse. EGA is the minimum resolution satisfactory on the IBM PC line of color adapters.

Graphics Adapter	Resolution (line x color)	Graphics Resolution (pixel grid x color)
Monochrome Text (Mono)	25 x 80	None
Monochrome Hercules Graphics (HGA)	25 x 80	720 x 348
Color Graphics Adapter (CGA)	25 x 80	320 x 200 x 16 color
Enhanced Graphics Adapter (EGA)	43 x 80	640 x 350 x 16 color
Video Graphic Adapter (VGA)	50 x 120	640 x 480 x 16 color
Enhanced Video Graphic Adapter	50 x 120	640 x 480 x 256 color
High Resolution Graphic Adapter	?	1024 x 1024 x 256 color

IBM PC Graphic Adapters

Table 7

Programming Languages

Virtually every development language written is running on a PC. There are compilers for Fortran, BASIC, Pascal, C, ADA, Prolog... . You name it, there is a compiler running on a PC today. In fact, Borland revolutionized software development by introducing Turbo Pascal on the PC. For less than $50 one could purchase a complete, fast, user-friendly Pascal language development system. This product has extended into other languages such as C, BASIC and Prolog. Other companies have followed suit and developed similar products to compete with Borland.

Purchasing a Computer System

With all of this technology how does one go about purchasing a computer? To start you must decide what you want the computer to do. Define your task in detail. Specify your software needs. Find the software or have it developed for your particular needs. Then go about selecting the hardware needed to implement the software. Try to avoid proprietary computer systems with proprietary operating systems. Choose a system that your colleagues use so that you can share the crying towel together as well as data files, application programs, etc.

You may then acquire both hardware and software in several different ways:

1) Purchase a complete system from a local vendor with full training and local support.
2) Purchase hardware inexpensively and contract for software to be developed as per your specifications.
3) Purchase hardware and software from discount sales stores. There will be little or no local support for these products.
4) Purchase components of the systems from mail order houses. You are obligated to assemble and test these pieces with little or no support and no system warranties.

Of these options option one is probably the most expensive with option four being the least expensive. However, if you consider your time or the time of a person to assemble the system, options three and four may not be so inexpensive. Unless you are very knowledgeable with computer hardware **do not** undertake option four. My experience has been that even if you follow the steps in the literature, putting your own system together, although very rewarding, it is very frustrating and time consuming.

The personal computer can be purchased with or without expansion capabilities. Usually this expansion capability is handled by peripheral slots located on the motherboard where interface boards and memory can be added. If you are going to need expansion capabilities or interface lab acquisition devices choose your hardware carefully.

Purchasing Software Products

Before buying any software product today, I suggest that you test drive it. One way of doing this is to develop a good relationship with one or more local software houses which let you try or borrow software. Another way to try software is to borrow a copy from a colleague or your local software store. Most vendors approve of this practice as long as you buy the program if you decide to use it for any task.

One source of inexpensive software for the PC is known as shareware or freeware. These programs are public domain programs available through local bulletin boards and local user groups. Shareware is offered by the author for use and sharing with others under the stipulation that if you find it useful that you send them a small fee which may range from $15 to $75. This price usually determines the level of support for the product. Shareware is a wonderful source of programs to get started including menuing programs, accounting, word processing, games, and many others to numerous to mention. Even AAPM has its software exchange at the annual meeting each year where users can exchange non-copyright software with fellow members.

Which purchasing path you choose depends upon your personal setting. In the hospital I find that options one and two are the most economical whereas home I use options three or four. Remember, when purchasing computer systems corporate policy may dictate your choice of hardware and/or software. Standardization within the corporation is not always bad. In fact, in most cases the advantages outweigh disadvantages because uniformity leads to uniform support and exchange of information amongst similar systems within the corporation.

System Security

The advent of the PC has introduced new problems. The greatest is software security. Although most software protection schemes on commercial software have been dropped by the manufacturers, security and thievery problems have not been eliminated. Remember, software is copyright protected and copying software other than public domain software is illegal. Usually, one copy of a program per computer or user is allowed. Backups are also allowed to protect the original archival master. Most copyright agreements require that each computer have its own purchased copy of the software. This is no different than providing a desk, book, pen or pencil for each employee. A major problem with copy protection still exists for those very high priced software packages such as computer aided design (CAD) and treatment planning software. Some devices are being used to protect the average user from using multiple copies, but even a good hacker can defeat these schemes. Where this really becomes a problem is in software sales outside of the United States. Maybe with the cost of these expensive software systems decreasing and their limited market, piracy will be reduced. At least within the United States and the medical physics community, the grapevine should afford the vendor some protection for treatment planning systems.

Another problem which has developed with the introduction of the PC is that the operating systems often lack security capabilities for your data and programs. This includes lacking the capability to backup your work on a regularly timed basis or securing entry into programs and files by only those people who should have access to these files. There are a few software packages on the market now which will enhance your operating system and provide the capability of doing automatic saves on your current work as well as programs which will protect or lock programs and data from undesired access, but this is very limited. In order to protect your programs and data in the meantime I suggest backing up your hard disk on a regular basis and your proprietary information and secure information must be protected by removing this data from the system on diskettes or tape after each work period. Loss of immediate data can often be protected by installing a non-interruptible power source. This also protects your hardware from line surges. Programs are also available for continuous silent back up of your on-line work.

A new problem developing in the PC environment is the dangerous and dreaded computer virus. This virus was originally found in public domain software and on bulletin boards, but is now being discovered infecting commercial software. Just as with any human virus, such as AIDS which can be terminal, the computer virus can be terminal to your system and data. Each PC owner must keep his/her system in contact with clean software, clean data and associate with other clean systems.

Utilizing the Personal Computer Capabilities

Most of the new dedicated data acquisition and data processing equipment has been designed to use with the personal computer. Usually, the systems use a standard PC compatible with special hardware installed in one or more of the peripheral slots. The graphic display and hardware peripherals are standard products. Typically, the graphics display will be of either EGA or VGA display types. In some cases a very high resolution screen is needed requiring the system to have two displays. Most vendors have stayed with standard hardware as much as possible to minimize the cost of updates in the future. This is a wise choice and facilitates standardization and interchange of data between software modules. Listed below are several examples of such systems.

1) Multichannel analyzer and storage oscilloscope - system adds an analog to digital converter board to a standard PC with proprietary software
2) Beam acquisition system - system adds a proprietary ADC and drive board to the PC along with proprietary software. Data format is standard, making data interchange with other programs easy.
3) Linear accelerator control system - system adds special interface board to a standard IBM-AT computer (Varian 2100C)
4) PC's are being purchased by many mini owners as intelligent terminals or stand alone graphic workstations (Sun Micro Systems, Apollo, etc.)

Communications

One large area of rapid growth is the interconnectivity of computers. Going beyond using the PC as an intelligent terminal on the mini is the desire to share total resources of all computers connected together. Local Area Networks (LAN) which will be discussed elsewhere are the most practical systems to share local resources. Exchange of data outside of your environment is done with local bulletin boards, commercial bulletin boards such as CompuServe, electronic mail (often called "E mail") and direct computer to computer interaction. Supporting this technology are modems which operate over regular telephone lines at speeds of up to 19,600 baud allowing rapid telecommunications of even image data. At 19,600 baud a system can transfer a 512 by 512 CT scan in compressed mode in 10-12 seconds. I expect rapid technological growth in this area.

Selecting Hardware

For those people interested in scientific applications where number crunching is involved I would recommend the following PC configuration. IBM 386 compatible with an 80387 math coprocessor operating at a minimum clock speed of 16 MHz. Added to this system should be a graphics display of EGA or preferably VGA quality, 2MB of fast memory, 2 serial ports, 1 parallel port, at least a 40MB fast hard disk, mouse, 24 pin dot matrix printer or laser printer. Cost of such a system can range between $6,000-12,000.

For those not interested in speed but a very user-friendly system I suggest the Apple Macintosh computer with a hard disk, mouse and a good printer. This system can run between $6,000 and $8,000.

At the high end of prices there is an increasing number of manufacturers of graphic workstations such as Apollo or Sun Micro Systems. These systems can range in price from $6,000 to $150,000.

Where Is Microcomputer Technology Going?

Within a year or so I expect to see a 64 bit microprocessor, parallel processing disk cache, memory cache, and CPU's operating at between 25 and 50 MHz clock speed. Peripherals such as large 100-500 MB optical disk storage and fast 1,024 x 1,024 bit mapped graphic screens in 256 colors will be common. Software operating systems are going to be the limiting factor. None of the PC 286 and 386 machines have operating systems which fully utilize their architecture. Even APPLE with the MAC II does not support the 32 bit architecture. Until new operating systems are developed, hardware performance will outstrip that of software.

References

[1]Cradduck, Trevor, personal communication, August 1988.

Dvorak, John C. "Hmm, Mac Or PC?", *DEC Professional*, Vol. 7, No. 2, 1988.

[2]Fox, Ron, "Why MIPS Are Meaningless", *BYTE*, Vol. 13, No. 6, June 1988, pp. 225-234.

[3]Grehan, Richard, Tom Thompson, Curtis Franklin Jr., and George A. Stewart, "Introducing the New BYTE Benchmarks", *BYTE*, Vol. 13, No. 6, June 1988, pp. 239-266.

Heintz, Philip H., "Using the Personal Computer in the Radiation Oncology Department", UCSF Annual Meeting, San Francisco, CA, March 1986.

Heintz, Philip H., "A Personal Computer Treatment Planning System", RAMPS, New York, New York, April 1987.

Appendix A[1]

Floating Point Test Program
Fortran

```
C      Mflops test prograam
C
         REAL*4 a(1000),b(1000)
         INTEGER*2 i,j,lu
         INTEGER*2 jtime, jtime1,jdelt
         INTEGER*2 jtimeh,jtim1h,k,m
         REAL*4 atime
C
         DO 10 i=1,1000
C      Initialization (not timed)
         a(i)=1.0+1.0/FLOAT(i)
         b(i)=1.0+2.0/FLOAT(i)
10       CONTINUE
C
         CALL GETTIME(k,m,jtime,jtimeh)
C
         DO 20 i=1,50
C      Actual computation
         s=0.0
         do 20 j=1,1000
         s=s+a(j)*b(j)
         s=s-b(j)/a(j)
20       CONTINUE
C
C      Timing and display of result:
C
         CALL GETTIM(k,m,jtime1,jtim1h)
         atime=jtime1-jtime+(jtim1h-jtimeh)/100.0
C
         WRITE(6,50) s
 50      FORMAT(' Test result s= ',F12.6)
         WRITE(6,100) atime
 100     FORMAT(' Time= ',F6.3,'seconds')
         END
```

December 1989

ENVIRONMENTAL CONCERNS

Jack L. Lancaster, Ph.D.
University of Texas Health Science Center
San Antonio, TX 78284

Introduction

Proper environmental control for a computer system extends well beyond the basic need of controlling the physical characteristics of the room in which the system is operated. To attain optimum use of a computer system the impact of the environment on the user, the machine-user interface, as well as the computer and associated hardware must be taken into account. In addition, the user must learn the proper methods for care and use of the system to guarantee that the computer and environment remain at an acceptable state.

User-Machine Environment

The science of adapting machines to people is called ergonomics. The goal of ergonomics is to provide a good working environment for the computer user based on the physical factors. A great deal of research has been devoted to the video display terminal (VDT) because so much of the user's time is involved with VDT interactions. Additional factors for concern are keyboards, chairs and tables, and alternate I/O devices.

Video Display Terminal - A VDT placed in improper surroundings can generate complaints such as headaches, backaches, and eyestrain (5). A microcomputer placed on an existing desk is not uncommon and ranks next to lighting as the most often committed ergonomic mistake (7).

The visual information perceived by the operator is derived from that of the display screen and is subject to the legibility of the screen characters. A number of factors relate to the successful transmission of information from the screen to the user.

Time Dependent VDT Display Characteristics (2) - Several phenomena dealing with oscillating brightness of pixels have been observed by users of VDT's. Of these it has been suggested that flicker may adversely affect visual comfort. Variations in luminance during one video refresh period can be perceived as flicker. Perception of this flicker is primarily dependent on the relationship between phosphor decay time and video refresh rate (higher video refresh rates and/or longer phosphor decay times reduce effects of flicker). Additionally, flicker perception is dependent on video polarity, screen luminance, room illumination, image contrast, as well as the variable degree of sensitivity between users. Jitter, which is the time variation in the position of display characters, is observed with some VDT's. This phenomenon is due to improper variations in the circuitry which deflects (or scans) the electron beam and may be the result of poor design. Jitter is in principle harmless but may be irritating and tiring to the eyes. Afterglow which is due to longer persistence phosphors will reduce the observable effects of both flicker and jitter but may interfere with screen displays of rapidly changing scenes.

Structure of VDT Screen Characters (2)- Fonts used for VDT representations are often very different from those encountered in normal printed text. A visible raster scan line decreases legibility. For dot matrix representations larger matrix sizes are desirable. One recommendation for character height for good optimum reading performance is 20-28 minutes of arc (3.8 mm @ 60 cm distance). Acceptable line spacing is determined by the eye saccade movement. The distance between adjacent lines should increase with increasing line length with a minimum of 2 degree return trajectory (end of line to beginning of next line). To meet this requirement vertical orientation of a full page screen or a two-column presentation can be used. Most displays fail on several of these recommendations.

VDT Screen Polarity and General Luminance (2) - A negative polarity text display emits light from the text and not from the background. Positive polarity is just the opposite, providing text displays which look more like normal printed material. Though there are good arguments for both types of displays most groups polled preferred the positive display mode. It is believed that the positive display mode is better when viewed in a room at a higher ambient light level which may partly account for this bias. Also, the positive display mode requires less adaptation for users who spend a great deal of time looking back and forth between screen text displays and printed materials.

VDT Workroom Illumination (2) - Computer screens have not been optimized for viewing using existing fluorescent lighting which has proven good for the pen and paper environment (7). To reduce screen glare and yet maintain useful levels of light at the VDT it is recommended that indirect lighting originating from behind the VDT be used with adjustable brightness. Task lighting is recommended where use of the VDT involves simultaneous viewing of printed materials. For those who work long hours at the VDT it is recommended that a break of 15 minutes be taken every 2 hours.

It is important to minimize reflections from the computer display screen. Proper positioning and tilting of the screen may be adequate, but in extreme cases solutions require the use of non-reflective polarizer type filters over the CRT screen (8). In very bad situations where the bothersome reflections originate from the user's own reflection it has been recommended that the user wear dark clothing (9).

The workroom illumination level desired by different users is often quite variable. When different users work at the same VDT at different times, it is desirable to provide a means for easy user adjustment of the illumination level.

VDT Visual User Interface (2) - Finally, to optimize the user screen visual interface, software and hardware should provide for screen layouts and interactions which have the following characteristics:

1. an orderly, clear, clutter free appearance;
2. an obvious indication of what is being shown and what should be done with it;
3. expected information where it should be;
4. a clear indication of what is related to what;
5. plain simple language;
6. a simple way of finding out what is in the system and how to get it out; and
7. a clear indication of when an action could make permanent change in the data or system operation.

The basic goal for developing better user interaction interfaces is to enhance user perception skills. The whole process of perception focuses on the ability to relate new experiences with old ones in a meaningful way. Perceptual skill may then be defined as a means to quickly and efficiently combine new experiences with old (6). A good example of the type screens which meet many of the visual interaction goals of the user interface are seen on the Apple MacIntosh computer system and with Microsoft Windows software on the IBM personal computer.

Keyboard - The I/O device most commonly used after the VDT is the keyboard. The common keyboards use the QWERTY layout of alphabetic characters which was designed for optimized use of the mechanical typewriter. Because so many users already had satisfactory typing skills this key layout has been retained even though it may not be optimum for today's electronic keyboards.

Since a majority of today's work with computers seems to be with word processing the keyboard is an especially important device. For those applications which involve extended typing tasks a keyboard with good feel and properly sized keys is important. Unfortunately, personal computers often are sold with keyboards with cramped key layout (portables) or improperly sized and positioned keys. Several companies have prospered by providing alternate keyboards with better ergonomic design.

The optimum keyboard height is a complicated problem in that it is related to chair height, table height, screen position, as well as individual physical differences and preferences. A detachable keyboard and adjustable chair height are needed to allow the user to select an appropriate location and comfortable height. Recently, it has been suggested that keyboards communicate with the computer using RF technology thereby eliminating the constraints imposed by cabling used on most current machines (11). IBM actually implemented an LED communication scheme on its unsuccessful PC Jr.

Chairs and Tables - The ergonomic rule for chairs and tables is that the chair height and back support be adjustable and that the table provide sufficient working area for the VDT and other materials associated with the work requirements of the user. Some specialized computer tables provide an additional adjustment for height as well as separate keyboard height adjustment.

Alternate Input Devices - There are a number of alternate input devices, each with good and bad points, but the one most commonly used is the mouse. The mouse is said to be ergonomically superior to the trackball and probably the best general purpose alternate input device for tasks such as desktop publishing (10).

Physical Room Conditions (3)

For most microcomputer systems used in medicine today no special computer room is required. They operate reliably as long as reasonable precautions are taken to provide clean power and no extremes in temperature, static electricity, or dust. However, additional considerations may be necessary to keep the physical room conditions suitable for users.

Temperature - The air temperatures which produce a definite deterioration of human performance are 27 C (80 F) for heat and 13 C (55 F) for cold (4). These limits should be avoided within the working environment.

Heat is generated by the working environment, people in the room, light fixtures, sun shining through windows, as well as by air conditioning units. In rooms with large windows, the use of drapes or other shades is recommended. Always avoid direct sunlight in the computer area.

Certain components such as removable hard disks are more sensitive to temperature variations and special operating procedures are necessary for optimum use (see hard disks section).

Humidity - The relative humidity within the computer room should be from 40% to 80% non-condensing. The ideal relative humidity is 50%, but most equipment will generally operate well within the range of 20% to 80%. The effect of humidity on user performance is somewhat temperature dependent, but most people become uncomfortable at the higher humidity levels.

When the relative humidity is below 40% static electricity becomes a danger. To minimize the effects of static electricity it is recommended that the computer room floor not be carpeted (especially using nylon carpet). Static electricity discharge within the vicinity of the computer can be damaging to sensitive components such as CMOS IC's. It is often a high voltage discharge between the user and some computer component that causes a problem. A rule-of-thumb is for the user to attempt to attain the same state of electric potential as that of the computer system before making contact with critical components. This can be accomplished by approaching the computer terminal slowly and touching everything with open hands to allow charge differences to equalize slowly.

Additionally, prolonged levels of high humidity are bad for most forms of equipment. The air conditioning system for most facilities tends to maintain the humidity at an appropriate level. However, if there is a problem with the central air conditioning equipment, humidity levels may rise above acceptable levels. If surfaces of equipment feel damp the humidity is excessive. A possible early sign of excessive humidity is the formation of a corona discharge around the high voltage leads to the CRT (a hissing sound may accompany this corona) or even an occasional arc accompanied by a loud popping sound. Both are indications that the humidity is too high and that corrective action should be taken.

Sound Level (Noise) - Noise is more likely to reduce the accuracy of worker's performance than to reduce the total quantity of work. When a task does not involve auditory signals (as with most computer work) the effects of noise on performance are much harder to assess. Nevertheless, certain general conclusions have emerged (4):

1. Steady noises without special meaning do not seem to interfere with human performance unless the noise level exceeds 90 dB and performance on the task lasts for at least 30 minutes.
2. Intermittent noise is much more disruptive than steady noise even with sound levels below 90 dB.
3. High-frequency components of noise, 1000-2000 Hz, seem to produce more interference with performance than low frequency components.
4. Complex tasks are more likely to be adversely affected by noise than are simple tasks.

The main recommendation for today's microcomputers is to keep the physical room conditions comfortable for humans, and the computer will do well.

Equipment (3)

Proper care and operation of the computer and accessories is important to assure uninterrupted error free operation of the system. Those components which need the most attention are the ones which are involved with I/O and mechanical devices and include disks and drives, tape and tape drives, and printers.

Floppy Disks - This medium is probably the main cause of failures in any computer system that uses it. Damage may not be detected immediately, and critical data may be lost. There are some good rules to follow when working with diskettes:

1. Respect the physical integrity of the diskette.
2. Don't bent or fold a diskette.
3. Don't touch the surface of a diskette. The oily chemicals secreted by the skin can permanently damage an area of a diskette.
4. Keep all sources of magnetic fields away from diskettes, including magnets as well as magnetized objects.
5. Maintain the proper working environment. Avoid heat, moisture and dust. Keep diskette in protective envelope when not in use.
6. Insert the diskette into the drive properly.

One of the most important defensive measures when using diskettes is to make a backup copy of critical data. The backup should be stored in a different location from the main copy. Diskettes should be labelled to include date of backup and a good description of the data category. Additionally, use only soft-tip felt pens when writing on the soft jacket type diskettes (5-1/4" or 8"). It is also recommended that a felt tip pen be used when labelling the hard jacket 3-1/2" diskettes. When diskettes are not in use they should be kept in their protective envelope.

Finally, there are several non-obvious locations near computers where harmful magnetic fields may be produced. One is near a magnetic coil used to ring older type telephones. Another is near the front of a color CRT where the degaussing coil is located (activated each time the CRT is turned on). Diskettes should not be placed near such sources of potentially damaging magnetic fields.

Diskette drive heads should be cleaned regularly to eliminate dust and other particulate accumulations. The frequency of cleaning depends on the environment, but once a year is generally sufficient. Special head cleaning kits are available for this task. When moving a disk drive unit be careful to avoid shock and vibrations that may cause misalignment of heads. There is software available to perform rather elaborate testing of the disk drive and report findings which can be used to correct problems relating to head alignment.

Hard Disks - There are several formats for hard disks. The removable disk cartridge is a single platter contained within a hard plastic protective cover. The removable disk cartridge can be either top or front loading. It is usually combined with a fixed disk in a single drive dual disk unit. Disk packs generally come with 5 or more platters on a common spindle in each package. Disk packs are removable but since they hold so much data they are usually left in place for long periods of time once installed. Winchester disks are a sealed form of the single or multiple platter design.

The main advantage of a hard disk is the large storage capacity. This large storage capacity leads to the main disadvantage that a large amount of data may be lost if damage occurs. Thus data from the large capacity hard disks must be backed-up frequently. Because of the large storage capacity of a hard disk, only two devices are recommended for back-up purposes: a removable hard disk cartridge or a magnetic tape. The removable disk cartridge is faster, but the tape will be less expensive. If the quantity of data to back-up or archive is small and the storage time not too great, the removable disk may be just as economical as tape.

Dust is more of a problem with disk packs and removable disk cartridges than with floppy disks. The read/write head cannot be allowed to come into contact with the hard surface of a disk platter. However, to produce the high data storage density needed for these disks the head must be positioned very close to the surface of the disk media. By design the heads literally fly above the surface of the platter (about 100 m gap). The head can be caused to crash into the surface if it collides with an object even as small as a smoke particle (about 300 m) on the surface of the platter. When the hard read/write head crashes into the hard surface of the platter both are usually damaged. To prevent such events from occurring the air which enters the disk drive during operation must be filtered. The filters must be checked and cleaned according to manufacturers' specifications. For protection when outside the disk drive platters, disk cartridges and disk packs are packaged in hard, sealed plastic containers which provide openings for read/write heads which are normally closed when not in use.

Because of the relatively high data density associated with hard disks they are more sensitive to distortions in data due to temperature. A good operating rule is not to use disks which have been brought to the computer from a location with a substantially different temperature (either lower or higher) without allowing sufficient time for temperature equalization.

Loss of electrical power is one of the worst potential dangers to the data as well as the disk unit. Well designed drives will retract the heads when power is lost and require operator intervention to be placed back into service. However, if the unit is writing data at the time of the power loss the affected disk file may not be recoverable.

Tape Drives - Magnetic tapes are less sensitive to shocks, dust, and magnetism than hard disks but they should still be treated delicately. Currently tapes are used primarily as back-up for data and software stored on hard disks. Commonly used tapes fall into two categories: (1) 9-track 1/2" tape on reels of various diameters and (2) 4" x 5" tape cartridges with 1/4" tape. Both types allow for tapes of various lengths to suit a broad range of user needs. The data density of the 9-track tapes can be as high as 6250 characters per inch while the 1/4" tape cartridges can attain up to 12,500 characters per inch.

Tapes should be stored in a clean environment at the proper temperature (59 to 122 F). To accomplish this tape reels should be stored in air-tight canisters and shelved vertically. Tape cartridges should be stored in their protective container in a vertical position also.

Periodically inspect the uniformity of the winding of the tape. If the tension of the tape drive is too low, tape slippage and cinching should be visually indicated. Print-through of data can occur if the tape is wound too tightly and the tape remains unused too long. Frequent rewinding or reusing of the tape solves this problem.

To attain the most reliable use of tapes make sure that the system is kept properly adjusted, the tape heads are cleaned regularly, and that only high quality tapes are used.

Printers - There are many types of printers available for use with computers. Because of the relatively large number of mechanical parts and the large motions needed the printer is prone to a higher failure rate. Fortunately, printer failure does not destroy data, but it can be very irritating and sometimes costly.

The printer should be positioned for stable operation, for unobstructed paper flow, and for easy operator access. It is usually best to install the printer on a stable table or stand to avoid excessive vibrations. With high speed printers where the head momentum changes rapidly at the end of each print line a dedicated pedestal or stand is a necessity. Adequate clearance for paper access should be provided (1) below for the fresh paper box, (2) behind for the printed paper, and (3) on one side for access to both.

Static electricity build-up on paper can occur where high speed printers are used in a low humidity environment. If such a condition exists static-discharge brushes can be installed at the exit of the tractor feed to drain off excessive charge so that the paper (fan fold) will fall correctly behind the printer.

While major maintenance and preventive maintenance are best performed by manufacturers and qualified service agencies, the responsibility for keeping the printer clean falls upon the user. When dust has accumulated it should be wiped off with a lint-free cloth, vacuumed, or simply blown away.

Proper paper selection is important for most printers, but more so for the newer laser type printers. To obtain the clearest, sharpest images, paper manufactured for photocopying should be used. Though laser printers work well with many types of paper, the following types are to be AVOIDED:

1. Those not meeting the specifications given by the manufacturer.
2. Extremely smooth or shiny paper, or paper that is highly textured.
3. Coated papers.
4. Letterheads using low temperature dyes or thermography.
5. Damaged or wrinkled paper, or paper with irregularities such as tabs, staples, etc.
6. Multipart forms or carbonless paper.

Certain supplies must be maintained for each printer to guarantee uninterrupted and high quality operation. These include paper, printer ribbons, toner cartridges, print wheels, and spare fuses.

The Electrical Environment

A computer must have a clean and stable supply of electrical power. Any interference with the supply of electricity to a computer system can disrupt its operation and/or cause problems that are hard to diagnose. Computers have an internal power supply which converts the line voltage (110V @ 60 Hz or 220V @ 50 Hz) into voltages required by various internal components. A typical power supply will provide +5 volts, +12 volts, and -12 volts with sufficient power capability to handle built-in components as well as a limited number of internal and external accessories. As a rule more expensive computers have higher quality power supplies which can better withstand fluctuations which inevitably occur with the electrical power supplied to the computer.

Electrical power provided by most utility companies is neither truly stable nor accurate. Even if supplied electricity is satisfactory most of the time there will be times when it may suddenly become unacceptable. There are a number of electrical power problem areas which can be readily identified and properly addressed.

Source Power Problems - The computer system should operate on its own electrical circuit connected directly to the main distribution box (a dedicated line) in order to prevent fluctuations which are caused by varying loads on locally shared lines. This is more important for larger computers such as file servers which handle a large quantity of critical data. In some cases it is even desirable to have powerful peripherals such as large disk drives connected to a dedicated line to minimize fluctuations in the distributed power when they are energized. With personal computers at home it is important that a device such as an air conditioner or electric iron not be connected on a common line. As a check of potential problems all electrical devices in the computer room should be turned on and off to verify that they do not affect normal operation of the computer system. At home it may be helpful to turn off the circuit breaker at the distribution box to determine what, if any, outlets share the same line. This method may also be useful in verifying a dedicated line within a building.

It is advisable to place a tag on the circuit breaker or main distribution box warning others within the building not to turn the power off without notifying the computer operator. This is especially true in large buildings when electrical problems in an adjacent area serviced by the same distribution box arise, and service personnel need to turn power off for servicing.

Low Frequency Fluctuations - These are variations which occur slowly (over many cycles of the sinusoidally varying supply voltage). Such variations are the result of a slow time varying load somewhere on the line. These can be caused by motors and some appliances.

Fluctuations are often correctable by the power supply since it can provide a constant output voltage over an input voltage range of approximate +/- 5% for slowly varying voltages. Whenever these fluctuations exceed the limits of the computer's power a line filter or line conditioner may be required to correct line problems. The line filter is the cheapest solution but is better suited to the higher frequency ranges of fluctuations. A line conditioner is a regulator that includes an isolation transformer so that it provides both isolation as well as regulation. The line conditioner is more expensive but usually a better choice for low frequency fluctuations.

Line Transients - Line transients are fast changes in the line voltage that occur within a few cycles or less. Transients may cause a voltage drop called a dip or notch, a voltage rise called a spike or surge, or both. They may be due to lightning, inductive kick-back from inductive switching equipment, or electromagnetic interference (EMI). Several sources of EMI are automotive ignitions, fluorescent lights, radio and television transmitters, general communications equipment, and some medical equipment. The higher amplitude transients or surges are usually due to lightning or power restoration following power outages. Most good power supplies include some form of surge protection, but devices called surge protectors can be placed in line with the power source to divert much of the effect of surges away from the sensitive computer equipment. Surge protectors are also available for data lines to prevent damage to sensitive components such as modems.

If the computer is affected by transients then it is best to find the source and eliminate it or move it further away from the critical component. A line isolator may be used on the troublesome component to keep the transients from getting onto the power supply lines.

If the source of transient interference is EMI it may be radiated or conducted into the critical components. Power lines are the main entry route of conducted EMI. Installation of power line filters with appropriate bandwidth and attenuation factors will generally alleviate this problem. Radiated sources of EMI may be due to improper or ineffective grounding or shielding or just poor routing of unshielded data cables. Rerouting cabling and providing shielding for long unshielded data cables can solve many of these problems.

EMI conducted through the main power lines can enter the system by two modes:

1. When EMI is found on both the hot (white) and neutral (black) wires of the power line and not on the safety ground (green) it is called common mode noise. Filters must be installed between white and green and between black and green to attenuate this mode of EMI noise.
2. When EMI is found on only the hot (white) or neutral (black) but not both it is called differential mode. A single filter can be installed between white and black to attenuate this mode of EMI noise.

Since both modes of EMI may be present, filters are usually installed between all three pairs of the power lines.

<u>Brownouts</u> - A brownout is a deliberate reduction of line voltage by the power company. In some areas of the country brownouts are common during periods of high electrical use (1). During a brownout the power company deliberately reduces line voltage to avoid an overload situation. The line voltage at the computer may drop as much as 8% or more which can cause problems in many computers and associated components. If brownouts are a problem a <u>power line conditioner</u> must be used to maintain sufficient line voltage.

<u>Power Failures</u> - If power is interrupted for more than a few milliseconds it is called a power failure. Many computers can sustain short interruptions, but most will fail as the duration of the failure increases. If power failures are common an **uninterruptible power supply** (UPS) is recommended. The UPS uses a battery to generate the needed electrical power for short periods (several minutes) during power failures. The UPS switches into action when power is lost and switches back out when it is restored. If the power is to be off for extended periods of time (longer than the UPS can sustain operation) the computer should be turned off. The UPS is desirable when it is essential that critical data must be preserved.

If operation is needed for extended periods of power failure a <u>motor generator</u> is more practical.

The following table summarizes the general features of electrical devices available to correct common electrical noise problems:

	F	T	E	B	PF
UPS	*	*	*	*	*
Line Conditioner	*	*	*	*	
Line Regulator	*	*		*	
Isolation Transformer	*		*		
Surge Protector		*			
Line Filter	*		*		

F - Low Frequency Fluctuations.
T - High Amplitude Transients (lightning, etc.)
E - Electromagnetic Interference.
B - Brownout.
PF - Power Failure.

This table is presented as a guide, and specific features may vary between manufacturers. In general, devices near the top of the table provide a broader spectrum of protection.

Summary

Environmental concerns were presented in terms of user-machine interface, physical room conditions, equipment, and electrical power. Many aspects of environmental problems have been studied and often good solutions have been found. Other problems are so variable that each site must seek its own solution. The methods presented here should prove useful to those who seek to attain a good working environment.

References

1. Dang, Gurcharn, **Electrical Noise: Causes, Effects, Solutions**, Topaz, Inc., San Diego, CA., 1987.

2. **Visual Display Terminals and Worker's Health**, WHO Offset Publication No. 99, World Health Organization, Geneva, 1987.

3. Zaks, Rodney, **DON'T! Or How To Care For Your Computer**, Sybex Inc., Berkeley, CA., 1981.

4. Bailey, Robert W., **Human Error In Computer Systems**, Prentice-Hall, Inc., Englewood Cliffs, NJ, 1983.

5. Corson, Richard G., "VDTs - New evidence indicates helpfulness over harmfulness", Data Management, Dec 1986, p23(3).

6. Bailey, Robert W., "IN DEPTH: HUMAN PERFORMANCE ENGINEERING, Looking Through the User's Eyes", Computerworld v17-May 2, 1983, pID1(5).

7. Berney, Karen, "The Cutting Edge", Nation's Business, V74-June '86, p86.

8. Tobias, Arthur. "Ergonomic Accessories Keep PC Users in the Pink", Today's Office, 22:8 pp 53-56, 1988.

9. Hatoff, Howard I., "Ergonomic Suggestions on the Proper Use of VDTs", Legal Economics, 14:1 pp 35-37, 1988.

10. Saffo, Paul. "Alternate Input Devices", Personal Computing, 12:5 p59(2), 1988.

11. Menke, Susan M., "Design expert proposes solutions to office sterility", Government Computer News, 7:1 p17, 1988.

II. APPLICATIONS IN DIAGNOSTIC RADIOLOGY AND NUCLEAR MEDICINE

IMAGE PROCESSING: BASIC CONCEPTS AND TERMINOLOGY

Gary T. Barnes, Ph.D.
Department of Radiology, University of Alabama Hospitals
University of Alabama at Birmingham, Birmingham, Alabama 35233

Karen Lauro, BS
CT Engineering Group, Picker International, Cleveland, Ohio 44118

ABSTRACT

Digital x-ray images are routinely processed to enhance diagnostic information and to suppress irrelevant detail, and also to extract quantitative information. The basic concepts and terminology of image processing as it applies to x-ray projection radiography are discussed and defined. In general, the processing of an image involves one or more point, local, and/or global operations. Clinical examples of linear and nonlinear gray-scale and algebraic point operations are presented. Examples are also given of local corrections, misregistration corrections, linear filtering, and nonlinear filtering.

INTRODUCTION

A digital image is a two dimensional array of quantized gray scale values or pixels

$$\mathbf{A(m,n)} \tag{1}$$

where **m** represents the pixel column location and **n** represents the pixel row location. For an M x N image matrix or array **m** can take on values 1, 2, ···, M; **n** can take on values 1, 2, ···, N; and for a K bit deep image matrix individual pixels can have integer values from 0 to $2^K - 1$ (i.e., 2^K shades of gray). Digital image processing involves manipulating one or more such images. The subject covers a broad range of techniques, and medical imaging applications generally involve the enhancement of diagnostically important (and the suppression of nondiagnostic) information. An exception to this generalization is image compression, which involves a reduction in the amount of memory required to store an image.

Operations of practical importance on a single image involve artifact suppression, gray scale manipulations, distortion corrections and edge enhancement. Important techniques involving two or more images are (time difference) subtraction, dual energy cancellation of bone or soft tissue structures, and the extraction of flow or organ function parameters. The processing of single image can be classified as either a point operation, a local operation or a global operation. A point operation utilizes a single point (pixel) of the initial or input image to obtain the corresponding point of the final or output image. A local or neighborhood operation utilizes several pixels in a limited area of the input image to obtain a point in the processed or final image. Likewise, the less common global operation utilizes information from essentially all, or almost all, of the input image to obtain a pixel in the output image. Frame processing techniques (i.e., algorithms involving two or more images) can also be similarly classified. The objective of this paper is to review these basic image processing concepts and their utilization in medical imaging.

POINT OPERATIONS I: GRAY SCALE MANIPULATIONS

Gray Scale Processing

An important class of image manipulations involves changing the gray scale values to better display diagnostically important information. This is analogous to manipulating the sensitometric curve of conventional x-ray film to optimize radiographic information content. However, it is a much easier process with a digital image.

Gray scale manipulations are more generally referred to as point operations because each pixel value in the input image is mapped into the output image in such a way that each output pixel's value depends only on the value of the corresponding input pixel. Such gray scale manipulations take a single input image into a single output image and are also known as contrast enhancement and contrast stretching. A given point operation is completely specified by a functional relationship that determines the mapping of the input gray levels to the output image gray levels.

The Gray Level Histogram

Prior to discussing gray scale manipulations, it is useful to define the gray level histogram of an image and review its properties. The histogram of an image (see Figure 1) is a plot of frequency distribution of the pixel gray levels, i.e., the number of pixels with a given gray level vs. gray level. It is a useful tool in image gray scale processing and can be employed to optimize the information displayed. It can also be employed to determine if the digitization process is linear.

Windowing and Leveling

Commonly used linear point operations are windowing and leveling. These can be characterized by the equation

$$B(m,n) = c \cdot A(m,n) + d \tag{2}$$

where **c** and **d** are constants determined by the window and level display controls. When **c** = 1 and **d** =0 (depicted by line C in Figure 2) the input and output images are identical. Increasing **c** (line A in figure 2) decreases the range (window), lowers the midpoint (level) and increases the contrast of pixel gray scale values displayed. Increasing **d** (line B in Figure 3) increases the level of the window. These point operations were introduced to the medical imaging field in 1972 by EMI on their Mark I CT scanner.

Non-Linear Gray Scale Mapping

Often to enhance diagnostically important information, it is desirable to employ a non-linear gray scale transformation. As illustrated in Figure 3 a non-linear operation modifies the shape of the gray scale histogram in a predictable manner and can be characterized by the equation

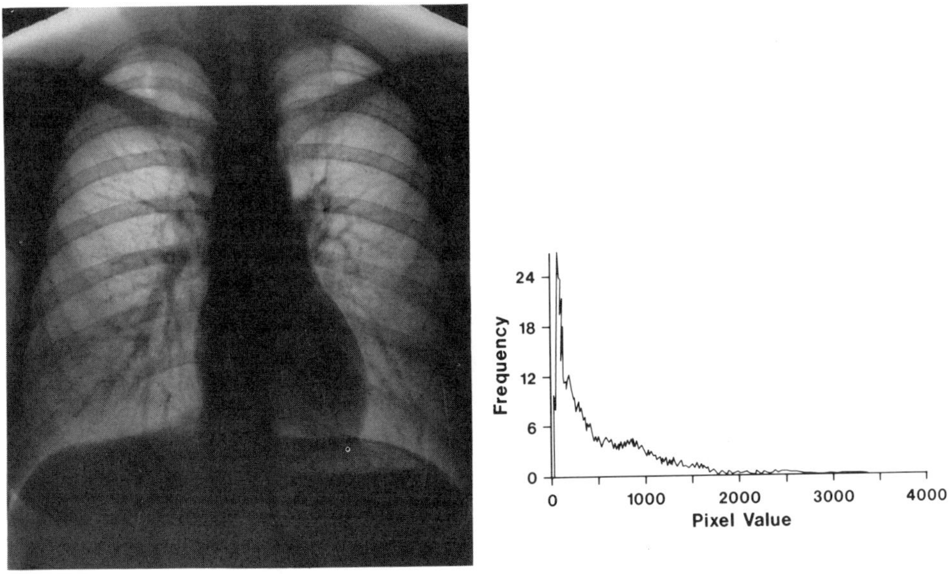

A B

Figure1. Unprocessed digital chest image (**A**) and its histogram (**B**). The image was obtained with the Picker prototype scanning slit dual-energy chest unit.

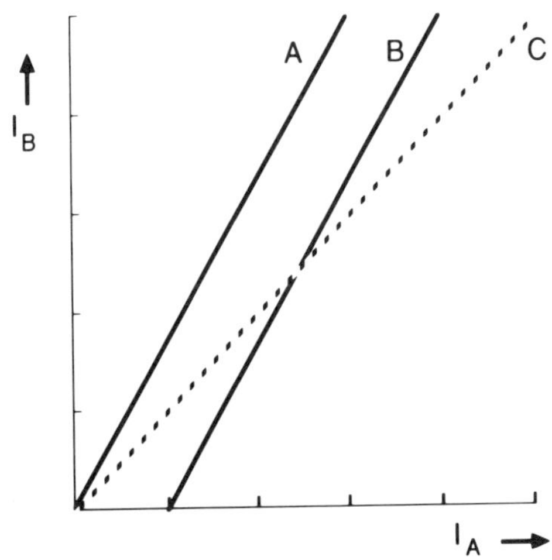

Figure 2. The linear point operations of windowing and leveling.

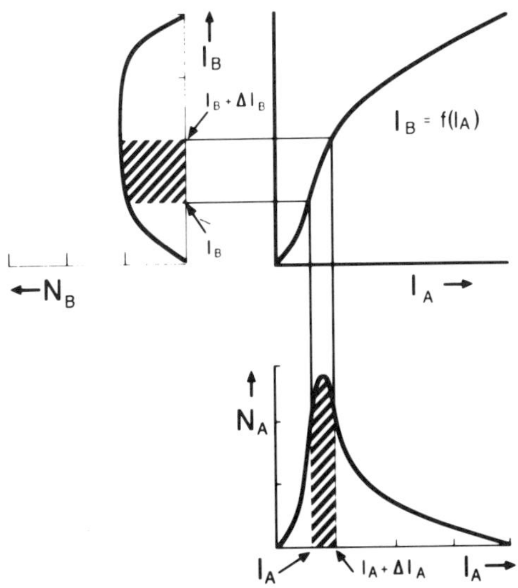

Figure 3. Effect of a nonlinear point operation output image histogram.

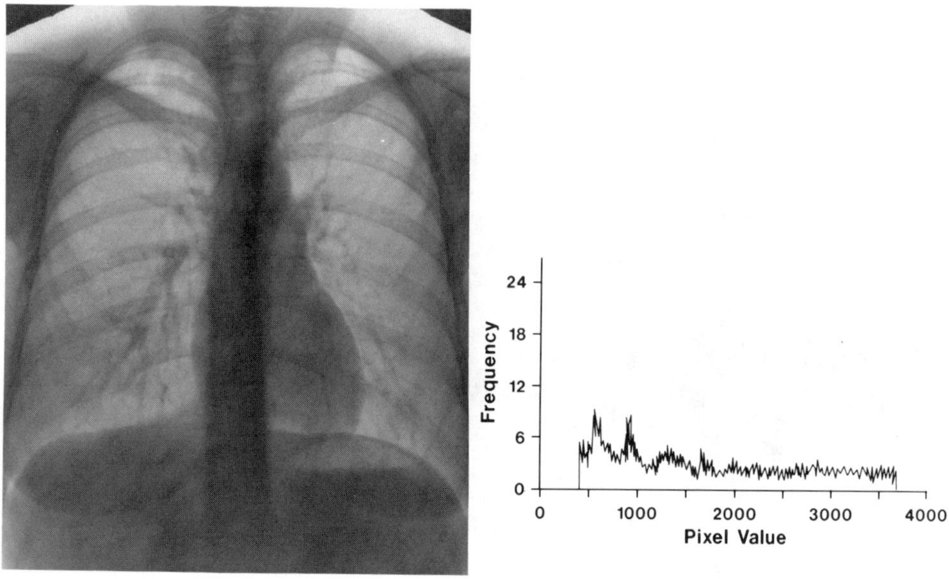

A B

Figure 4. Result of histogram equalization processing on the digital chest image of Figure 1a (**A**) and on the image's histogram (**B**).

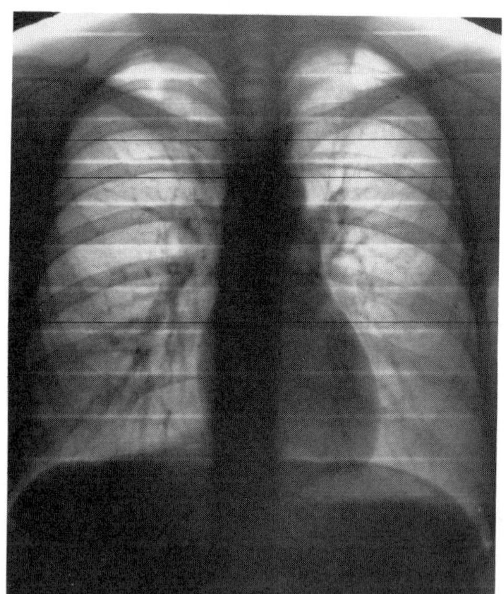

Figure 5. Digital chest image of Figure 1a prior to applying offset and gain corrections.

$$N_b \cdot \Delta B = N_a \cdot \Delta A \quad (3)$$

where N_a is the number of pixels in the input image with gray scale values between **A** and **A** + Δ**A**. N_b and Δ**B** are similarly defined.

One common operation of this type involves flattening or equalizing the histogram. Assuming N_{ave} equals the average number of pixels per unit gray level, the operation that accomplishes this is

$$\Delta B = (N_a/N_{ave}) \cdot \Delta A \quad (4)$$

Shown in Figure 4 is an example of histogram equalization applied to the chest image of Figure 1. In comparing the two images (Figs. 1 and 4) it is apparent that a wider range of anatomy (i.e., mediastinum as well as lung) is displayed with the histogram equalized image. However, low contrast lung detail is less apparent. Since this information is important clinically, histogram equalization is not advantageous in the chest. It has proven useful in situations where one does not have small, low contrast detail of clinical interest, such as CT of the mediastinum.

Offset and Gain Corrections

Often in acquiring a digital image there are systematic errors introduced which create noticeable and disturbing artifacts. Such artifacts depend on the method of image acquisition and the fact that they are systematic allows one to process and improve the image. For example in a CT scanner or a slit radiographic unit where one has an array of detector elements significant differences in leakage (or dark current) and gain occur between elements. Unless these problems are corrected, the image conveys little or no diagnostic information. For a slit radiographic device in which a given row is obtained by sampling a given detector element during a scan the correction takes the form

$$B(m,n) = G(m) \cdot [A(m,n) - L(m)] \quad (5)$$

where **L(m)** is the leakage or dark current offset vector, **G(m)** is the gain vector and **B(m,n)** is the corrected image. Typically the dark current offset vector is determined just prior to an x-ray exposure and the gain vector by a calibration scan of air or a known thickness of tissue substitute.

Illustrated in Figure 5 is an example of the chest image shown in Figure 1 which was acquired with a prototype scanning slit unit and not corrected for detector-to-detector offset and gain variations. The difference in the two images is striking and illustrates the fundamental importance of this type of processing.

POINT OPERATIONS II: ALGEBRAIC MANIPULATIONS

Addition and Subtraction

Addition and subtraction are important point operations in medical imaging that involve two or more images, i.e, frame operations. In these cases a pixel value in the final image is the sum or difference of the corresponding pixels of the input images

$$C(m,n) \quad = \quad A(m,n) \ + \ B(m,n) \tag{6}$$

$$C(m,n) \quad = \quad A(m,n) \ - \ B(m,n) \tag{7}$$

A common example of addition is the averaging of rapidly acquired images in digital subtraction angiography. Provided patient motion is not a problem, the average of N images

$$B(m,n) \quad = \quad N^{-1} \sum_{i=1}^{N} A_i(m,n) \tag{8}$$

results in an image in which the noise level is reduced by a factor of $\sqrt{N}$ compared to one of the initial images. This occurs because the noise fluctuations vary randomly from image to image and add incoherently (i.e., tend to cancel out); whereas, the anatomical structures of interest are the same in each image of the sequence and add coherently. An example of averaging is shown in Figure 6.

Most medical physicists are aware that subtraction is routinely used in digital fluorographic (or fluoroscopic) procedures to obtain an image of vascular structures. Typically several frames are acquired before and after the injection of contrast media. The images acquired before the injection are averaged to obtain a low noise mask which in turn is subtracted from a postinjection image acquired when the contrast media is passing through the blood vessels of interest. As shown in Figure 7 subtracting the mask leaves only the radiopaque blood vessels. Usually contrast media is present in several of the postinjection images and these can in turn be averaged or preferably their weighted sum obtained to further reduce the noise level in the final image. In the latter case each image has a weight based on the amount of contrast media present in the blood vessel(s). Of interest is that with modern equipment the subtraction is so perfect that virtually all background structures and bony landmarks are absent. This is clinically undesirable and often the preinjection mask is added to the subtracted image with a fractional weight to obtain an image with high contrast vasculature and low contrast bony landmarks. An example of this is illustrated in Figure 8.

Dual-Energy Techniques

The manipulations performed in dual-energy imaging are in essence point operations. The first step is to generate two images of the patient, identical in every respect except that one is obtained with an x-ray spectrum with a low mean energy and the other is obtained with an x-ray spectrum with a somewhat higher mean energy. These are then decomposed point by point into water and bone images whose intensity values are proportional to the overlying masses per unit area of soft tissue and bone(1).

$$w_b(m,n) \quad = \quad f\,[S_{lo}(m,n),\ S_{hi}(m,n)] \tag{9}$$

$$w_s(m,n) \quad = \quad g\,[S_{lo}(m,n),\ S_{hi}(m,n)] \tag{10}$$

where $S_{lo}(m,n)$ and $S_{hi}(m,n)$ are the pixel values of the low and high energy images and

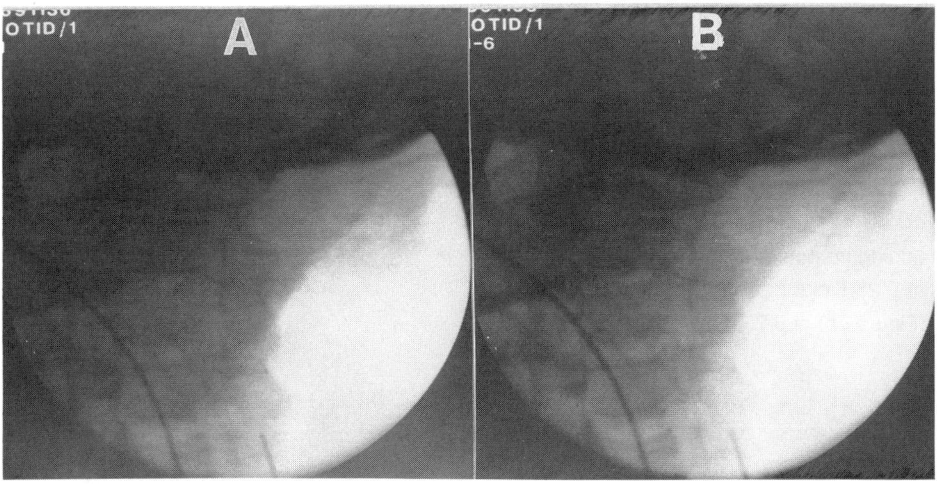

Figure 6. Comparison of a single frame (**A**) and an average of six frames (**B**) of a preinjection cerebral DSA mask.

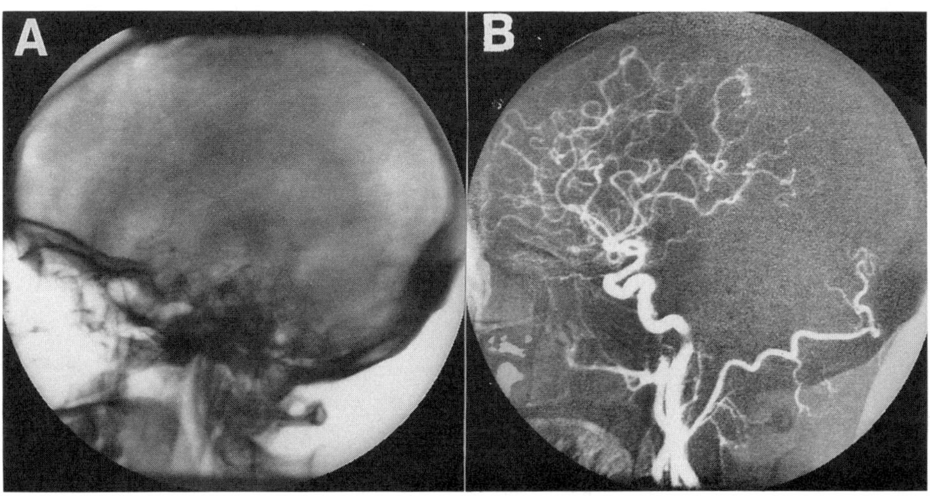

Figure 7. Postinjection DSA image before (**A**) and after subtraction of mask (**B**).

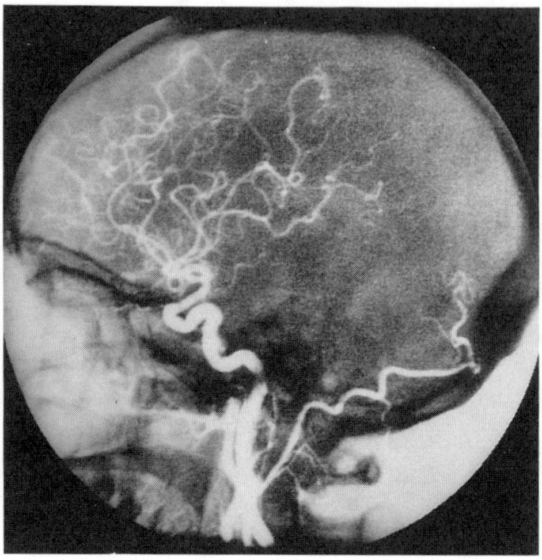

Figure 8. High contrast fluorographic cerebral arteriogram obtained by addingthe postinjection image (Figure 7a) at 0.25 weight and the subtraction image (Figure 7b) at 0.75 weight.

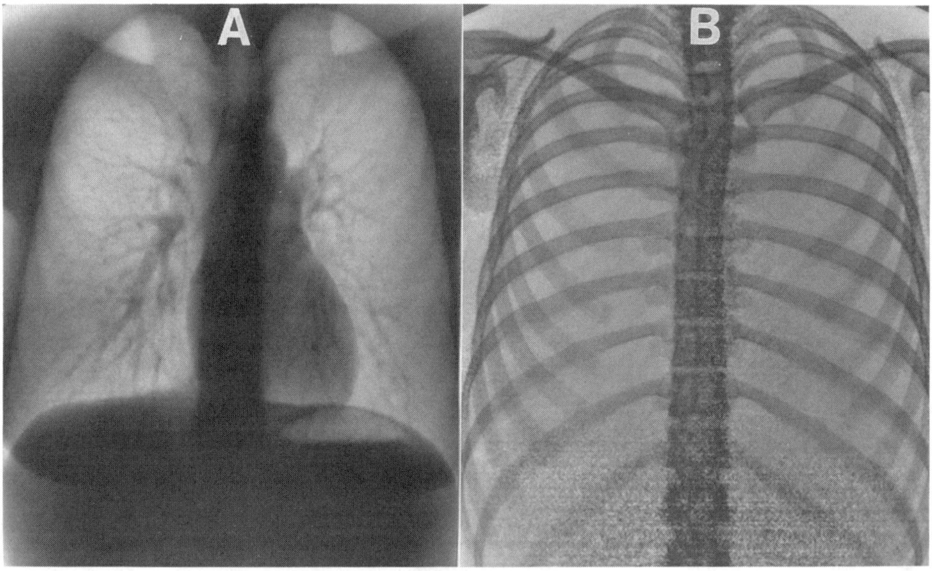

Figure 9. Soft tissue (**A**) and bone (**B**) digital chest images of our reference patient (see Figure 1a) obtained with Picker prototype dual-energy chest unit.

with different masses per unit area of soft tissue (water) and a bone equivalent material. A bone cancelled (soft tissue) image is then generated by adding a fraction ($\approx$ 0.55) of the bone image to the initial soft tissue (mass per unit area) image to fill in the voids or "shadows" left by the removal of bone. In addition a normal or single energy image can be obtained by adding the original low and high energy images together. Bone cancelled and soft tissue cancelled images (see Figure 9) can then be presented to radiologists simlultaneously along with the standard single energy image. Of practical interest is that the bone cancelled image shows only soft tissue structures - the ribs and spine are completely invisible. The bone image, on the other hand, shows the ribs and spine but no soft tissue structures.

The obvious advantage of this type of point operation is that it separates the two types of anatomical information present and provides the radiologist with simpler images to view. Thus, a nodule which may be hidden by the ribs in a conventional chest film is clearly visible on the soft tissue image. Recent results indicate that a dramatic improvement in pulmonary nodule detection rates is realized with dual-energy techniques(2). In addition, dual-energy chest radiography reveals nodule calcifications with greater sensitivity and accuracy than other noninvasive techniques(3), an important capability since the presence of calcium is a primary radiographic criterion for benignancy.

LOCAL OPERATIONS I: GEOMETRIC CORRECTIONS

Spatial Distortion

Often undesirable spatial distortions or changes occur when images are acquired. When it is clinically advantageous, these distortions can be corrected employing digital processing techniques. In such cases the final image is obtained from the original image employing a two stage process. The first step is to map the input image pixels to their location in the output image, and then utilize this spatially shifted image to obtain the final image. In general the first step will result in fractional pixel shifts and interpolation is required to determine the value of the pixels in the final image. The simplest interpolation scheme is to use the nearest neighbor value. This, however, is usually not satisfactory in medical imaging because gray scale values often change dramatically between adjacent pixel locations. A more desirable approach is to employ bilinear interpolation of the four nearest neighbor pixel locations. More sophisticated interpolation schemes using cubic spline, sinc or similar functions can also be employed but rarely offer any practical advantage (and have the disadvantage of requiring more computational time) because of the inherently noisy character of medical images.

Practical examples of this type of local operation are correcting for motion misregistration in DSA, correcting for misregistration in dual-energy radiography employing a sandwich detector, and correcting for image intensifier distortion in digital fluorography. The improvement that can be realized by employing a patient motion misregistration correction is illustrated in Figure 10. Misregistration between the front and back elements of a dual-energy sandwich detector results in imperfect bone and soft tissue cancellation and introduces noise. Likewise, when an image intensifier is used "pincushion distortion" occurs. In the latter two examples, the spatial aberrations or distortions are systematic and can be quantitated by imaging a periodic pattern. By knowing what the calibration image should look like, one can then determine the pixel shifts that are necessary to remove the

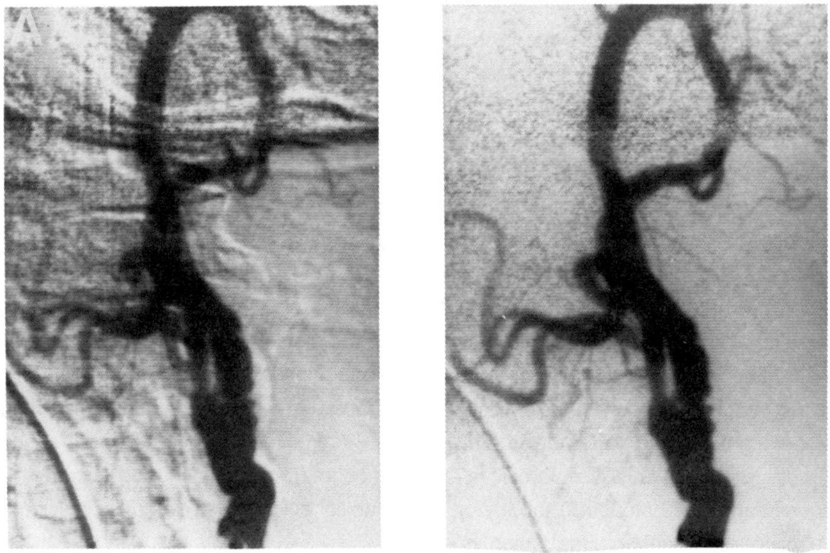

Figure 10. Comparison of DSA images that were obtained without (**A**) and correcting for (**B**) patient motion misregistration.

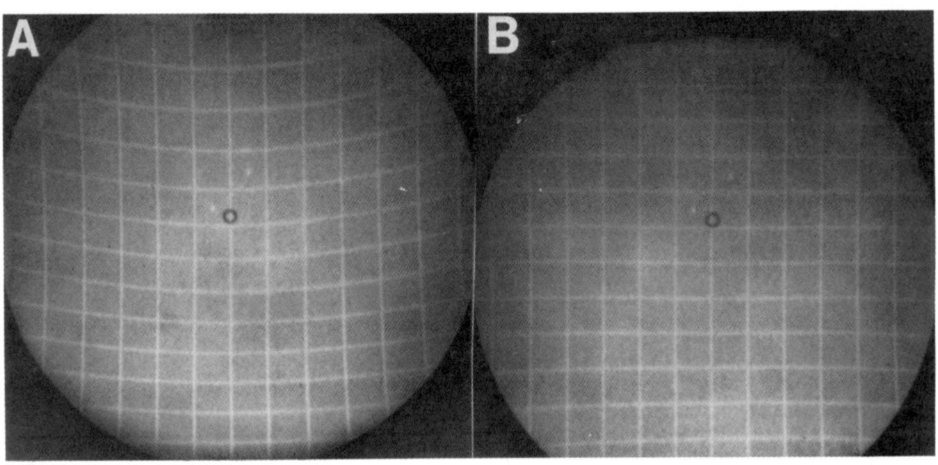

Figure 11. Fluorographic image of grid phantom obtained with the central ray angled at 28.4° to the plane of the image intensifier, i.e., the x-ray tube was moved toward the top of Figure. Unprocessed (**A**) and corrected (**B**) for geometric distortion. Courtesy of D. P. Chakraborty (4).

distortion. The improvement that can be realized when one corrects for pincushion distortion is illustrated in Figure 11. Such corrections are essential in applications requiring precise spatial information from a conventional fluoroscopic imaging chain (i.e., electro-optical image amplifier tube and TV).

LOCAL OPERATIONS II: FILTERING

Linear Filtering

An imaging system is said to be linear if the image pixel values double when the input x-ray photon fluence is doubled. Likewise, when one is spatially filtering an image it is considered to be a linear process if, when the pixel values of the input image are changed by a constant factor, the corresponding pixel values of the output image change by the same factor. That is, if the output image **B(m,n)** is obtained by spatially filtering the input image **A(m,n)**

$$\mathbf{A(m,n)} \rightarrow \mathbf{B(m,n)} \tag{11}$$

the process is said to be linear if

$$\mathbf{k \cdot A(m,n)} \rightarrow \mathbf{k \cdot B(m,n)} \tag{12}$$

where **k** is a constant.

Linear systems analysis has an established theoretical basis and is commonly employed in the study of electrical circuits, optical systems and also x-ray imaging systems. With the exception that discrete rather than continuous mathematics is employed, the same theoretical framework can be employed to study the linear filtering of digital images.

Convolution - Smoothing and Edge Enhancement

The operation of convolving one function with another is an essential concept in image processing and is best understood by considering the one dimensional case. Mathematically, the convolution of the function or kernel **g(m)** with the image **A(m)** is represented by the expression

$$B(m) = \sum_{i=-(N-1)/2}^{(N-1)/2} g(i) \cdot A(m+i) \tag{13}$$

where **B(m)** is the output image and **N**, for reasons of symmetry, is an odd integer. If the kernel extends over 5 pixels then

$$\begin{aligned} B(m) = {} & g(-2) \cdot A(m-2) + g(-1) \cdot A(m-1) + g(0) \cdot A(m) \\ & + g(1) \cdot A(m+1) + g(2) \cdot A(m+2) \end{aligned} \tag{14}$$

For simplicity and notational ease, $*$ is often employed to denote the convolution operation. With this notation Equation (13) reduces to

$$B(m) = g(m)*A(m) \tag{15}$$

Extending Equations (13) and (15) to two dimensions, one obtains

$$B(m,n) = \sum_{i,j=-(N-1)/2}^{(N-1)/2}\sum g(i,j)\cdot A(m+i, n+j) \tag{16}$$

$$B(m,n) = g(m,n)*A(m,n) \tag{17}$$

The degree and type of filtering, i.e., smoothing or edge enhancement, depends on the shape of the kernel. For example, if a rect function kernel (Figure 12a) is convolved with an image (Figure 12b), the result is a smoothed or blurred image (Figure 12c). How one obtains the final blurred image is conceptually illustrated in Figure 13. Of interest is that if one subtracts the blurred image from the original image, only the edges remain (see Figure 14). Likewise, if the kernel has a positive central peak (Figure 15a) and negative side lobes, the output image will only contain information about the edges of the input image (Figure 15c). Noteworthy is that Figure 15c was obtained with one operation while Figure 14d required two operations. This edge image can in turn be added to the original image to obtain an image in which the edges are enhanced (Figure 15d). This image could also be obtained with one operation if the positive central peak of the Figure 15a kernel is increased by one, i.e., the kernel has unity area. In addition to illustrating the fundamental concept of convolution and the effects of smoothing and edge enhancing kernels, there are also two salient points in these examples that should be emphasized. First, in digital image processing the kernels are usually symmetric (i.e., incorporates an odd number of pixels) and second, if the kernel sum is unity (i.e., has unit area), the average pixel values of the input and output images will be the same.

In two dimensions a representative smoothing or blurring kernel is

$$\begin{matrix} +1 & +1 & +1 \\ +1 & +1 & +1 \\ +1 & +1 & +1 \end{matrix}$$

Convolution with such a kernel would also usually incorporate division by nine so, as noted above, the average large area pixel values in the output and input images would be the same. The effect of a 3x3 smoothing kernel on a digital chest image is shown in Figure 16a and can be compared to the unprocessed image (Figure 1a). Parenchymal detail as well as noise is suppressed. A kernel that enhances edges is

$$\begin{matrix} -1 & -1 & -1 \\ -1 & +8 & -1 \\ -1 & -1 & -1 \end{matrix}$$

This is often referred to as a "Laplacian" kernel or filter because it approximates an

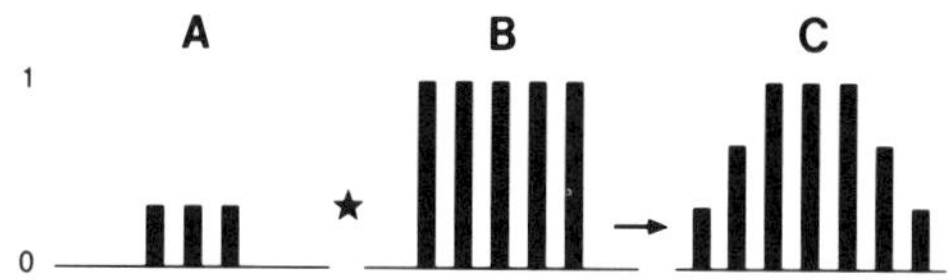

Figure 12. Convolution of 3 pixel unity area rect function kernel (**A**) with 5 pixel rect function image (**B**) yielding image (**C**).

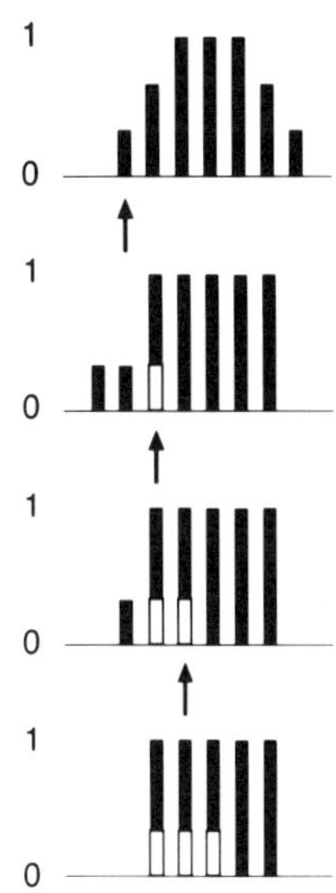

Figure 13. Illustration of the convolution process. The top line represents the resultant function (i.e., Figure 12**C**). At any pixel location the value of the process is determined by the overlap area of the kernel and the function it is operating on as the kernel is moved from left to right. The overlap area is shown in white in lines 2, 3, and 4 for three adjacent positions of the kernel. The arrow above the lines points to the pixel of the final function that the overlap corresponds to.

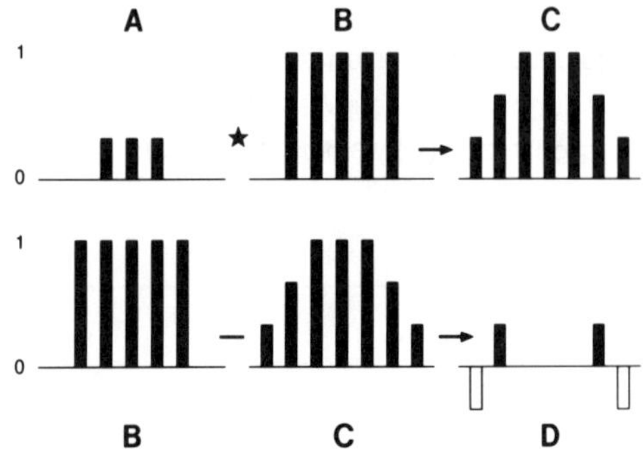

Figure 14. Convolution of unity area rect function kernel (**A**) with image (**B**) yielding "smoothed" or "blurred" image (**C**). An edge image (**D**) can subsequently be obtained by subtracting image **C** from image **A**.

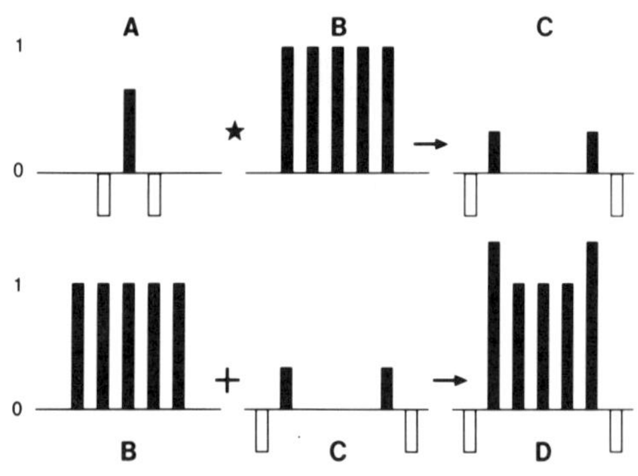

Figure 15. Convolution of zero area positive central peak, negative side lobe kernel (**A**) with image (**B**) yielding edge image (**C**). Adding this image to the original image results in edge enhanced image (**D**).

unoriented second-derivative operation. The result of applying it to our reference chest image is shown in Figure 16b. It in essence leaves only the high spatial frequency edge information and is often used as "edge detector" in image analysis algorithms. As can be appreciated by comparing it with the unprocessed image, other than its use as an edge detector, it has limited clinical utility. A Laplacian kernel is useful, however, when it is added to the original image (Figure 16c). In practice this can be combined into a single operation with the kernel

$$\begin{matrix} -1 & -1 & -1 \\ -1 & +9 & -1 \\ -1 & -1 & -1 \end{matrix}$$

As one would expect from comparing these four images, Figure 16c is preferred by radiologists.

An alternative and commonly used linear high pass filtering approach is "blurred mask subtraction." As the name states one convolves the input image with a blurring kernel (i.e., a 50x50) and then subtracts the result (often fractionally weighted) from the input image. The blurred image contains only low spatial frequencies and the subtraction of this low frequency image from the original image suppresses the low and thereby enhances the final image's medium and high spatial frequencies. An example of this two step approach (i.e., low pass filtering followed by the algebraic operations of subtraction and addition) is shown in Figure 17a, 17b and 17c. As illustrated in Figure 17d adding a Laplacian filtered image further enhances the high spatial frequency information. The advantage of this approach over that used to obtain Figure 16c is that the only the very low spatial frequencies are suppressed, the medium frequency content is maintained and the high spatial frequencies are enhanced. As can be appreciated by comparing Figures 16c and 17d this allows one to display a wide range of anatomical information as well as emphasize fine detail in a single image. Of practical interest is that the same result can be achieved by convolving the original image with a single, appropriately designed kernel.

Convolution and the Fourier Transform

The discrete Fourier transform of an image $\mathbf{g(l\,\Delta p)}$ is given by

$$G(f) = \sum_{l=0}^{N-1} g(l\,\Delta x)\, e^{-j2\pi f\, l\, \Delta p} \tag{18}$$

$$f = k/N\,\Delta p, \qquad k = 0, 1, 2, \cdots, N-1$$

where $\mathbf{N}$ is the number of pixels comprising the image and $\mathbf{\Delta p}$ is the size of the pixels. Likewise, the image can in turn be determined from its Fourier transform

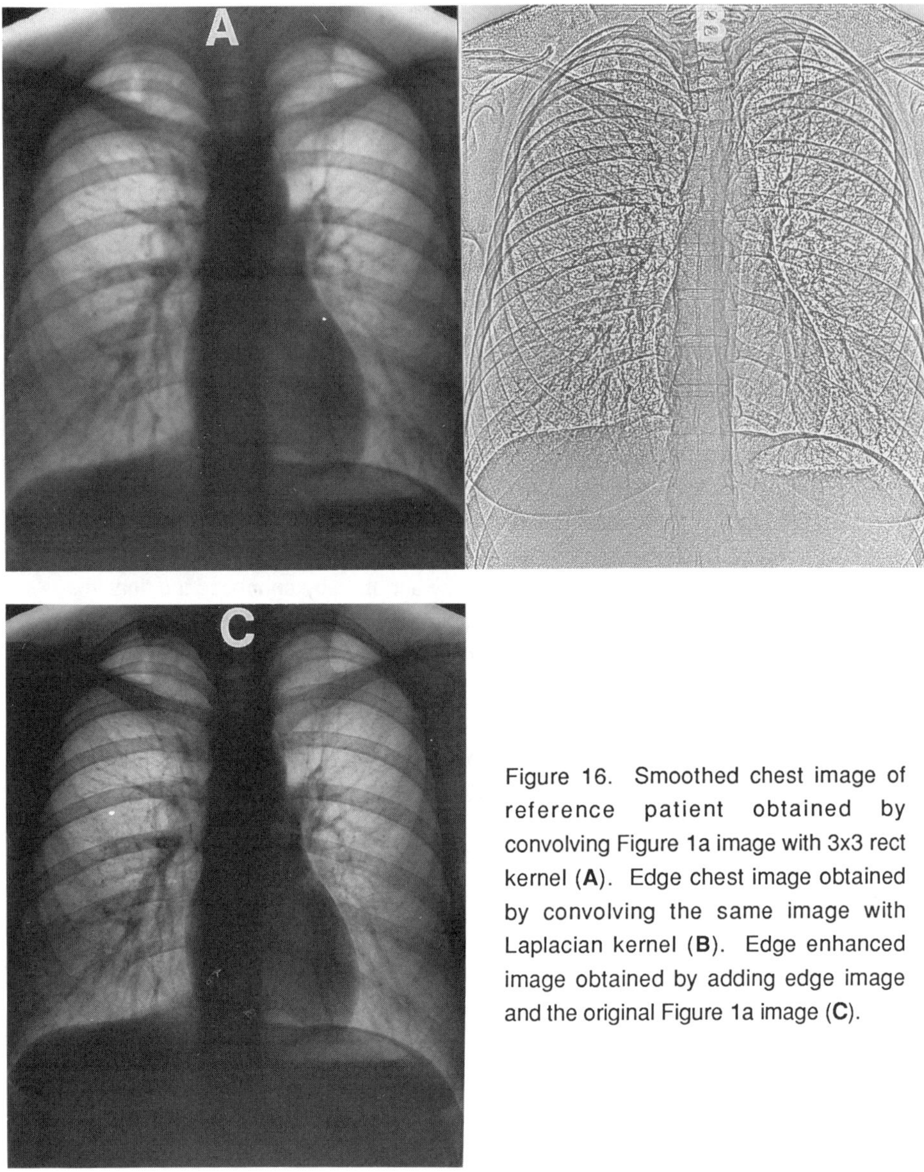

Figure 16. Smoothed chest image of reference patient obtained by convolving Figure 1a image with 3x3 rect kernel (**A**). Edge chest image obtained by convolving the same image with Laplacian kernel (**B**). Edge enhanced image obtained by adding edge image and the original Figure 1a image (**C**).

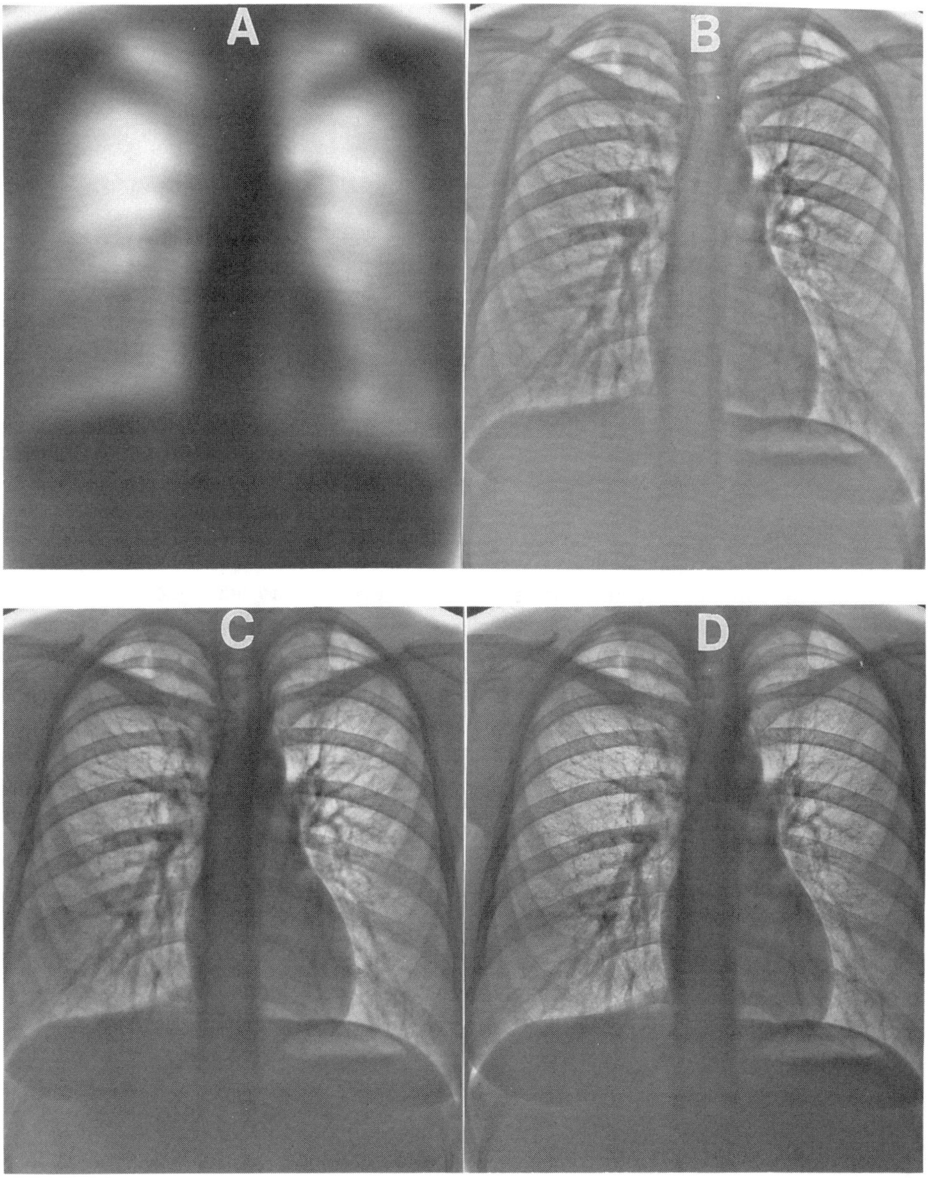

Figure 17. Smoothed chest image obtained by convolving Figure 1a with 50x50 rect kernel (**A**). Edge image obtained by subtracting smoothed image from the original image (**B**). Blurred mask subtraction image obtained by linear combination of **B** and Figure 1a (**C**). Edge enhanced blurred mask subtraction image obtained by adding **C** and Laplacian edge image (Figure 16b) (**D**).

$$g(x) = N^{-1} \sum_{k=0}^{N-1} G(k/N\ \Delta p)\ e^{j2\pi k\ x/N\ \Delta p} \tag{19}$$

$$x = l\ \Delta p, \quad l = 0, 1, 2, \cdots, N - 1$$

The functions **g(x)** and **G(f)** are known as a Fourier transform pair, and for any **g(x)**, **G(f)** is unique and vice versa.

A useful property of Fourier transforms is that the transform of the convolution of two functions is the product of the individual Fourier transforms of the two functions in the frequency domain.

$$g(x)*h(x) \quad \rightarrow \quad G(f)\cdot H(f) \tag{20}$$

This implies that one can process an image by first Fourier transforming it to frequency space, multiplying it by the transform of the kernel, and then transforming the result back to the spatial domain. Although this can be done, it usually is not in practice because of the excessive amount of computer time required. An exception to this generalization often occurs in the processing of nuclear medicine images where one is dealing with small matrices and reasonably short Fourier transform execution times on an array processor. The large matrices employed in digital radiography dictate that one employs the convolution operation.

Implicit in Equations 18 and 20 is that the frequency response of a filter is given by its Fourier transform. Plotted in Figure 18 are the Fourier transforms of the kernels employed to obtain the one dimensional processed images in Figures 12, 14 and 15. As one would expect the response of the smoothing kernel falls off at higher spatial frequencies and because of this it is referred to as a low pass filter. The behavior of the Laplacian filter is quite different and has a response only at higher spatial frequencies, e.g., a high pass filter. Adding the original image to the Laplacian filtering operation enhances the high spatial frequencies and also maintains the low spatial frequencies. The ability to analyze the frequency response of a filter is of fundamental interest and also useful in its design. In practice one often uses a high pass filter that compensates for MTF falloff.

Non-Linear Filtering

A high pass linear filter enhances the high spatial frequency content of the low as well as the high pixel count regions. This often poses a problem in the low pixel count regions because there is little information and considerable noise which becomes more apparent with a linear high pass filter. An alternative and often more pleasing result can be obtained by scaling the degree of edge enhancement with pixel count. An example of this type of adaptive filtering on our reference chest image is shown in Figure 19. This image was obtained by a linear combination of the original unprocessed input image and the Laplacian filtered blurred mask subtraction image (Figure 17d). At the high pixel count regions (i.e., lung) the edge enhanced blurred mask subtraction image is given a weight of one and the original unprocessed image a weight of zero. In the low pixel count regions (i.e., below the

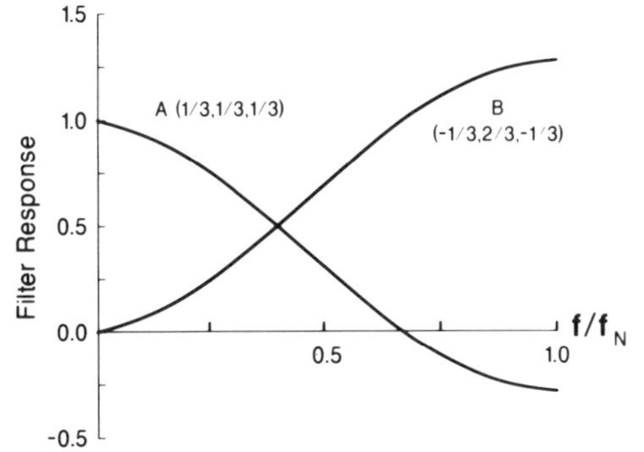

Figure 18. Fourier transform of Figure 12a and 14a rect function kernel (**A**), and Figure 15a Laplacian kernel (**B**).

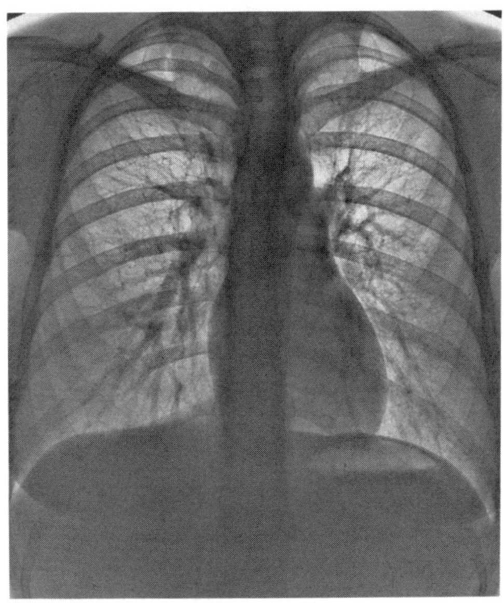

Figure 19. Adaptively filtered reference chest image obtained by linearly combining Figure 1a image with Figure 17d image. In the lung field (i.e., the high pixel count region) the Figure 17d image had a weight of 1.0 and the unprocessed image a weight of zero. In the subdiaphragmatic region (i.e., the low pixel count region) the Figure 17d image had a weight of zero and the unprocessed image a weight of 1.0. In between these two regions the weight of the processed image increased and the weight of the unprocessed image decreased linearly with pixel count.

diaphragm) the Laplacian filtered blurred mask subtraction image is given a weight of zero and the unprocessed image a weight of one. In between the weighting of the Laplacian filtered blurred mask subtraction image increases linearly and the weighting of the unprocessed image decreases linearly with pixel count of the original image. In comparison with our previous best (Figure 17d) the degree of parenchymal detail is similar and the noise in the subdiaphragmatic region and mediastinum is less. Needless to say, this image is preferred by radiologists over the other types of filtered images of this patient we have studied.

DISCUSSION

There are several important practical applications of point and local operations in medical imaging. Discussed above are the gray scale point operations of windowing and leveling, histogram shaping and correcting for equipment nonlinearities. We would anticipate that in the digital radiology departments of the future these operations will be completely automated so that no intervention is needed between the acquisition and displaying (filming) of an image. One only has to look at the current CT sections in most radiology departments to realize the potential savings in manpower that would result. In widespread use are the algebraic point operations of averaging and subtraction.

Local geometric operations can be employed to correct for image distortions. Local filtering operations can be utilized to enhance either the low, medium or high spatial frequencies of an image. Moderate-to-high pass filters (i.e., convolution with an edge enhancing kernel or blurred mask subtraction) are generally preferred in clinical practice. The optimal degree of filtering is not known at this time and is likely to vary from exam to exam. In general radiologists prefer to suppress much of the low spatial frequency (quasi-DC) and to enhance the high spatial frequency (small detail) information. Probably more important than the filtering employed is that it is not varied from day-to-day or week-to-week. Radiologists' pattern recognition depends on image consistency and if the images appear different from day-to-day, they lose their base of reference and have a tendency to over or under call.

In addition to enhancing the diagnostic information content of an image, digital image processing can be employed to obtain quantitative information from one or more images. Functional parameters of clinical interest are routinely attained in nuclear medicine. Examples are cardiac ejection fraction, segmented wall motion, thallium and kidney function studies. Examples in other areas are quantitation of blood vessel per cent stenosis (DSA) and pulmonary nodule calcium content (dual-energy chest radiography). These and other applications will be discussed in more detail in subsequent presentations.

Feature extraction and the development of algorithms for computer aided diagnosis are research areas that hold considerable promise. The general strategy is to analyze the physical characteristics of the features of interest and then use image processing techniques (i.e., edge detection, gray level thresholding, etc.) to isolate these features in the image of interest. The characteristics of the isolated features are in turn compared to reference data. Areas of current interest are lung nodule detection, microcalcification detection and analysis in mammography, and interstitial disease. The results to date are provocative and techniques being developed could have significant impact on the practice of medicine.

BIBLIOGRAPHY

R. N. Bracewell, ***The Fourier Transform and Its Applications*** (McGraw-Hill, New York, 1978).

E. O. Brigham, ***The Fast Fourier Transform*** (Prentice-Hall, Englewood Cliffs, New Jersey, 1974).

K. R. Castleman, ***Digital Image Processing*** (Prentice-Hall Inc., Englewood Cliffs, New Jersey, 1979).

W. K. Pratt, ***Digital Image Processing*** (John Wiley & Sons, New York, l978).

REFERENCES

1. L. A. Lehmann, R. E. Alvarez, A. Macovski, and W. R. Brody, Med. Phys. **5**, 659 (1981).
2. L. T. Niklason, N. M. Hickey, D. P. Chakraborty, E. A. Sabbagh, M. V. Yester, R. G. Fraser, and G. T. Barnes, Radiology **160**, 589 (1986).
3. R. G. Fraser, N. M. Hickey, L. T. Niklason, E. A. Sabbagh, R. F. Luna, C. B. Alexander, C. A. Robinson, A. A. Katzenstein, and G. T. Barnes, Radiology **160**, 595 (1986).
4. D. P. Chakraborty, Med. Phys. **14**, 249 (1987).

AN OVERVIEW OF DIGITAL RADIOGRAPHY

Stephen Balter, Ph.D.

Cornell University Medical College, New York, NY 10021
Philips Medical Systems North America Company

ABSTRACT

Digital radiography is the application of digital image handling and processing techniques to transmission radiography. Various detectors are used to capture the modulated x-ray beam and transform it into a digital image. The separation of image capture, processing, and display permits optimization of each component of the imaging chain. This paper will discuss some of the special considerations related to digital projection radiography as well as some of the more general considerations associated with the processing and display of any digital radiological image.

IMAGE ACQUISITION

A projection radiograph is formed using one of the three principal geometries shown in Figure 1: Projection of a cone of radiation onto a two-dimensional detector, projection of a fan of radiation onto a one-dimensional detector, and projection of a line of radiation onto a point detector. For the fan and line beam modes, appropriate means are used to translate the beam-detector assembly relative to the patient. At the end of the image acquisition process, a two-dimensional transmission image has been formed and detected. The useful information is carried by spatial and temporal modulations of the detected signal. In the case of scanned systems there may be a second temporal modulation of the signal which is superimposed on the primary modulations. As the solid angle of the instantaneous x-ray beam is decreased the effects of scatter are reduced. This increases the Signal-to-Noise Ratio (SNR) in the modulated x-ray beam. However, the effective efficiency of the x-ray tube is decreased. This eventually leads to increased geometric or motion unsharpness.

Presently, there are no line beam systems used for clinical projection radiography. A fat pencil, or a narrow cone, is used in the scanned equalized system developed by Plewes, et al.[1,2] A novel multiple beam system was recently described by Vlasbloem.[3] Linear arrays of detectors, used with a fan of radiation, are evolving from the fan beam geometries used with third and fourth generation CT systems. An example of a system under development is the digital chest unit developed by Tesic et al.[4,5] Its geometry is shown in Figure 2. Each detector in this class of system usually has its own signal amplifier and conditioner. Individual or multiplexed analog to digital converters are assigned to each channel. By placing energy sensitive detectors in series, it is possible to construct a system which produces energy dependent images in near real time.[6] Fan beam systems (either single or dual detector) have potential imaging problems including a relatively low spatial resolution due to the large number of separate detectors required and a kymographic distortion of the image resulting from anatomical motion during the scanning interval. A profound advantage of fan beam systems is that they are, in practice, the only ones that routinely use slot geometry for radiographic scatter suppression.

Important parameters of the modulated x-ray beam include its absolute intensity and intensity range, its signal-to-noise ratio, and its spatial resolution. An ideal detector will transform all of the information content of the modulated beam into either the analog or digital domains.[7,8] Suboptimal conversion anywhere in the system can cause an irreversible loss of information content. Subsequent image processing cannot improve the image quality or information content beyond that in the detected x-ray beam or compensate for irreversible losses. Image processing may, however, influence the perceptibility of clinically relevant structures.[9-15]

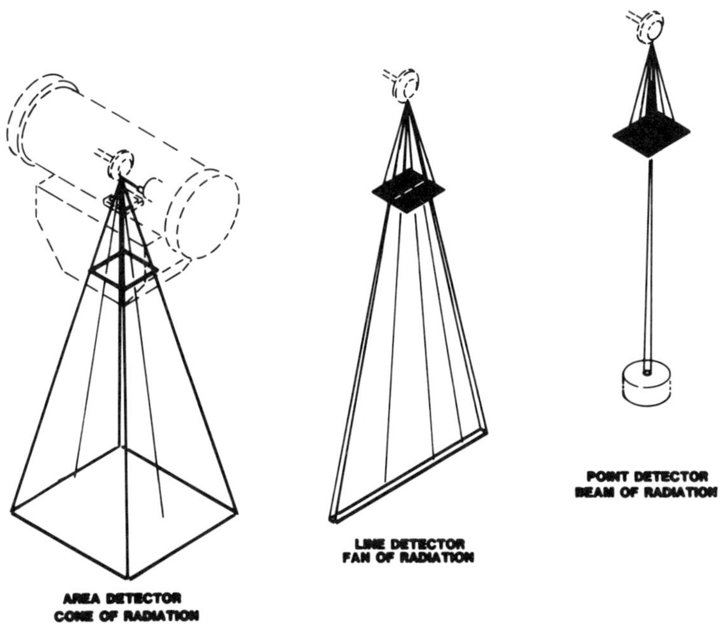

Figure 1. Radiographic Projection Geometries: Cone, Fan, and Line.
As the solid angle decreases, scatter decreases, acquisition time increases.

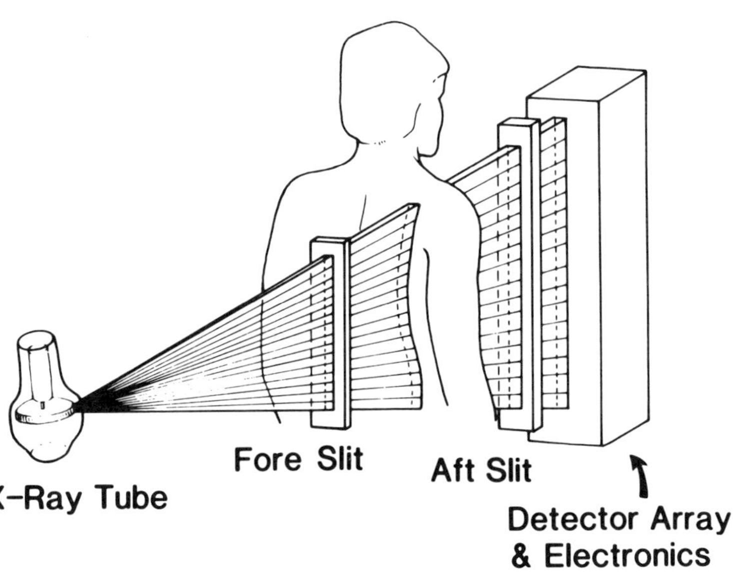

Figure 2. Experimental Digital Chest Radiographic Device.
Note the use of fan beam geometry. (Courtesy Picker Medical Systems)

Figure 3 illustrates the effects of improper ranging in the analog domain. If potentially useful information is captured in either horizontal region of this curve it is irretrievably lost. A similar data loss can also occur in a digital system. This is illustrated in Figure 4. The rounded toe and shoulder of analog transfer functions ordinarily give regions of limited image capture which are typically not present in digital systems. In particular, the shoulder region of film curves is accessible to (and used by) radiologists using bright lights. The abrupt black clipping of digital images can be very confusing.

There are engineering and practical limitations which cause the digital image to retain less total information than that present in the modulated x-ray beam. Some are related to the intrinsic nature of specific image conversion technologies. Others result from the desire to minimize the number of bits needed to represent an image. The loss of "information" in an absolute sense may or may not affect the clinical information content. Each situation must be evaluated on its own merits. Thus it is to be expected that an optimized digital imaging system will handle images of widely varying matrix sizes and bit depths.

If a full size chest radiograph is represented by 50 micron pixels (corresponding to a limiting resolution of 10 lp/mm) 60 Megapixels would be required. A simple logarithmic digitization of the x-ray intensity in each pixel, with allowances for the real dynamic range of the chest, and with guard bits to give some measure of exposure latitude, brings the total to around 1,000 Megabits per image. Such a representation is too large for practical purposes. Compromises are sought in the size of the acceptable matrix size[16] and pixel bit depth. The matrix size should be selected based upon the required system resolution and the required field of view. Acceptable matrices range from 128^2 to 4096^2, with 512^2 to 2048^2 being most common. The bit depth should be selected based on an acceptable signal-to-noise ratio. If the quantization level is too coarse relative to the structure and noise content of the final image then pseudocontouring will result; if it is too fine, then bits are wasted on recording statistical fluctuations. Clinical images require between 8 and 14 bits for image data.

Image intensifier - video chains are the radiation transducers which are used as the front end of most Digital Fluorographic (DF) systems. The most common DF systems are those specialized for the performance of Digital Subtraction Angiography (DSA).[17] The block diagram of a generic DF system is shown in Figure 5. A single- or multi-mode image intensifier is used to convert a radiographic image into optical format; the video chain converts the image into a serial analog electrical signal; finally an A/D converter produces the digital image. There are numerous feedback loops in such a system which affect image quality. These include circuits which directly influence the radiographic factors such as automatic brightness control and automatic exposure termination. Other systems control the optical and electrical characteristics of the image. These include servo control of the system's optical apertures and automatic gain control in the video chain. More complex loops provide coupling between the video and radiographic systems. The purpose of this design is to transform all of the useful information into the working range of the analog-to-digital converter. Control of the optical intensity also permits the operator to select a working dose level which provides an appropriate level of quantum mottle.

The use of multimode image intensifiers presents the first application of the principle of zooming the image so that the clinically important field of view fills the entire aperture. This process yields improvements in spatial resolution (including "smaller" pixel sizes) as the field of view is reduced (Figure 6). Many clinical fluoroscopic and fluorographic applications do not simultaneously require fine spatial resolution and a large field of view. In these cases, the appropriate use of image intensifier zoom (and corresponding radiation beam restriction) to the smallest relevant field of view maximizes available spatial resolution of any given image matrix while minimizing scatter degradation of the SNR. Even though the input dose in the center of the field increases with decreasing field size because of less geometric minification within the image intensifier, if the optics are correctly set, the patient's integral dose is independent of zoom mode.

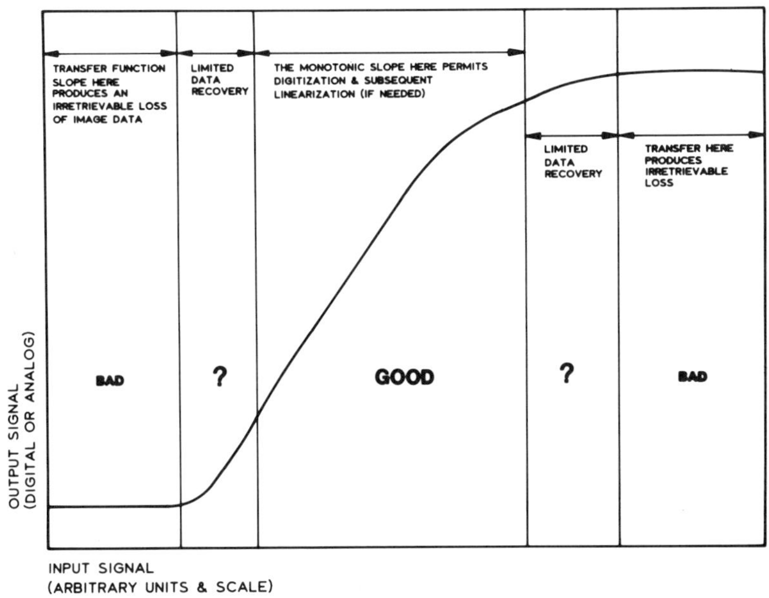

Figure 3. Gray Scale Transfer Function for an Analog Imaging System. Note the characteristic rounded toe and shoulder.

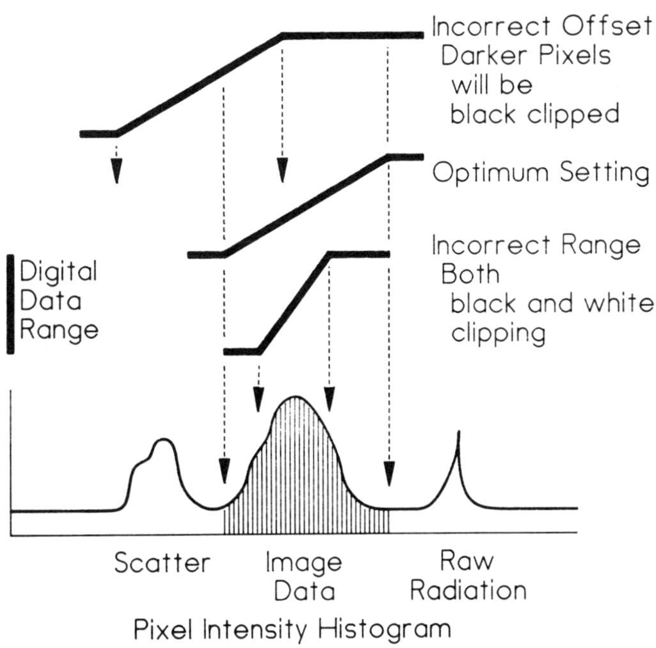

Figure 4. The Intensity Histogram of an Analog Image. Improper A/D conversion can result in considerable irretrievable data loss.

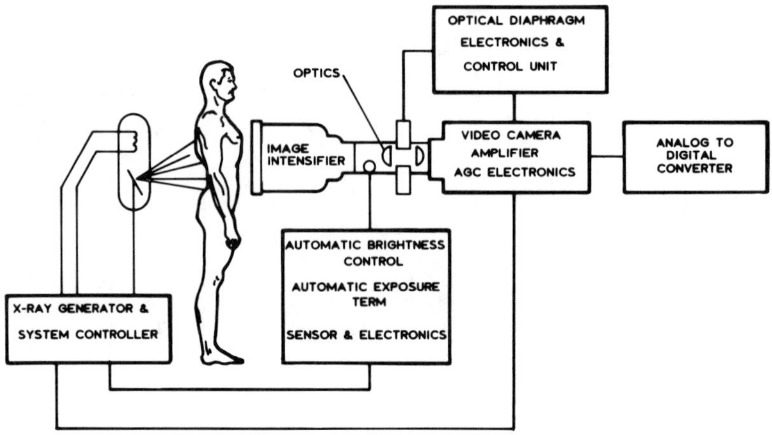

Figure 5. Block Diagram of a Fluorographic System.
Note the elaborate use of feedback and control elements.

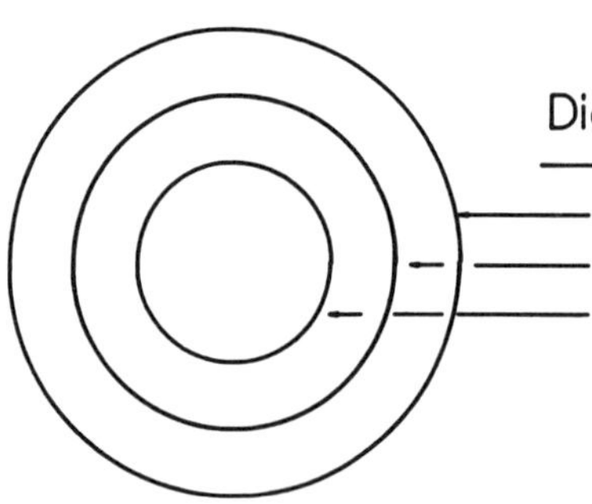

Nominal Pixel Size

Input Diameter	Matrix Size 256	1024	4096
350	1.37	0.34	0.09
250	0.98	0.24	0.06
150	0.59	0.15	0.04

All values are in millimeters

Figure 6. Multimode Image Intensifier Field Sizes.
Corresponding Digital Fluorographic Pixel Sizes.

The usual system specification of a Digital Radiographic (DR) system includes a 1,000:1 SNR. This implies that a 10 bit pixel depth is required to capture and handle all of the available information in the modulated x-ray beam. The required bit depth may be less when scatter and quantum noise factors are taken into consideration.

The most common radiographic image capture modality is the film-screen system. This system carries the entire burden of radiographic image capture, display, archive, and communication in most departments. In its ninety-year life, film has evolved to do all of these tasks reasonably well. Specialized screen-film systems can achieve a great deal of optimization in niche environments. General purpose systems require more compromises. The principal decision is that of selecting a system that will give enough contrast for easy interpretation while preserving enough latitude to encompass both the dynamic range of the body part being imaged as well as providing a margin for exposure errors. As previously shown, portions of a film which are over- or under-exposed will lead to an irretrievable information loss. The useful dynamic range can be extended by reducing the contrast of the film. This makes the image more difficult to visually search and interpret. If a film-screen system is to be used only as the input to a film digitizer (a new task for radiography), then a very low contrast transfer characteristic is appropriate.

Most clinical departments restrict the number of available film screen speed combinations so that confusion and resulting exposure errors are minimized. Therefore, the "departmental standard" film-screen system must possess resolution and noise properties for the most demanding examination. The resulting image is likely to have more resolution and require more than the minimum necessary dose for most examinations. The gray scale transfer function will be that compromise needed for viewing all examinations reasonably well.

Energy storage detectors are comprised of sheets of material having the property that when they are irradiated, a portion of the signal is stored as a two-dimensional electron distribution. The system of computed radiography (CR) based upon photostimulable phosphor imaging plate technology developed by the Fuji Photo Film Company[18] is an excellent example of a radiographic imaging system in which primary image capture is separated from other aspects of imaging. This separation of functions permits separate optimization of image capture and image display parameters.[19-22] Conceptually, the benefits of such a separation are available for photofluorographic systems as well as for all other systems in which the image exists in either electronic or digital form.[23,24]

The functional cycle of the photostimulable detector is shown in Figure 7. Europium dopants in the phosphor trap about half of the signal carrier electrons released in the crystal during irradiation. Subsequent irradiation of the phosphor using light releases a portion of this stored energy as blue fluorescence. The image plate is read out by scanning with a pencil laser beam and the resulting fluorescence is collected and digitized. An example of the resulting image is shown in Figure 8. The raw digital image obtained in this manner may then be processed to produce alternate representations of the data. A sample image pair, of a difficult radiographic projection, is shown in Figure 9. The remaining images in this paper were all obtained and processed with a computed radiography system.

The dose response characteristics of the photostimulable phosphor are linear over a dose range of more than four orders of magnitude. With proper setting of the CR system, subject contrast is captured with the same efficiency at "low", "correct", and "high" doses. Figures 10 to 12 illustrate this point using a skull phantom. These images were obtained using detector exposures of 60, 300, and 3000 microroentgens. The overall contrast and density of the resulting radiograph are preserved over this range. The noise characteristics of the low dose image may make it unacceptable for many purposes. At high doses, system noise and structure mottle dominate x-ray quantum noise. Once this effect occurs, further increases in dose will not lead to direct improvements in image quality. The stability of contrast and density in this dose range does, however, yield a clinically important improvement in image uniformity.

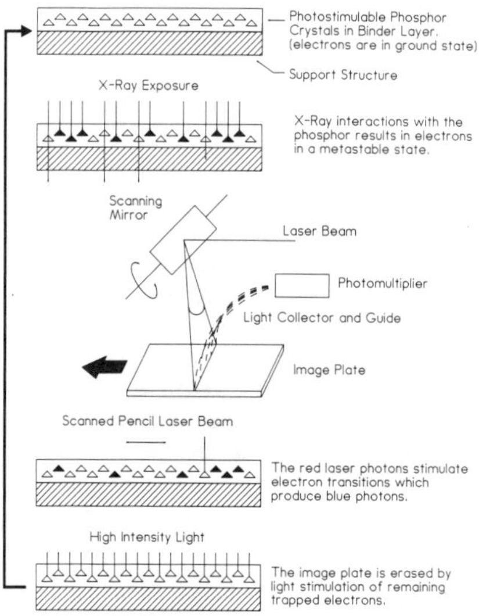

Figure 7. CR Stimulable Phosphor Image Plate Working Cycle. Pixel size is determined by the laser spot size and sampling rate.

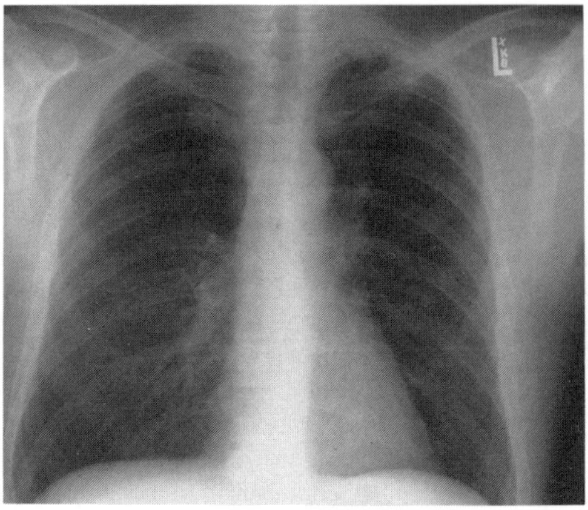

Figure 8. Neutral Digital Radiograph. Nominal 2100 x 1800 pixels, corresponding to 2.5 lp/mm. Linear Grey Scale Transfer. No Unsharp Masking.

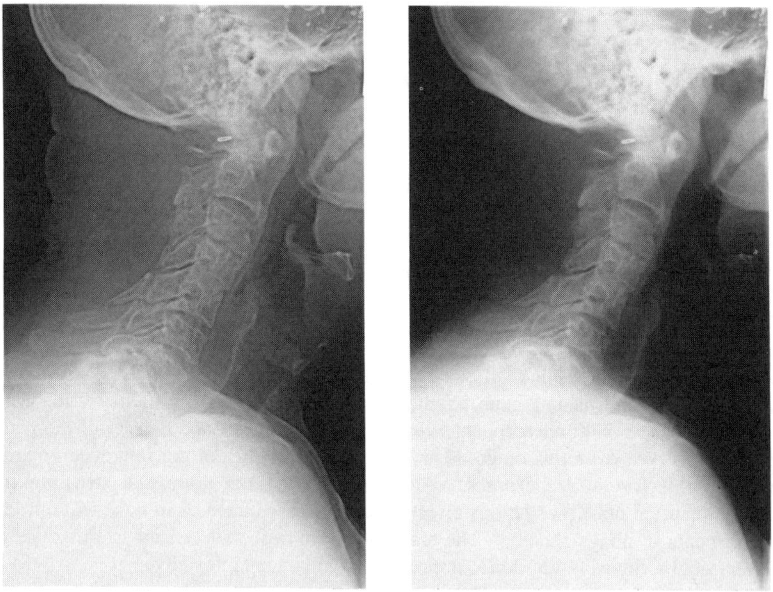

Figure 9. Typical digital image pair produced by the CR system. Different image processing operators were used on each side.

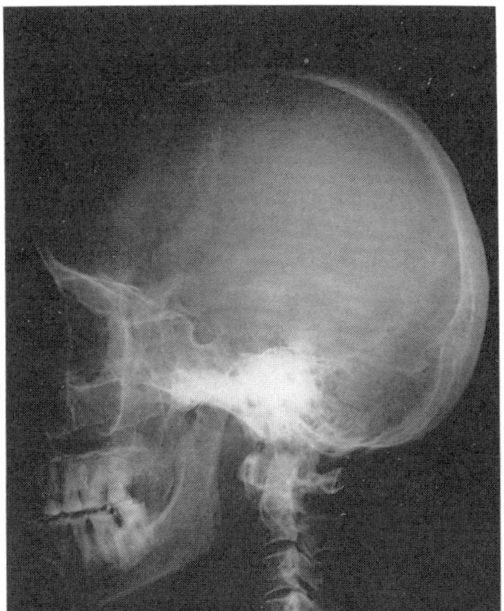

10a

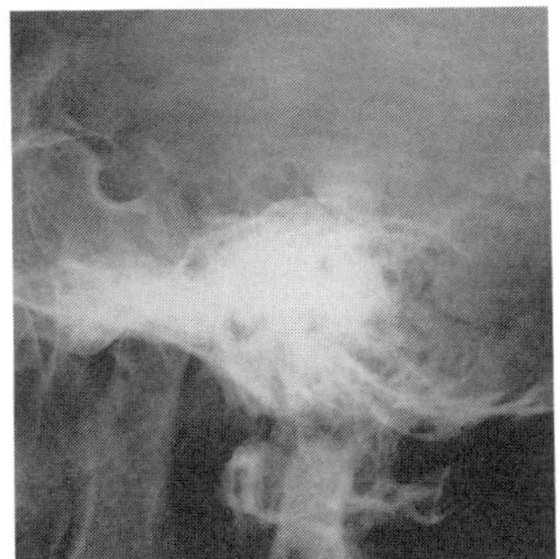

10b

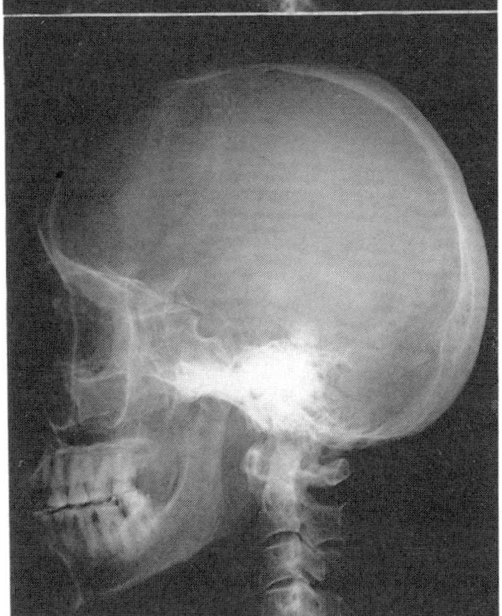

11a

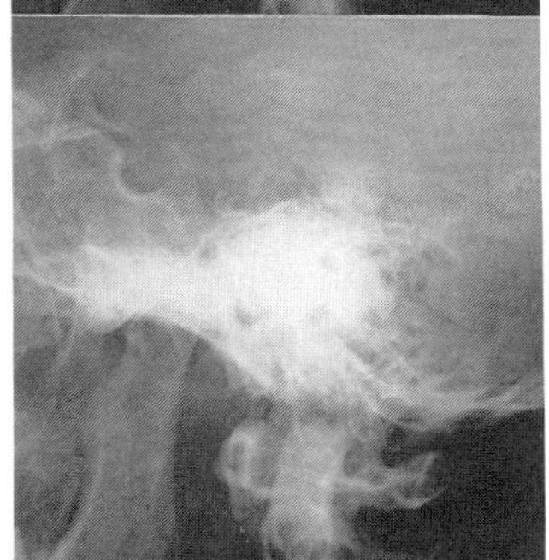

11b

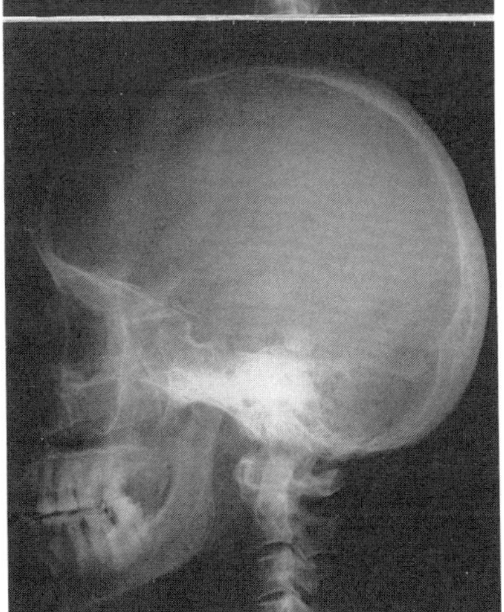

12a

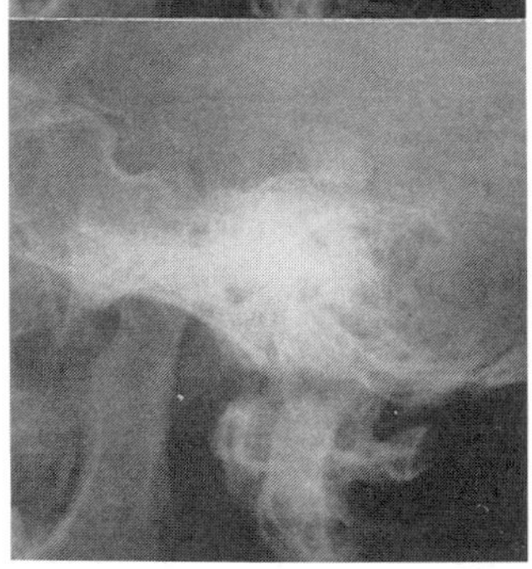

12b

Figure 10. Digital Radiograph of a Phantom.
Image Plate Exposure 3,000 microR.
System and Structure Noise Visible.

Figure 11. Exposure 300 microR.
The same image plate was used.
Normal Clinical Dose Range.

Figure 12. Exposure 60 microR.
Below the normal clinical working range.
X-Ray noise dominates this image.

IMAGE HANDLING

Image handling and image processing are two distinct entities. An image handling system physically moves or archives images without transforming their properties. An image processor takes an input image and produces a transformed output image. Most complete imaging systems have both handling and processing components. For example, when an image is acquired, it undergoes both analog and digital processing (some of which is unintentional) as it is moved from the detector, through the system, in and out of archives, and finally to an analog display device. A few specific considerations related to the design of a digital radiography system and its internal image handling will be discussed below. In the clinic, the tens of images associated with a typical patient must be collected, archived, communicated, and displayed. The implementation and technology of Picture Archiving and Communications Systems (PACS) are discussed elsewhere.[25-27]

The raw information content of individual images is very high in comparison to most data sets handled by computerized means. A single digital image ranges from 0.25 MByte in a digital fluorograph to about 10 MByte in a high resolution digital radiograph. This has several implications on system design. For example, a DR system needs high internal bus and processor speeds. Digital fluorographic systems often have the most critical speed requirements because image capture and processing must occur at rates ranging from about one image per second (fluorography) to thirty images per second (cinefluorography). Digital fluorographs are often viewed and clinically used immediately after acquisition. Digital radiographs are acquired at rates ranging from a few per second (serial changers) down to a few per hour. Digital radiographs seldom need to be viewed in less than a few seconds after capture.

Speed at reasonable cost places constraints on the nature of the image structure and those processing algorithms which are to be employed. Image matrix size and pixel bit depth are reduced so as to move images quickly through a system with fixed bandwidth.[28] The elimination of unimportant bits should terminate just short of the point at which the clinical purposes of the examination are compromised. Because medicine is a conservative art, clinicians may doubt the integrity of the image if image quality is reduced without safety margins. The use of high speed image capture and a subsequent slower processing cycle is acceptable provided that the original data capture does not degrade the image and that the complete cycle is fast enough to meet the clinical needs of the examination.

The question of data compression is almost as important to the design of digital radiographic systems as it is to the design of PACS systems. Appropriate image compressors are needed to efficiently store digital flourographic and cinefluorographic images. For example, the raw data bandwidth needed for cinefluorography can exceed ten megapixels per second for periods exceeding ten seconds. Image compression algorithms are either totally reversible or produce some irreversible "loss" of image information. Digital radiographs can be reversibly compressed by around 2.5:1 and irreversibly compressed to a much greater degree.[29,30] At present, many physicians are unwilling to work clinically with any image that has undergone an "irreversible" compression. This may only be due to an unfortunate semantic consequence associated with the word "irreversible" and not a real limitation of appropriate data compression. The difference between the original and the compressed-decompressed image can be regarded as an additional noise source in the imaging system. A reversible algorithm is free from compression noise. An irreversible algorithm will add some degree of additional noise to the image. If compression noise makes a small contribution relative to other noise sources in the complete imaging system (including x-ray quantum noise), it is likely to be unnoticeable.

IMAGE PROCESSING

The processing of images in a stable and reproducible manner is a major technical and clinical justification for digital radiography. In addition, digital tools permit transformations which cannot be achieved in the analog domain. With appropriate processing, one can obtain an improvement in the perceptibility of clinically meaningful structures and possibly improve the accuracy of the diagnostic process. However, image processing when carried to an extreme will disorient both radiologists and clinicians. In any radiographic setting, there is a need to build strong bridges between the observer's concept of the patient's anatomy and pathology and the particular visual representation of the observer's mental construct in the image.[31-35]

One must also consider that radiologists have a finite amount of time for the routine interpretation of radiographs. The image processor should use prestored algorithms for the computation of optimum output images in most cases. The processing of most images should be completed prior to viewing. As an example: A single observer reading screening chest radiographs in almost any format will miss about 30 percent of the positive findings. In part, this is due to the low prevalence of disease in screening populations. The false negative rate can be reduced to around 10 percent by having two radiologists interpret the films separately. It is to be carefully noted that the present use of radiological images is not a solitary vice: Images are discussed between residents and attendings, radiologists and referring physicians, and in clinical conferences. This methodology while expensive in physician time maximizes the transfer of important information from the images to the responsible clinician.

For interactive image processing to be economically viable it must perform as well in extracting information about the patient as double reading, without doubling reader time. There are special cases, such as trauma, in which unlimited observer time is justified. In these cases, interactive image processing, and other computerized aids, may be of vital importance.[36] For this reason, digital radiology facilities should have at least one fully equipped work station available for such tasks.

MULTI PIXEL TRANSFORMATIONS

The histogram of pixel count vs pixel value in a digital radiograph is a complicated structure. Histogram analysis can be used to find the regions of valid clinical data within a much wider preliminary search range and thereby maximize the bit depth of the A/D converter. A representation of such a histogram is shown in Figure 4. Histogram equalization techniques are a way of mapping the digital representation of an image so as to make maximum use of the available output gray scale.[37,38] The image resulting from a full histogram equalization is presently considered too different from the conventional appearance of a radiograph to be useful in routine practice. Global and constrained histogram equalization is beginning to be used in the digital processing of radiotherapy port films.[39,40]

Convolution operations can be used to modify the Modulation Transfer Function (MTF) of a digital system. In principle, many different convolution filter shapes can be applied to an image. In practice, the need for rapid image throughput currently limits the choice to variants of the square kernel unsharp mask algorithm for clinical use. Other desirable filters will no doubt be implemented as image processor bandwidth increases.

It is worthwhile to look at the unsharp mask operator in some detail. Each pixel in the image becomes the "target pixel" in turn. An average is taken of all of the pixels in a square region of the original image surrounding the target pixel. A weighted sum is then constructed by combining this average with the target pixel value. The size of the averaging region and the weighting factor are commonly adjustable factors. As requirements for image processing speed increase the size of the averaging region usually decreases due to image processor bandwidth limitations. Thus digital fluoroscopic systems may average over a maximum mask region of only a few pixels while digital radiographic systems (where speed is less critical) permit use of a maximum mask region of more than 100 pixels on a side. The selection of the appropriate unsharp mask size and weighting factor is highly task dependent. Figures 13 - 16 illustrate the influence of unsharp mask kernel size on the appearance of a chest radiograph. The weighting factor used for these examples is larger than those used clinically.

NOISE REDUCTION

Algorithms exist for minimizing the perception of noise in digital images. Most of these algorithms are spatial frequency filters. They are effective only in those cases in which the spatial frequencies of clinically useful signal and those of noise can be separated. For example, an algorithm which replaces single pixel noise spikes with an average of nearest neighbors is effective because the spike's signal intensity is located in those spatial frequencies corresponding to a region of low MTF of the total imaging system.

The visual effect of noise is enhanced by the unsharp mask operator. In practice, the application of unsharp masking is adapted to the potential interference of enhanced noise on viewing conditions. Examinations such as neonatal chest images are often computed with little or no unsharp masking because enhanced noise matches the frequency content of certain types of interstitial lung disease. If one processes chest examinations while searching for nodules, one may accept a great deal of visual noise in return for the enhanced visualization of nodules. The same algorithm would be inappropriate when the clinical question is interstitial disease because the unsharp mask operator increases the visualization of unwanted anatomy.

SINGLE PIXEL OPERATIONS

The ability to arbitrarily map the digital representation of a pixel to an arbitrary gray scale value is a key tool available in a digital radiography system. It is possible to produce digital images with gray scale characteristics that are chemically and physically impossible using either film or pure analog computational technologies. For example, one can produce the same subjective representation (on the output of a fixed parameter hard copy or soft copy display digital system) that is produced by any given radiographic film. However, the capability of tracking a variable CRT image so that its corresponding hard copy is an exact subjective match is not within the state of the art.

Users of digital imaging systems should be aware that gray scale transformations are more subtle than the simple window and level controls found on CT scanners. It is possible to transform the data so that the "level shift" occurs along either the horizontal or vertical axis. Only in the case in which the transfer function is linear with a slope of one will these two transformations have the same effect on the image. As shown in Figure 17, if a system has a sigmoidal gray scale transfer function, a simple vertical level shift changes the brightness (density) of an image but does not affect the relative contrast of different structures. A horizontal shift (corresponding to a mAs change in radiography) affects both local brightness (density) and local contrast. Figures 18 - 20 illustrate the effects of changing both global contrast and density shifts on the appearance of the same image. Here again, the demonstrated images use exaggerated shifts for teaching purposes.

Figures 21-24 are examples of empirically optimized displays for chest and abdominal radiography. The profound effect of using incorrect display functions are also shown in this series. It should be obvious that the selection of digital display functions needs to be done with at least as much care as the selection of a film for conventional radiography.

Figure 13.

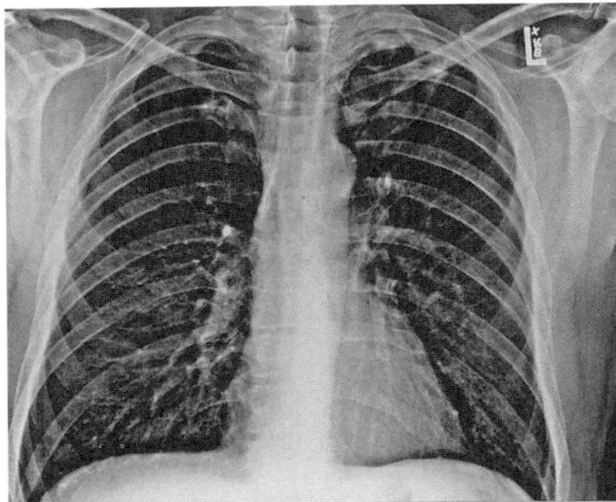

Figure 14.

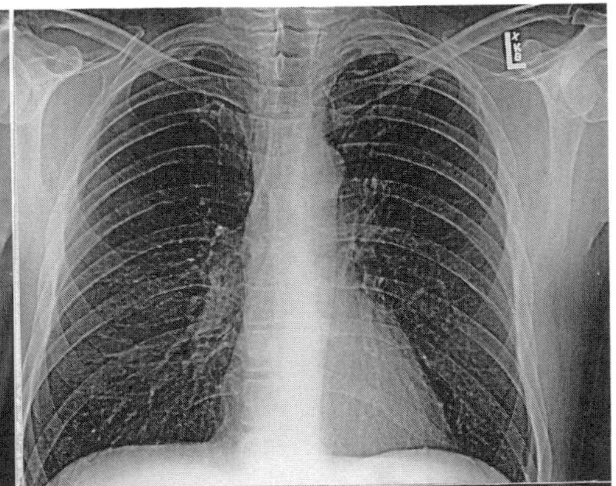

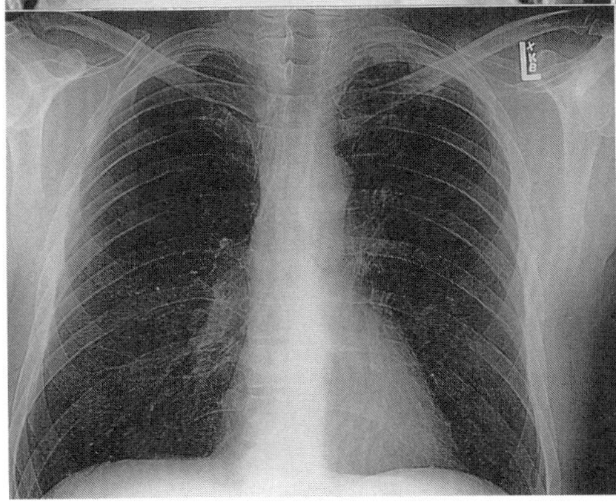

Figure 15.

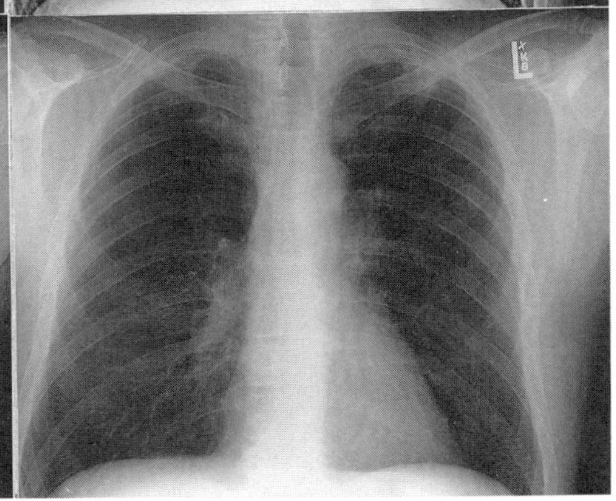

Figure 16.

The Effect of Unsharp Masking
on a digital chest radiograph.

Figure 13. Kernel = 99 x 99 pixels
Figure 14. Kernel = 41 x 41 pixels
Figure 15. Kernel = 21 x 21 pixels
Figure 16. Kernel = 3 x 3 pixels

Pixel size = 0.2 millimeters.

The weighting factor is 6
for all four images.
This is greater than required
for most clinical applications.

Figure 17.

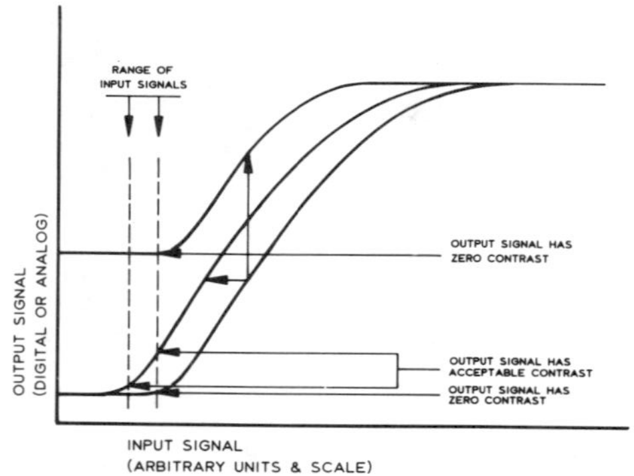

Figure 18.

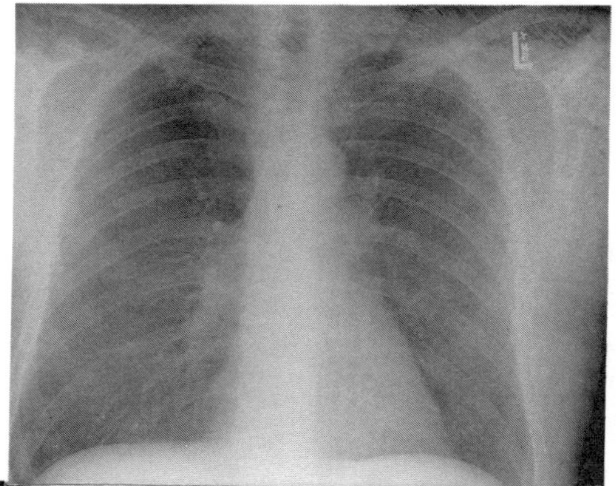

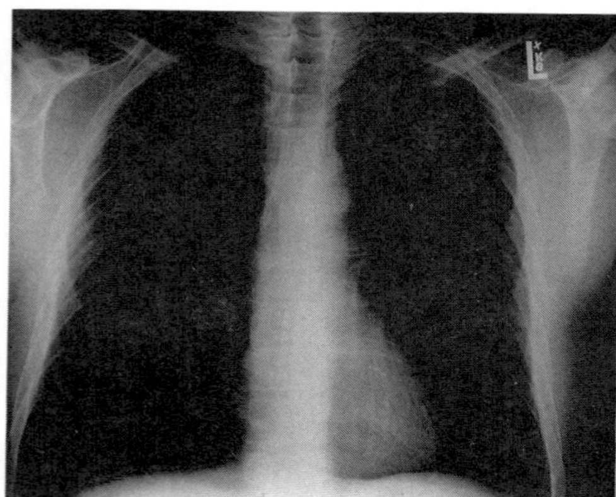

Figure 20.

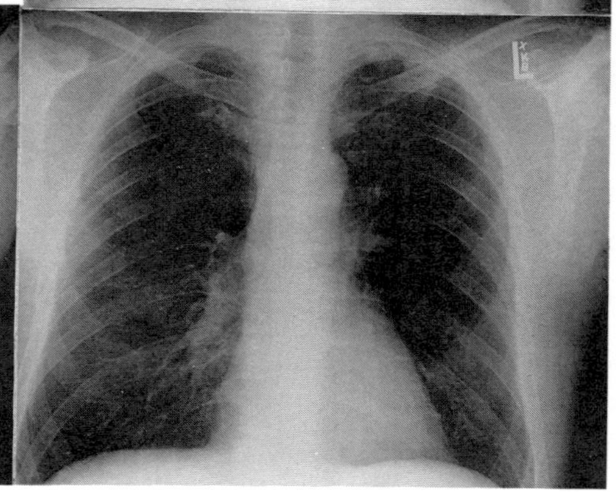

Figure 19.

Figure 17. The effects of horizontal and vertical density shifts on local contrast and density.

Figure 18. Sample radiograph with lowered contrast, density horizontally shifted toward low values.

Figure 19. Same radiograph with clinically optimized contrast and density.

Figure 20. Same radiograph with increased contrast, density horizontally shifted toward high values.

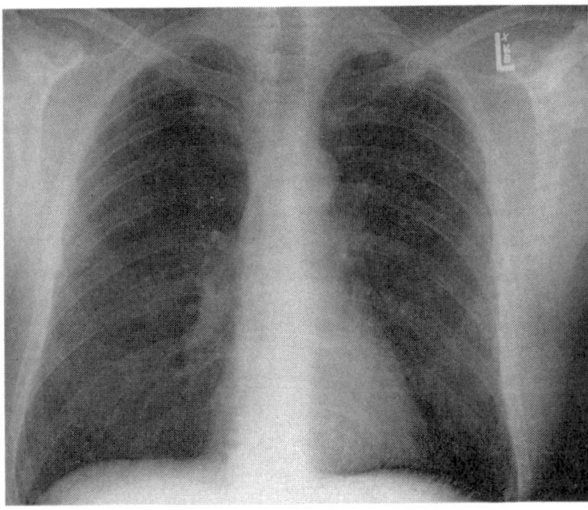

Figure 21. CR Chest Image. Displayed using clinically optimized chest imaging parameters.

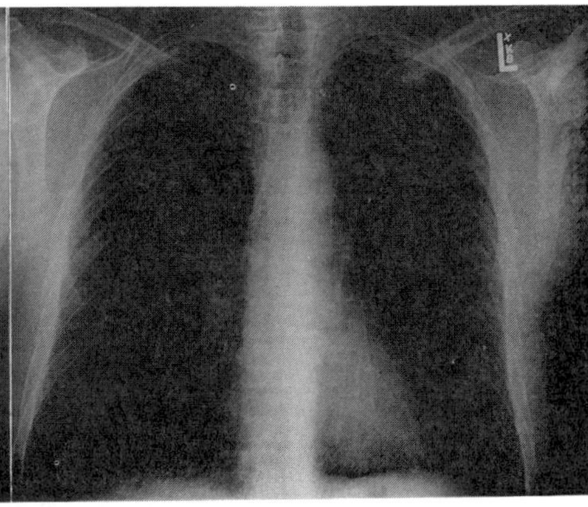

Figure 22. CR Chest Image. Displayed using clinically optimized abdominal imaging parameters.

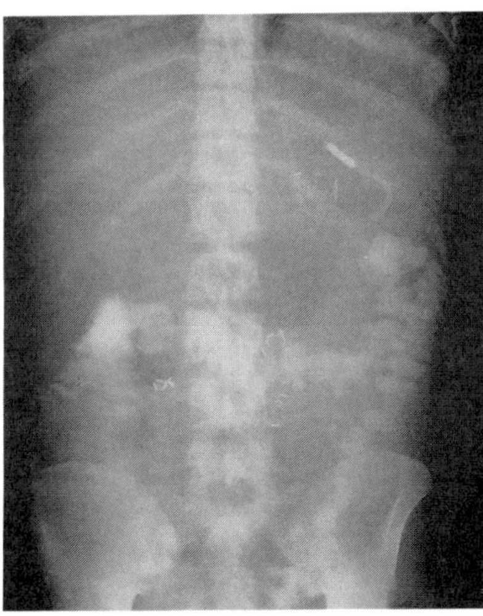

Figure 23. CR Abdominal Image. Displayed using clinically optimized chest imaging parameters.

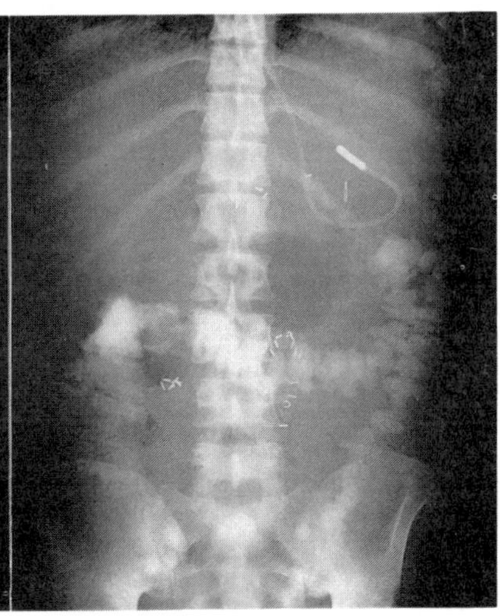

Figure 24. CR Abdominal Image. Displayed using clinically optimized abdominal imaging parameters.

DIFFERENCE IMAGING

Because of the stability of digital image processors, the first routine application of digital radiography was the subtraction of angiographic fluorographic images. Subtraction imaging demonstrates the difference between a reference image and a given image. These images are intended to show temporal or energy differences between the two original images. Since all changes between the two images are demonstrated, radiographic exposure conditions as well as image processing parameters must be controlled so as to minimize unwanted contributions to the difference image.

The difference image has unusual properties in relation to either of the two original images. These properties include a reduction of global signal intensity and an increase in both global and regional noise intensity. The subtraction process removes common mode information, therefore there is less total signal in the difference image than in either original. Stochastic noise is increased by the subtraction process because it is not common mode in the two images. It is therefore summed by the subtraction process. In a situation in which the original images are quantum noise limited, a ghost of the subtracted structures is seen in the resultant image.[41] Subtraction images are of value only when the conspicuity of clinically needed information is improved relative to the original image and the amplified noise in the subtraction. Since subtraction images represent contrast-amplified small differences between the two original images, they are susceptible to digital contouring artifacts. Digital contouring will occur when the digitization steps are too coarse relative to the difference signals.

The stability of digital radiographic systems opens up the possibility of routine energy difference imaging. In such systems, different x-ray spectra are used to acquire two images. The differences between such images may be exploited to obtain representations in which bone or soft tissue components are suppressed.[42] The presence or absence of calcium in a structure is of considerable diagnostic importance[43,44]. The method for performing such studies using stimulable phosphor image plates was recently described by Ishigaki et al.[45] Figures 25-26 are examples of an image pair produced by Ergun et al.[46] with bone and soft tissue suppression. Additional corrections have been made to these images for regional beam hardening, and for noise and scatter suppression. These images represent a single exposure of approximately 500 microR to the image plate cassette.

CONCLUSION

Digital radiography may be defined by the presence of an image in digital form. This operationally separates different imaging tasks into different system components. It is therefore possible to separately optimize image capture, PACS, the extraction of diagnostic information, and image display. This opens up new fields of investigation and development. The end result of such efforts should be the evolution of new digital imaging concepts which maximize diagnostic yield per unit of dose, cost, and physician time.

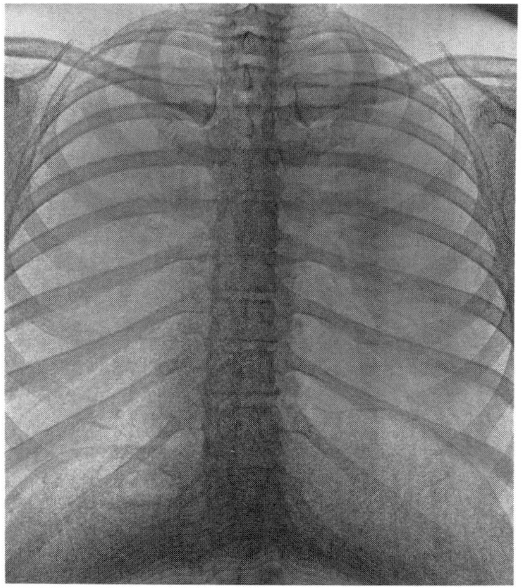

Figure 25. Soft Tissue Suppressed.

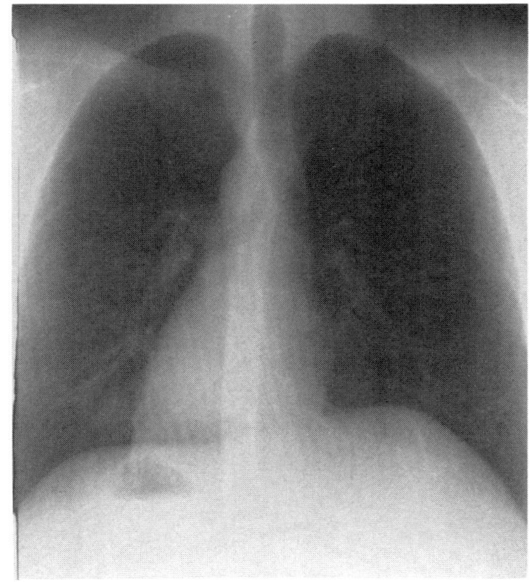

Figure 26. Bone Suppressed.

Example of an Energy Subtraction Image Pair.

Additional corrections have been applied for beam hardening, scatter supression, and noise reduction.

BIBLIOGRAPHY

1. Plewes DB. A scanning system for chest radiography with regional exposure control: theoretical considerations. Med.Phys 10:654-663; 1984
2. Wandtke JC. Improved pulmonary nodule detection with scanning equalization radiography. Radiology 169:23-27; 1988
3. Vlasbloem H, Schultze Kool LJ AMBER: A scanning multiple beam equalization system for chest radiography. Radiology 169:29-34; 1988
4. Tesic MM, Sones RA, Morgan DR Single - slit digital radiography: Some practical Considerations AJR 142:697-702; 1984
5. Barnes GT, Sones RA, Tesic MM Digital chest radiography: Performance evaluation of a prototype unit. Radiology 154:801-806; 1985
6. Fraser RG, Hickey NM, Niklason LT, et al. Calcification in pulmonary nodules: Detection with dual - energy digital radiography. Radiology 160:595-601; 1986
7. Geiger ML, Doi K Investigation of basic imaging properties in digital radiography. 3. Effect of pixel size on SNR and threshold contrast. Med.Phys 12:201-208; 1985
8. Geiger ML, Ohara K, Doi K Investigation of basic imaging properties in digital radiography. 9. Effect of displayed gray levels on signal detection. Med Phys 13:312-318; 1986
9. Davis GW, Wallenslager ST Improvement of chest region CT images through automated gray - level remapping. IEEE Trans.Med.Image:MI 5:30-34; 1986
10. Goodman LR, Foley WD, Wilson CR, et al. Pneumothorax and other lung diseases: Effect of altered resolution and edge enhancement on diagnosis with digitized radiographs
11. Oestmann JW, Kushner DC, Bourgouin PM, et al. Subtle lung cancers: Impact of edge enhancement and gray scale reversal on detection with digitized chest radiographs. Radiology 167:657-658; 1988
12. Rogowska J, Preston K Jr, Sashin D Evaluation of digital unsharp masking and local contrast stretching as applied to chest radiographs. IEEE Trans.Biomed.Engin: BME 35:817-827; 1988
13. Sakuma H, Takeda K, Hirano T, et al. Plain chest radiograph with computed radiography: Improved sensitivity for the detection of coronary artery calcification. AJR 151:27-30; 1988
14. Samarasekara MG, Hudson FR A technique for image density matching in radiotherapy planning radiographs. BJR 61:697-699; 1988
15. Sherrier RH, Chiles C, Wilkinson WE, et al. Effects of image processing on nodule detection rates in digitized radiographs: ROC study of observer performance. Radiology 166:447-450; 1988
16. Fujita H, Geiger ML, Doi K Investigation of basic imaging properties in digital radiography. 12. Effect of matrix configuration on spatial resolution. Med.Phys.15:384-390; 1988
17. Mistretta CA, Crummy AB Diagnosis of cardiovascular disease by digital subtraction angiography. Science 214:761-765; 1981
18. Sonoda M, Takano M, Miyahara J, Kato H Computed radiography using scanning laser stimulated luminescence. Radiology 148:833-838; 1983
19. Fuhrman CR, Gur D, Good B, et al. Storage phosphor radiographs vs conventional films: Interpreters' perceptions of diagnostic quality. AJR 150:1011-1014; 1988
20. Kangarloo H, Boechat MI, Barbaric Z, et al. Two - year clinical experience with a computed radiographic system. AJR 151:505-608; 1988
21. Kogutt MS, Jones JP, Perkins DD Low dose digital computed radiography in pediatric chest imaging. AJR 151:775-779; 1988
22. Merritt CRB, Matthews CC, Scheinhorn D, Balter S Digital imaging of the chest. J Thorac Imag 1:1-13; 1985
23. Balter S On the work of the radiologist - separation of image capture from image display. Acta Radiol. 29:257-265; 1988
24. Goodman LR, Wilson CR, Foley WD Digital radiography of the chest: Promises and problems. AJR 150:1241-1252; 1988
25. Cho PS, Huang HK, Tillisch J, Kangarloo H Clinical evaluation of a radiologic picture archiving and communication system for a coronary care unit. AJR 151:823-827; 1988
26. Mezrich RS The implication of PACS for radiology practice. AJR 151:828; 1988
27. DeSimone DN, Kundel HL, Arenson RL, et al. Effect of a digital imaging network on physician behavior in an intensive care unit. Radiology 169:41-44; 1988

28. Cannavo M Compressed data transmit faster without degrading image quality. Invest.Radiol. Aug 88:85-88
29. Lo SC, Huang HK Radiological image compression: Full-frame bit-allocation technique. Radiology 155:811-817; 1985
30. Lo SC, Huang HK Compression of radiological images with 512, 1,024 and 2,048 matrices. Radiology 161:519-525; 1986
31. Berbaum KS, El-Khoury GY, Franken EA, JR Impact of clinical history on fracture detection with radiography. Radiology 168:507-511; 1988
32. Berbaum KS, Franken EA, JR, Dorfman DD, Barloon TJ Influence of clinical history upon detection of nodules and other lesions. Invest.Radiol 23:48-55; 1988
33. Cowen AR, Hartley PJ, Workman A The computer enhancement of digital gray-scale fluorographic images. BJR 61:492-500; 1988
34. MacMahon H, Metz, CE, Doi K, et al. Digital chest radiography: Effect on diagnostic accuracy of hard copy, conventional video, and reversed gray scale video display formats. Radiology 168:669-673; 1988
35. Swensson RG The effects of clinical information on film interpretation: another perspective. Invest.Radiol 23:56-61; 1988
36. Getty DJ, Picket RM, D'Orsi CJ, Swets JA Enhanced interpretation of diagnostic images. Invest.Radiol 23:240-252; 1988
37. Dhawan AP, Buelloni G, Gordon R Enhancement of mammographic features by optimal adaptive neighborhood image processing. IEEE Trans.Med.Image MI 5:8-15; 1986
38. Braunstein EM, Capek P, Buckwalter K, et al. Adaptive histogram equalization in digital radiography of destructive skeletal lesions. Radiology 166:883-885; 1988
39. Balter JM Personal Communication; 1988
40. Wilenzick RM, Merritt CRB, Balter S Megavoltage portal films using computed radiographic imaging with photostimulable phosphors. Med.Phys 14:389-392; 1987
41. Balter S, Ergun D, Tscholl E, et al. Digital subtraction angiography: Fundamental noise characteristics. Radiology 152:195-198; 1984
42. Alvarez RE, Mascovski A Energy-selective reconstructions in x-ray computed tomography Phys.Med.Biol 21:733-744; 1976
43. Ishigaki T, Sakuma S, Ikeda M One-shot dual-energy subtraction chest imaging with computed radiography: Clinical evaluation of film images. Radiology 168:67-72; 1988
44. Asaga T, Chiyasu S, Mastuura H, et al. Breast imaging: Dual-energy projection radiography with digital radiography. Radiology 164:869-870; 1987
45. Ishigaki T, Sakuma S, Horikawa Y, et al. One-shot dual-energy subtraction imaging. Radiology 161:271-273; 1986
46. Ergun D, Mistretta CA, Bystrianyk RT. Dual-energy subtraction technique for computed radiography. Presented at the 1988 annual meeting of the AAPM, San Antonio; 1988

MICROPROCESSORS IN "SMART" CONTROL CONSOLES

Ernst Ammann, Dr.-Ing.
Kurt Schwarzmann, Dipl.-Ing.
Siemens AG, Medical Engineering Group
Erlangen, West Germany

ABSTRACT

This course presents an overview of actual microprocessor applications currently used in diagnostic X-ray equipment. A few highlights were selected to understand why the user and the physicist benefit from "smart" control consoles. For instance:

In mammography, the microprocessors contribute significantly to the uniformity of image quality and to a reduction in radiation exposure.

In angiography, the microprocessor provides a collision free operation which speeds up the procedure and reduces medical risk.

In generators, digital and hybrid feedback systems ease the operation during fluoroscopy, spot filming, and other recording modalities. These systems tend to reduce both cost of health care and radiation exposure.

There is absolutely no need to shift from medical physics to computer science. Conceivable trends in service and quality control procedures indicate clearly that the activities of a medical physicist in diagnostics are in many ways supported by microprocessors in "smart" control consoles.

INTRODUCTION

Why do we use microprocessors in diagnostic X-ray equipment? We use them because this technology has improved image quality and has proven to be cost efficient. Only this technology can satisfy the need for higher accuracy of diagnostic and therapeutic results. Let us select three main highlights to verify this:

1. mammography
2. tables and safety features, and
3. digital and hybrid feedback systems.

Conceivable trends will support the introductory statement.

MAMMOGRAPHY

Control actions by the technologist

A unit for routine and screening examinations is presented in figure 1. It provides extensive capabilities for special examinations including magnification technique and biopsies. The need for this procedure asks for top image quality, short examination time, simple operation, reproducibility, and reliability.

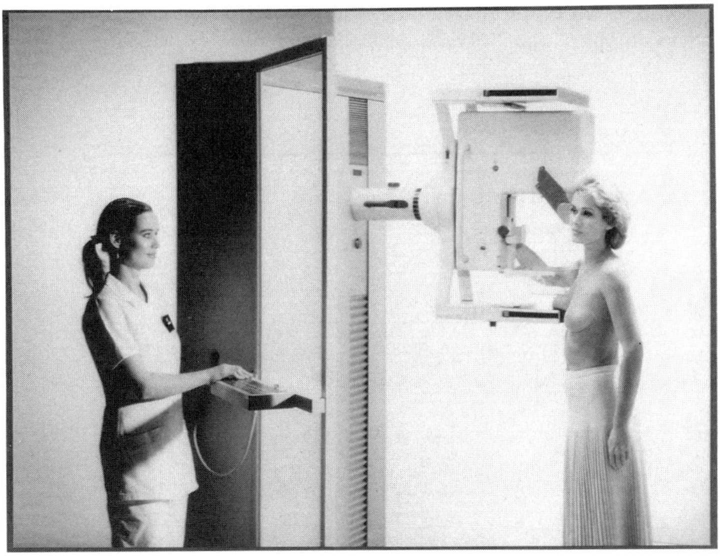

Figure 1: MAMMOMAT 2, a unit for routine and screening examinations of the breast

The steps to receive a mammogram or a film of a phantom for quality control purposes are very much the same.

Simple operation means selecting the appropriate

- cassette size to reduce costs of film and chemicals,
- mode of operation such as magnification technique, with or without grid to reduce unnecessary radiation,
- compression force, and
- program or technique factors.

Structure of the microprocessor system of the MAMMOMAT 2

Figure 2 shows the multiprocessor system with an external and internal bus structure. A master (M) controls the tube voltage, mAs switching, and four slave (S) processors. The first one controls technique factors and indications. The second one the selection of large or small focus, filament and tube current, and high speed starter. The third one handles the automatic exposure control system with on-line transparency calculation, film/screen speed, and density. The fourth one controls the compression force, k-edge filters, grid movement, and service functions.

Only a microprocessor with its performance can calculate and compare values on-line in such a way that the actual data are exactly the same as the selected ones. This improves reproducibility of a system but only if all relevant data are checked. Radiation can be turned off if the selected data do not comply with preprogrammed limits. Error messages can be stored in memory to support the service procedure.

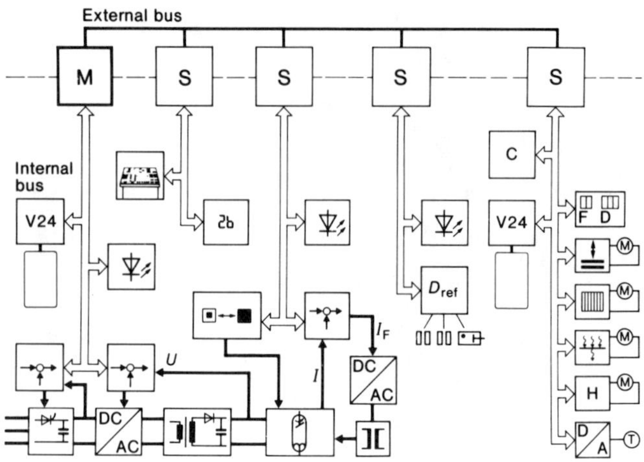

Figure 2: Microprocessor system of the MAMMOMAT 2 with a master (M) and four slave (S) processors. V24 interface to pocket terminal.

Closed loop control system for the tube voltage U, tube current I, filament current I_F, compression motor, grid drive, change of filter, height adjustment, and automatic exposure control system D_{ref}

Patient positioning and program selection

Engineered software should relieve the technologist from several control functions and from making any mistakes. Taking care of the patient and the procedure is the most important issue. Meeting regulations, safety standards, and good image quality requirements are far too complicated nowadays for one human being to keep in mind at all times.

Figure 3 demonstrates the automatic response of the system after each control action:
Experience and appropriate conditions including regulations are built into the system.

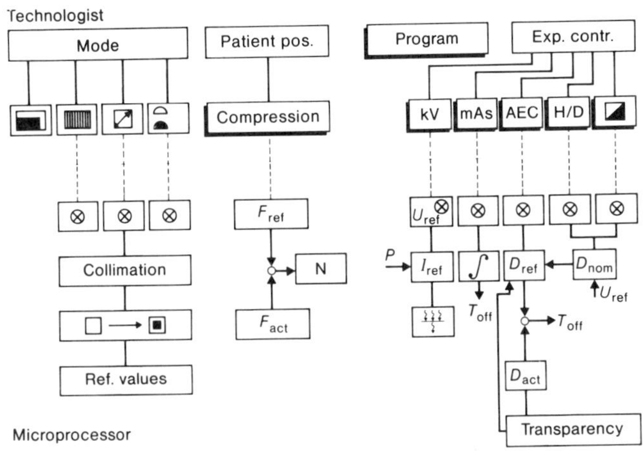

Figure 3: Control actions by the technologist and system response by the microprocessors of the MAMMOMAT 2. A single program button contains the reference values for the compression force F_{ref}, kV, mAs or automatic exposure control, high speed or definition of screen, and optical film density.

If the technologist selects the mode of operation - let's say magnification technique - the microprocessor performs multiple tasks. It indicates the proper mode and changes the collimation automatically. It switches the focus to the small one and selects the program for the exposure. It calculates and sets new reference values for the filament current depending on the tube voltage and on the output of the focus - because a tungsten anode provides a higher output than a molybdenum anode. In addition, it carries out many more activities.

Only one individual program button activated by the technologist sets all parameters automatically by the microprocessor system. Just try to follow the microprocessor's activities in detail as shown in figure 3. What a relief for the technologist. Nothing can go wrong anymore or may be forgotten.

Compression

Image quality highly depends on an appropriate compression of the the breast. It is cumbersome for the technologist to activate the compression device manually. Patient's pain is not a sufficient feedback for a reproducible force in follow-up examinations. Therefore, microprocessor technology provides a motorized compression.

The technologist has both hands free for accurate positioning of the patient and activates the compression by means of a footswitch. Repositioning becomes unnecessary and discomfort to the patient is minimized.

The principle function is explained in figure 4. The reference force is set with a potentiometer. The value is converted from an analog to a digital one. Now the microprocessor can handle the indication of the force value. When the footswitch is activated the compression motor is driven until the actual force applied is the same as the preselected one. Then the compression movement automatically stops. Force and thickness are measured and displayed. The very same parameters can now subsequently be applied throughout the follow-up examinations.

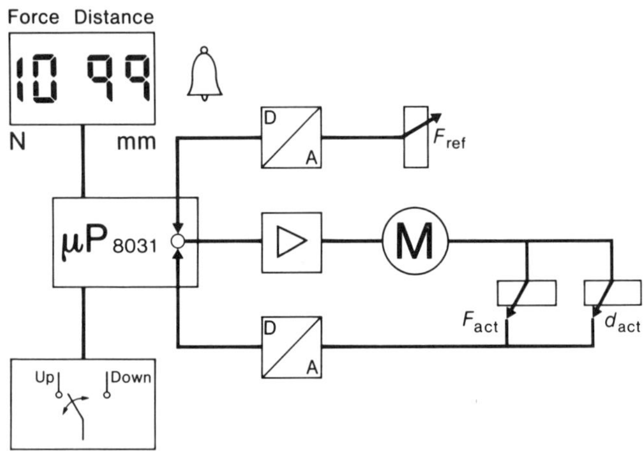

Figure 4: Closed loop control system of the motorized compression device

X-ray exposure

1. Dynamic grid movement

In early years of mammography film without screens was used. Extensive scatter radiation reduced the contrast of these mammograms. The arrival of low dose film-screen combinations reduced the radiation exposure at least by a factor of 10. Therefore, the application of a moving grid was indicated to reduce scatter. Now the mammograms show higher contrast.

Figure 5 shows the microprocessor controlled motor drive for the grid.

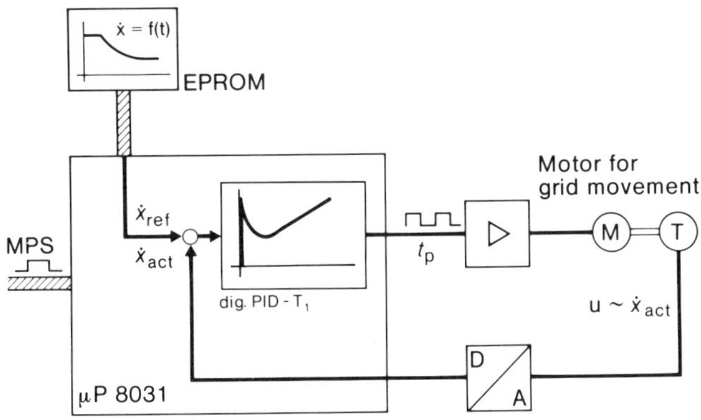

Figure 5: Dynamic grid movement of the MAMMOMAT 2

The speed $\dot{x}_{ref}$ of the grid is a function of time, stored in an EPROM (=Erasable Programmable Read-Only Memory). This is necessary to avoid grid lines on the mammogram because the exposure time depends on the transparency of the breast and other system parameters.

The speed of the grid during exposure has to be maintained within a very narrow tolerance range independent of the position of the bucky device (cranio-caudal, medio-lateral-oblique or lateral). The only way to solve this problem is the application of a feedback system which measures the actual speed and have it controlled by a sophisticated closed loop PID system (see the section entitled: Digital and Hybrid Feedback Systems).

2. Automatic exposure control (AEC) with transparency calculation

Without going too far into details of radiation physics and sensors, the dose rate measured behind the film-screen combination is not proportional to the optical film density anymore. In fact, there is a quite complicated relationship between a constant value of optical film density and the signal of the radiation sensor depending on the beam quality as well as density and thickness of the breast.

The technologist used to guess either mAs values or density correction points for the conventional AEC system. The nominal dose D_{nom} for the exposure is initially set by the system configuration and the technique factors.

Only the design of a special double detector with a microprocessor controlled AEC system (figure 6) could eliminate the guesswork of the technologist.

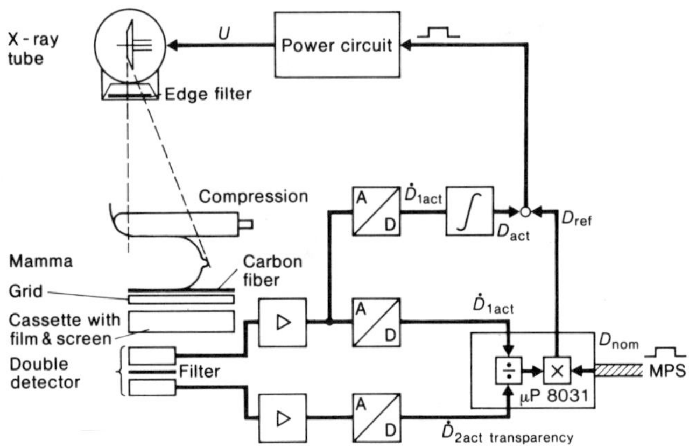

Figure 6: AEC system of the MAMMOMAT 2, an automatic exposure control system with calculation of the transparency to correct on-line the reference dose value D_{ref}. The signals of the double detector are proportional to the actual dose rate values $\dot{D}_{1act}$ and $\dot{D}_{2act}$. The MPS-bus system provides the nominal dose value D_{nom}

This system corrects during the exposure the reference dose D_{ref} by calculating a correction factor on-line from the two detector signals.

This procedure needs an extensive capacity to process large amounts of data for achieving a highly accurate exposure control.

Such an AEC system avoids any repeated exposures, provides uniform good image quality, and saves a lot of time and expenses.

<u>Optimization of parameters</u>

Figure 7 presents the structure of the software for both service engineer and technologist. Their tools are different: a pocket terminal for service and a key for the user.

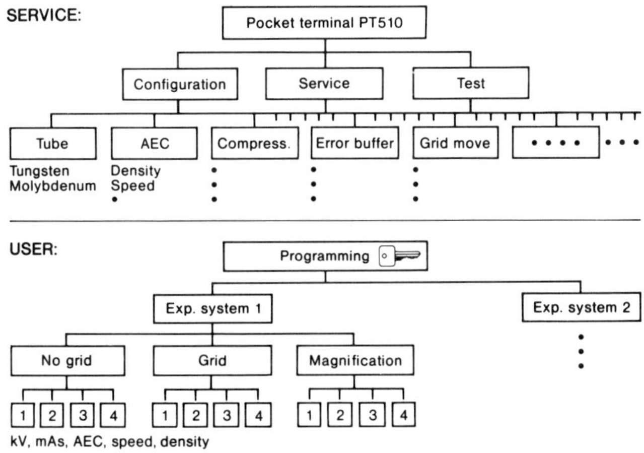

Figure 7: MAMMOMAT 2 software structure. The service engineer uses the pocket terminal for configuration, service, and test of the system.
The programming process is very easy:
All selected parameters are locked in by the user - turning the key

1. System configuration and service

The service engineer uses a pocket terminal to initially set up the configuration of the unit during installation. There is a checklist with more than 70 paragraphs to be answered during the configuration process. The system will be programmed with such data as the type of x-ray tube (molybdenum or tungsten anode), the sensitivity of the particular film, screen, and processor system to achieve the appropriate average optical film density of 1.5. Particular values can be changed if the physician prefers individually different ones.

All this information is very important whenever taking a mammogram, however, it needs not be remembered by the technologist. The smart microprocessor system will do it for them.

The same support is provided for the medical physicist or the service engineer. In addition, he is supported by service software, for instance an error buffer. He can also call test routines to check the reliability of a data transmission line.

2. User programming

There are exposure parameters which are different for each mode of operation and for each system used. The technologist is again relieved by the microprocessor from thinking of all these parameters.

The thinking has to be done only once during the programming process. All data is locked in with the turn of a key. Programmed data is preindicated whenever selected. This is very reliable support to avoid any mistakes. One single button recalls all relevant data for a correctly exposed mammogram. It uses the same values precisely and exactly time after time.

Can the program be changed by the user? Sure, but in a very safe way: with the key! Can somebody be misled by a changed program? Actually no, because there is still a preindication of data. One top of that, the technologist can override any parameter.

Trend

Can anybody imagine how much time, money, and radiation is saved with such a smart control console? Remember: Microprocessor hardware and software accomplish nothing unless the engineers' ingenuity puts everything correctly in place. This means first of all knowledge and then a lot of engineering time. As time and money are always limited, conceivable trends in mammography may be:

- Relief from cumbersome documentation.
- Calculation for positioning of biopsy needle.
- Calculation of the parenchyma dose.
- Pattern recognition as support for the diagnosing physician. This is a dream for years to come.

TABLES AND SAFETY FEATURES

Mammography gave us an overview of what a microprocessor system can do for us. Now let us turn to a new issue.

Angiography: Collision free operation

The capability of multidirectional projections are used in angiography and interventional procedures. There are quite a lot of possible collisions between parts of the moving equipment and other unmoved parts.

Figure 8 shows such a system which avoids a collision with the patient.

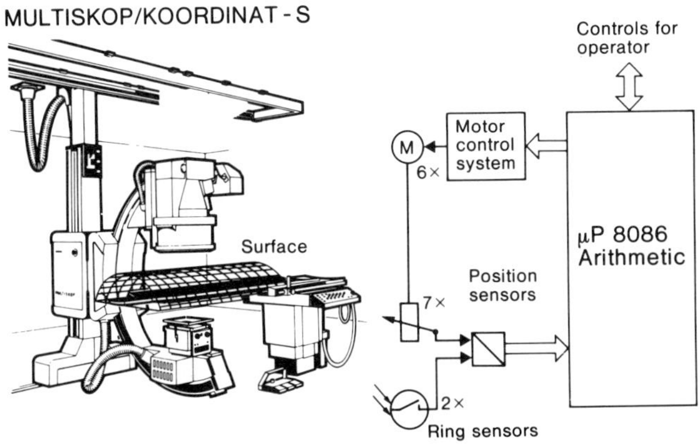

Figure 8: A special procedure room with collision free operation

On the other hand, it is very important to reduce the medical risk by providing fast movements and a support system to the operator. It enables them to work fast and safe in a specific area outside the patient's surface. It is a typical job for a computer system to handle a large amount of data for each and every projection and position.

Table 1 indicates criteria for a collision free operation and why we should and can provide it.

Table 1

CRITERIA	WHY
Adjustable isocenter	<u>Easier operation under stressful conditions</u> Individually adapted to the working position of every operator and to different medial procedures
Linear movement in 3 coordinates (x, y, z) Angulation around 2 axes	
Hard-collision: equipment with - equipment - floor	<u>Reduced medial risk</u> - High speed movement shortens examination time
Soft-collision: equipment with - tabletop - patient surface with safety margin	- Operator concentrates on patient and medical procedure rather than on avoiding a collision or an injury.
Immediate stop prior to a collision	<u>Money</u> - Reduced expenses for equipment repair
Indication of collision free movements to the operator	- Reduced risk of law suits from patient injury
Operation in override mode	
System configuration during installation	<u>Technology</u> Ability of
Internal test routines	- sensors and solid state components - safe and speedy data transmission
Redundant safety circuits	- powerful arithmetic microprocessors and software

Safety precautions

There is no such thing as 100% safety in life. However, there are several precautionary measures that provide a better, technically controlled safety.

Operator's attention to the patient and medical procedure is usually more important than to avoid a collision, especially in areas which the operator cannot see, i.e., under the table.

Figure 9 demonstrates a safety circuit.

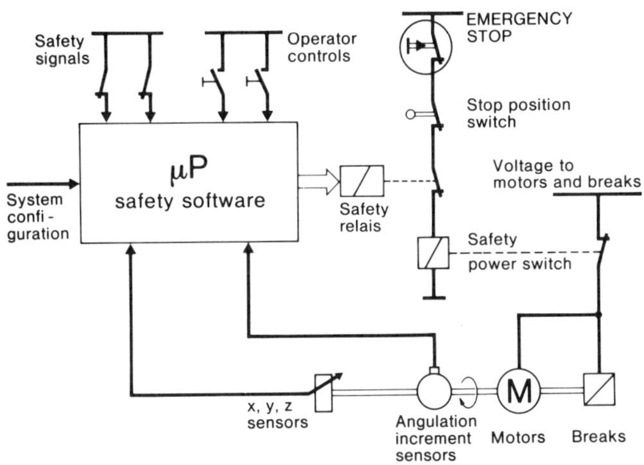

Figure 9: Safety precautions for moving parts of the table and the stand with the x-ray tube and image intensifier system

A variety of safety signals and of operator controls feed the microprocessor system which triggers a safety switch via software. An emergency stop switch with easy access for the operator is provided on top of all these controls.

The safety software can provide error codes to the user and the service engineer. This may reduce downtime and service expenses.

Further applications

1. Tables for angiography and R/F rooms

Positioning and movements are highly sensitive remote controlled. The memory has individual positions stored. The operator can recall them by a touch of the fingertip. The physician can speed up the procedure and therefore saves time which in turn reduces medical risk and increases patient throughput.

2. Spot film device

The cassette moves microprocessor controlled from parking to exposure position and vice versa. Fast yet smooth transition reduces mechanical oscillation of the spot film device. Therefore, the transition time is reduced to 0.6 seconds from fluoroscopy to exposure mode.

3. Universal collimation system

The positive beam limitation (PBL) system works with both rectangular and iris shaped collimator leaves. The achieved tolerance is less than 0.5% of 100 cm source image distance (SID).

Both a wedge shaped and a contoured semitransparent diaphragm improve image quality in angiographic procedures like digital subtraction angiography (DSA).

Different filters may be applied to reduce radiation exposure. They may also be helpful in producing dual energy images.

The universal collimation system is remote controlled by microprocessor technology and linked by fiber-optic cables. The positioning is highly sensitive to the object's shape. The operator can recall preprogrammed positions from memory.

4. Electronically controlled tomography

R/F-tables with tomo attachment used to be mechanically coupled by a tomo bar between the x-ray tube assembly and the bucky tray. Nowadays a microprocessor controlled tomo attachment does not need the tomo bar anymore. This tomography mode can be employed by the operator's fingertip.

Again, this is a fantastic application for a microprocessor shortening the procedure and relieving the technologist from cumbersome work.

DIGITAL AND HYBRID FEEDBACK SYSTEMS

Systems with and without feedback were mentioned in previous paragraphs. The next three sections will deal with the advantages and limitations of closed loop control systems.

Closed loop control system

Feedback on a natural or technical system provides an undisturbed flow of events. A closed loop control system, such as driving a car, is a manually controlled process by a human being who monitors the process and enforces any necessary change manually.

Figure 10 shows an analog controlled system. There are four main parts:

- a controlled system with output x,
- a monitor for the output x,
- a comparison with a reference value w, and
- a controller with an output y to change the output x of the controlled system.

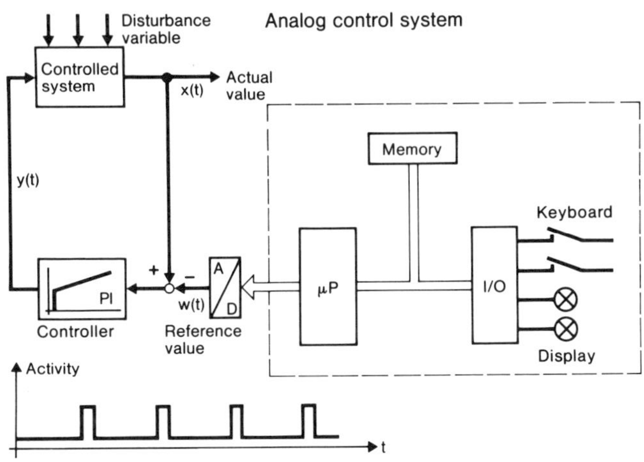

Figure 10: Analog closed loop control system with an activity-timing-diagram of the microprocessor μP

The reference value w exists in many cases in a digital form. Relevant parameters - we talked about them already - are tube voltage, time, dose or mAs, and so on. The circuitry consists of a microprocessor, a memory for codes and data, and an input/output interface which serves a keyboard and display.

The activity of the microprocessor in such a system is far below its capability or power. Therefore, the job of the analog controller and the comparison with the reference value can be shifted to the microprocessor. Figure 11 now shows this direct digital control system.

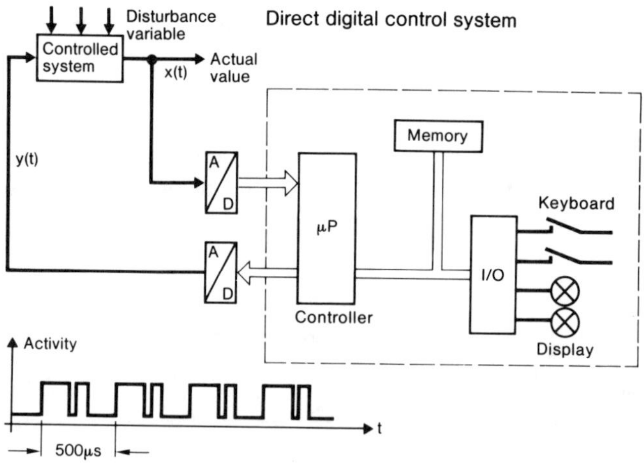

Figure 11: Direct digital closed loop control system with an activity-timing-diagram of the microprocessor μP

It needs fewer components than the analog one. In addition, the idle state of the microprocessor is now used for this job. A typical cycle time for an 8086 microprocessor is 500 μs. It will definitely be a limit for the speed of the control and consequently for both the accuracy and/or the equilibrium of the controlled system.

A typical example is the closed loop control for the tube voltage. The required speed of this system exceeds by far the power of a microprocessor. Therefore, this closed loop control system is designed as a combination of an analog controller supported by a microprocessor. It changes parameters of the analog controller depending on the tube load and on other parameters of the power circuit.

Let's have a review of the analog and digital closed loop control systems shown in figure 12.

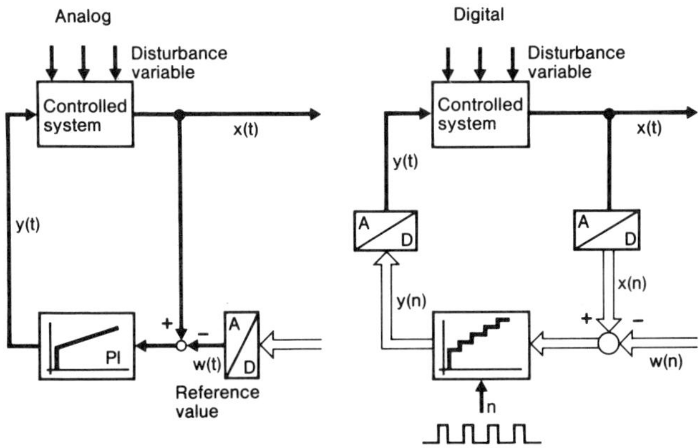

Figure 12: An analog and a digital closed loop control system

What has changed? The analog controller works continuously while the digital controller works step by step:

... n-1, n, n+1, and so on. n represents a value in a memory address at a specific time.

Only different properties of a controller can guarantee the stability of the output x of a complex system such as x-ray equipment.

Figure 13 explains an analog controller built with an operational amplifier and a network of passive components (resistors, capacitors) as feedback. The configuration of the network determines the property of the controller:

P	for proportional response,
PI	for proportional and integral response, and
PID	for proportional, integral, and differential response.

A digital controller provides additional advantages. The drift of the operational amplifier is completely eliminated by the microprocessor.

The same is valid for the tolerance of all the passive components and their parasitic properties within the particular layout.

The algorithm for the controller can directly be applied to the microprocessor system. An example of a PI procedure is given in figure 13. The algorithm is stored as a machine code in an EPROM and is executed in a loop every time with absolute accuracy. The value of each specific address will be rewritten by each step.

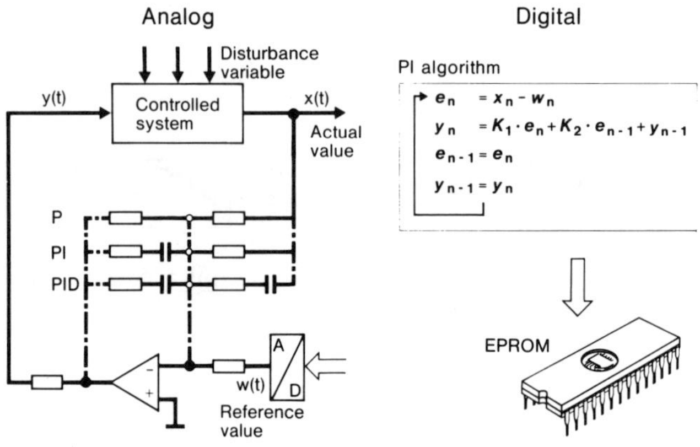

Figure 13: The controller of a P, PI, or PID analog system and the PI algorithm of a digital controller stored in an EPROM

Heat unit calculator

A high contrast fluoroscopy mode improves image quality during percutaneous transluminal coronary angioplasty (PTCA). The cut out of figure 14 displays the temperature distribution of the focal spot over the examination time. The anode disk has a lower permissible temperature than the focal spot. The temperature is a function of the exposure time and output.

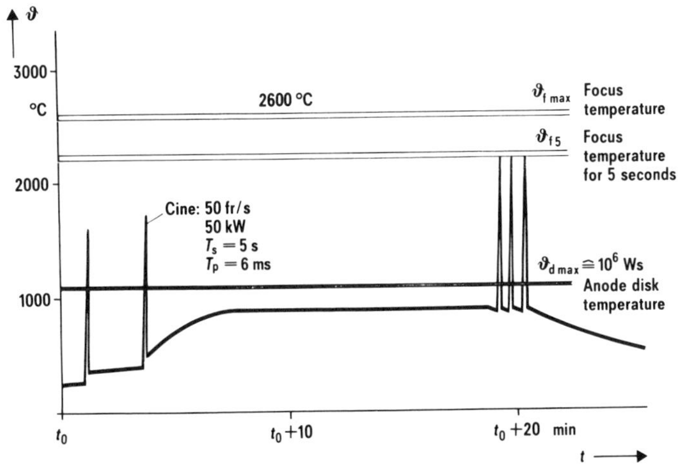

Figure 14: Temperature-timing-diagram during a percutaneous transluminal card angioplasty (PTCA) in a high contrast fluoroscopy mode of 2 kW

A heat unit calculator indicates the thermal situation of the tube housing assembly. This information supports the decision making process of the operator. He can change examination parameters if necessary: for instance, a lower frame rate or dose rate will reduce or avoid any waiting time. Therefore, the heat unit calculator can assist in providing an undisturbed procedure.

Figure 15 indicates the main parameters which have an impact on the thermal state of the focus, the anode disk, and the tube housing assembly.

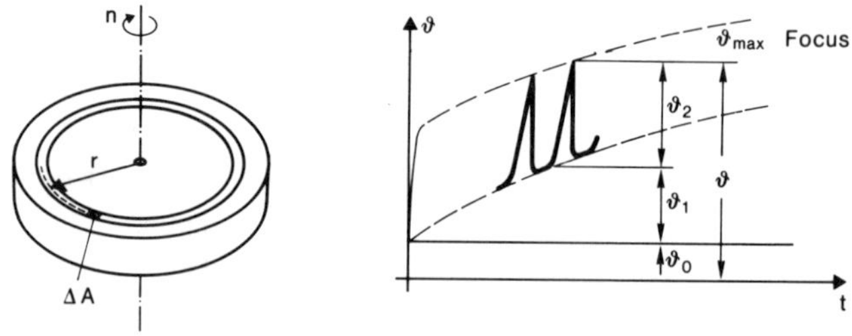

$$\vartheta = \vartheta_o + f(t, A, c_m) \cdot U \cdot I + f(n, \Delta A, c_m) \cdot U \cdot I$$

Heat storage capacity and cooling curve of anode and tube housing assembly

Waiting time calculated with Intel's 80188 + 8087

$$T_w = f(\vartheta, \vartheta_{max})$$

Figure 15: Heat unit calculator displays the waiting time T_W

Only a powerful microprocessor such as the Intel's 80188 and 8087 has the performance to handle the tremendous amount of ongoing changes of the input data on-line. Actual values must be compared with changing reference values and the waiting time must be calculated and displayed to show whether the thermal state of the tube has reached its limits.

Fluoroscopy and spot film exposures: POLYMATIC system

POLYMATIC is a microprocessor controlled system for several fluoroscopy modes and spot filming. It simplifies and standardizes user preferred techniques. Figure 16 and 17 show the principle of the POLYMATIC system.

The control console of the generator offers up to 49 programmable functions, each available at the touch of two buttons for routine diagnostics, tomography, spot film camera, and digital radiography techniques.

Figure 16 demonstrates the simple procedure of selecting an organ or examination technique. All parameters stored in a single function are displayed. The operator can override parameters if desired. He can also change them and lock the new ones in by using the key.

This is a very efficient way to achieve constant image quality. It eliminates repeated exposures, reduces radiation exposure, expenses, and it saves a lot of the patient's and staff's time.

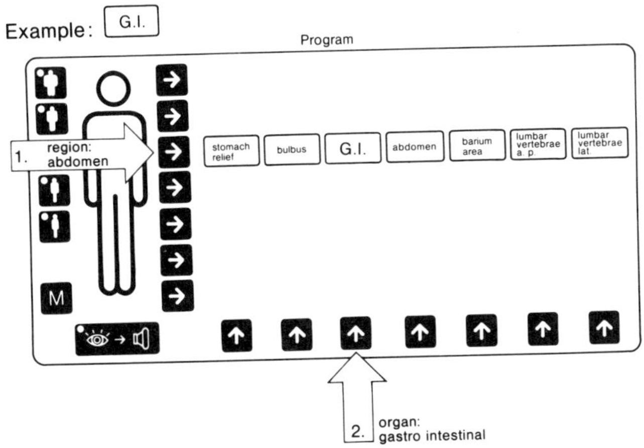

Figure 16: POLYMATIC system with standardized, user preferred techniques

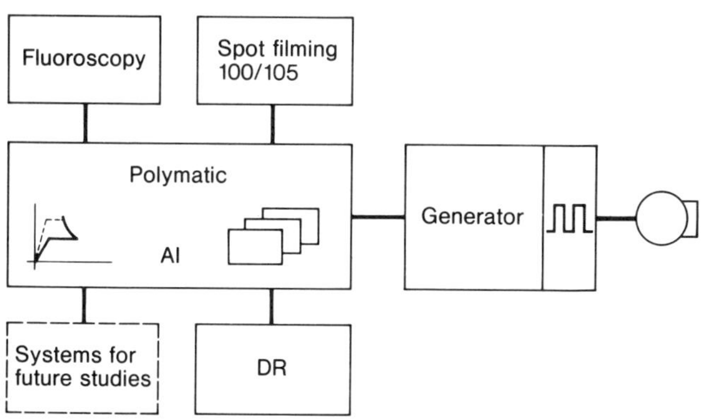

Figure 17: POLYMATIC system controls parameters relevant to good image quality and lowest possible radiation exposure

Figure 17 presents the POLYMATIC system. it consists of all imaging parameters of the x-ray system relevant to good image quality and lowest possible radiation exposure. In addition, it controls all fluoroscopy modes via the automatic brightness control (ABC) and stores tube voltage and current for calculation of the water equivalent value of the examined object on-line. This information is the basis to create technique factors automatically for exposures such as spot films with cassette or 100/105 mm camera, digital radiographs, or other studies to come in the future.

Operator's experience is fed into the system and it may well be considered a first step in the direction of artificial intelligence.

Now, how does this sophisticated microprocessor system work?

Figure 18 depicts the procedure. The fluoroscoped object of a patient provides its water equivalent value which corresponds to the object's transparency. This value is used to create information for an exposure. It is dependent on a number of system parameters and on the operator's expectation as to what an excellent image is. Contrast and resolution determine the technique factors and lead to reference values for tube voltage and focus.

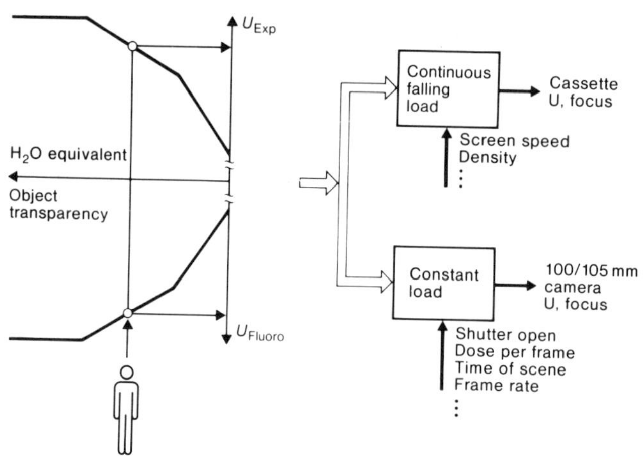

Figure 18: POLYMATIC system

The microprocessor system has stored knowledge and experience of hundreds of man-years' work because contrast requirements are different for each object, examination technique, and operator. One example under specific circumstances is shown in figure 19.

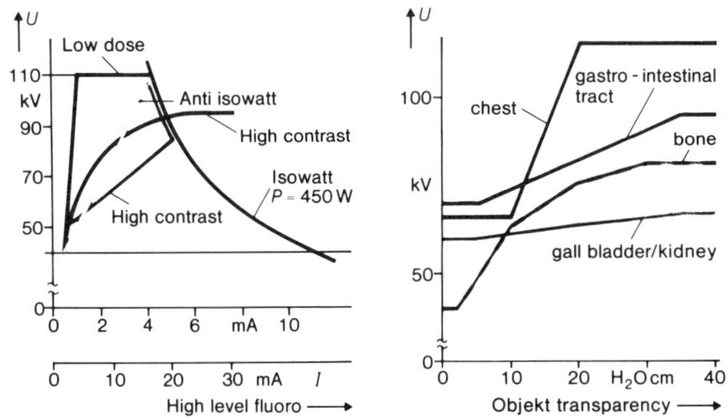

Figure 19: POLYMATIC system is based on different, to the object and examination technique, related tube voltages (U) for fluoroscopy modes shown in the left diagram; and for exposure modes shown in the right diagram.

Whenever the fluoroscopy mode is applied one has to compromise on radiation exposure and image quality (contrast, signal to noise ratio, et al). In case of motion blurring, pulsed fluoroscopy should be used with progressive scans and gap filling techniques to receive a flicker free TV image at frame rates of 30 frames/sec or even less. "Tail cutting" further reduces unnecessary radiation exposure.

The operator determines a particular fluoroscopy mode depending on the upcoming procedure. He may very well change the mode several times during the procedure. The decision remains his responsibility. In any case, the water equivalent value will be transferred to a reference tube voltage according to the examination technique, the object, and several other conditions for an excellent image.

There are absolutely no further control steps necessary by the operator. However, there is still the option of manual technique or only overriding a particular parameter. Reprogramming can easily be done by the operator. The system provides room for the user to implement her and his own experience as well.

The calculation of tube voltage and focus now takes into account the technical limitations of the installed system, such as shortest exposure time, different output of the focus, frame rate, and so on. The result of using the

POLYMATIC system is a remarkable increase in reproducibility of image quality.

The components of diagnostic x-ray equipment are shown in figure 20.

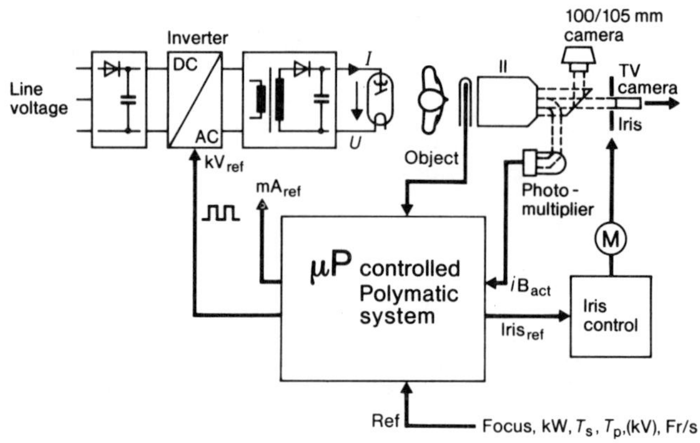

Figure 20: POLYMATIC system and components of diagnostic x-ray equipment

The line voltage supplies current to the DC primary power circuit, the inverter with its high frequency output, and the high voltage transformer which after rectification provides the tube voltage to the x-ray tube. The brightness of the image intensifier (II) output phosphor is sensed by a photomultiplier. The current $i_{B\ act}$ is proportional to the dose rate at the II input screen.

Reference values (Ref) are preprogrammed during installation in agreement with the operator's individual requirements such as optical film density and dose per frame. The microprocessor controlled POLYMATIC system is ready to support the operator during each and every procedure.

Another factor may be of interest in using microprocessor technology: Costs. Figure 21 shows well known x-ray control systems such as SIRCAMATIC for 100/105 mm spot films, CINEMATIC for 35 mm cinefilms, VIDOMATIC for TV images, and DIGIMATIC for digital subtraction angiography.

These components have a limited flexibility due to the extensive hardware involved. Cost of material exceeds engineering costs. High technology has changed this:

Microprocessors and memory chips are less expensive than conventional hardware. However, the cost of engineering software undergoes a dramatic increase because the capabilities of high technology components provoke the ever rising use as well as improved acquisition and processing of information. This means a shift of cost from material to engineering.

The advantages of a radiation exposure control system like the POLYMATIC surpass all existing systems by far.

It needs less space, generates less heat, reduces maintenance expenses, increases reliability, and provides more room for individuality.

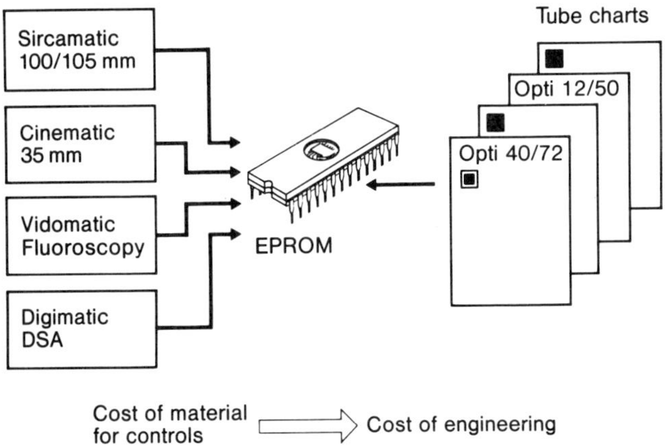

Figure 21: Conventional x-ray control systems for 100/105 mm and cine techniques, fluoroscopy mode and digital subtraction angiography (DSA) represent cost of material. The POLYMATIC system represents cost of engineering

CONCEIVABLE TRENDS

Service and quality control

Table 2 shows a guess of what can be done. How software controlled microprocessors in "smart" control consoles can contribute to a solution and why - after all - it should be done.

Table 2

Software Controlled Microprocessors in "Smart" Control Consoles

WHAT?	HOW?	WHY?
System configuration	Individually adaptable	Improved Diagnosis Reduced Radiation Exposure Saved Time and Money Easier Work
Surveillance of parameter tolerances	Built-in sophisticated sensors Closed loop control systems	
Statistics/Documentation	Networking/Printer	
Discrimination of information according to significance	On-line data comparison	Reduced Radiation Exposure Saved Time and Money Easier Work
Key data information on "what was going on BEFORE..."	Dynamic memory	

Regulations demand manual record keeping of medical and quality control procedures. The amount of data increases. Therefore, regulations may partially force us to automatic documentation of data. In addition, reliability and efficiency of human resources impel us to turn over cumbersome workload to the microprocessor.

Printouts of system and examination data are already available for some microprocessor controlled systems because the software and hardware had been developed. However, so far standardized interfaces to printers and networks exist only in a very limited and incomplete way: connectors, cables, and signal levels.

The cost of software engineering and testing increases tremendously. High quality standards are only assured if tests are performed thoroughly within a complete diagnostic system.

These facts diverge user's wishful thinking from the reality of conscientious engineering and manufacturing.

Figure 22 shows a conceivable trend.

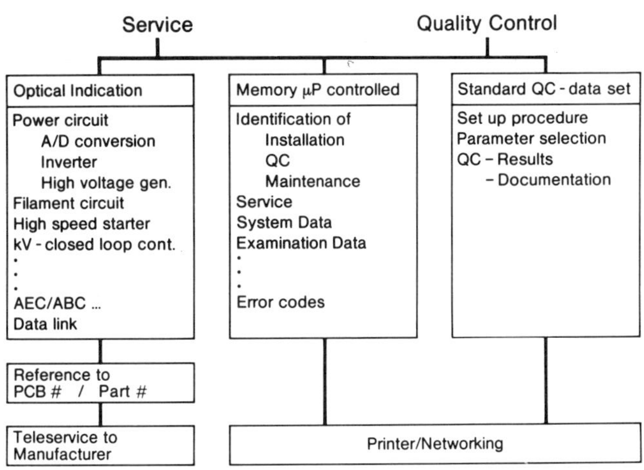

Figure 22: Conceivable trends for service and quality control

<u>Diagnostic x-ray equipment</u>

Figure 23 indicates three main roads of image receptors to digital radiography: Image intensifier, film, and reusable memory screen. Each of these three image receptor systems has a very similar x-ray source system which contains an appropriate x-ray tube, a beam limiting device, and a generator with image optimizing capabilities.

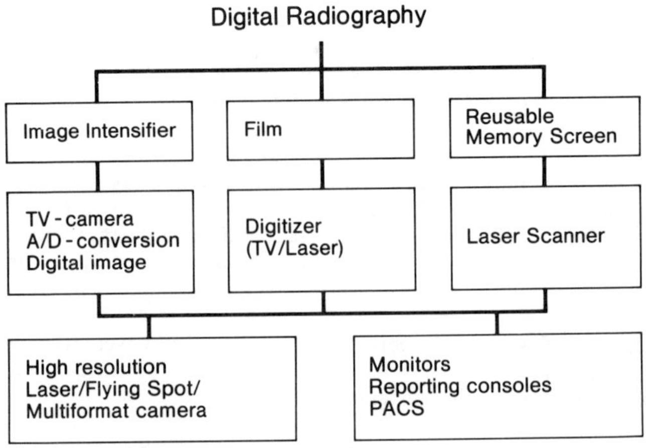

Figure 23: Trends for diagnostic x-ray equipment

CONCLUSION

Mammography, collision free operation in angiography, digital hybrid feedback systems, and conceivable trends for service and quality control were picked from a huge variety of x-ray equipment for diagnostic and interventional procedures.

In the introduction we asked the following question:

Why do we use microprocessors in diagnostic x-ray equipment? The answers are:

To improve image quality for a better diagnosis.

To apply dose according to the ALARA* principle which reduces radiation exposure to staff and patient.

* International Commission on Radiation Protection (ICRP) report #26 of 1977: ALARA means As Low As Reasonably Achievable and does not mean a renunciation of diagnostic information.

To relieve the staff from cumbersome workload.

To shift time from technical to medical procedures and patient.

To curb cost of health care by having

- more and accurate information for reject-repeat analysis
- support for troubleshooting
- higher uptime of equipment
- less service/maintenance expenses
- exchange of diagnostic information.

To cut a long story short: We keep in mind what Joel Gray said in his book on quality control: "Three D's":

Diagnosis
Dose
Dollars

Finally, we have to ask ourselves if all this is a first step to "INTELLIGENCE". Maybe - because user and engineering experience is implemented via software and therefore available to every user. Or - is all this a first step to "SMART" controls? Definitively yes - and there is much more to come. We look forward to an exciting future.

References

Joel E. Gray et al, Quality Control in Diagnostic Imaging, University Park Press, 1983

M. Säbel, F. Willgeroth, H. Aichinger and J. Dierker, X-ray spectra and image quality in mammography, Electromedica 54 (1986), 158-165

H. Aichinger, J. Dierker, P. Marhoff, Image Quality and System Dose with Pulsed Fluoroscopic Mode in Digital Subtraction Angiography, RSNA 86, Chicago

E. Ammann, Update on Technology of X-ray Generators, Refresher course RSNA 86 and amended RSNA 87, Chicago

Bildgebende Systeme für die medizinische Diagnostik, Herausgeber: H. Krestel, SIEMENS, 1988 (to be published in English 1989)

Computers in CT / US / MRI

Osman Ratib, M.D., Ph.D.

University of California at Los Angeles
Los Angeles, California 90024

INTRODUCTION

This paper discusses the utilization of computers in three common digital imaging techniques in diagnostic radiology. Computerized x-ray tomography (CT) was introduced in the early 1970s and it has become a standard radiological imaging technique. Ultrasound scanning (US) was originally designed as an analog imaging device, but digital technology has become an integral part of the instrumentation. Magnetic resonance imaging (MRI) was introduced in the early 1980s and has rapidly gained an increasingly popular position in diagnostic radiology.

What these three techniques have in common is the fact that they provide cross-sectional images of the body referred to as tomographic images. Each one, however, relies on a different physical mechanism for the generation of the images.

Common characteristic:
- **Tomographic images**

Physics of image acquisition:
- **CT: x-rays (Transmission)**
- **US: Ultrasound (Reflection)**
- **MRI: Magnetic fields (Resonance)**

CT: The scanning mechanism of a CT scanner consists of an x-ray tube, an array of detectors, and a gantry, which controls the relative position of the x-ray tube and the detector system. Tomographic images are generated from the collection of x-ray profiles (projections) from a transverse (cross) section of the body.

US: In two-dimensional ultrasound imaging the images are obtained by reconstructing a cross-sectional view of the patient by way of detecting the amplitudes of acoustical reflections (echoes) that occur at the interface of tissues with different acoustical properties.

MRI: Magnetic resonance imaging devices form images of objects through probing the magnetic moments of nuclei, employing radio-frequency (RF) radiation and strong magnetic fields. Information concerning the spatial (anatomical) distribution of nuclear magnetization in the sample is determined from RF signal emission by these stimulated nuclei.

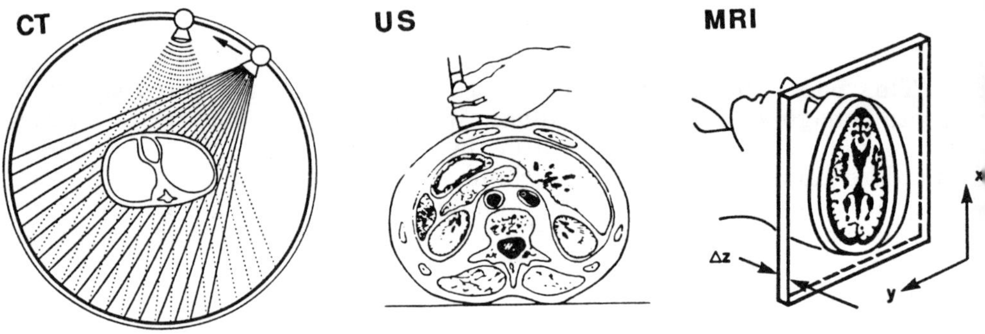

Figure 1 Schematic representation of the three imaging modalities.

The advantages and disadvantages of each of these three imaging modalities are summarized in the table below. MRI and US have the clinical advantages of using non ionizing radiation which is less hazardous to the patient than CT. In CT the total number of images is limited by the dose of radiation that the patient is allowed to be exposed to.

Advantages	*Disadvantages*
Computed Tomogaphy	
• High spatial resolution • Excellent for bone structures	• Slice orientation set by the gantry position • Poor discrimination between soft tissues • Ionizing radiation
Ultrasound	
• Portable and convenient • Non invasive • Free orientation of slices	• Poor spatial resolution • Limited penetration (air cavities)
Magnetic Resonance	
• Free orientation of slices • Non ionizing radiation • Excellent separation of soft tissue	• Long acquisition time

HARDWARE ARCHITECTURE

CT: Figure 2 shows a general schematic diagram of a fourth generation x-ray CT scanner. It is composed of a gantry which houses the x-ray tube, the detectors, and signal processing/conditioning circuits; a front end preprocessor unit which performs some data corrections and data reformatting, an image data buffer memory; a controlling computer; a high speed computational processor; a hardware backprojector unit; and a video controller used to display the resulting CT image onto a CRT monitor. The gantry unit houses the x-ray tube source which rotates about the patient. Beam collimators determine the beam thickness and angular extent of the fan-shaped x-ray beam. They are adjusted in accordance to the operator's selection of scan circle size and slice thickness. The gantry unit also houses a special detector known as a reference air detector used to calibrate the CT value of air.

Once the x-ray photons have passed through the patient and the flux has been converted to electrical signals by the detectors, the signal is processed by a series of analog signal conditioning circuits. Next comes a logarithmic amplifier to convert x-ray transmission data (I/Io) into attenuation(μ) and thickness(x) data. The next stage involves analog multiplexing of the signals from all detectors and serially feeding these signals from different detectors into analog-to-digital converters. Multiplexing is necessary in order to parallel process signals from the detection stages by the analog to digital converters (ADC), requiring serial input. The signal is then sampled and digitized to typically 12 to 16 bits.

Once the signals have been digitized, initial data correction is performed by the front end preprocessing unit. Some of the corrections to the data will include subtraction of the air reference detector signal to normalize the attenuation data, obtaining local averages of detectors to determine if any detectors are outside a predetermined standard deviation (which helps locate bad detectors), and corrections due to dead time losses (i.e., detection response time losses) by the individual detectors. The front end preprocessor may also provide the function of reformatting the data from what are known as source fans representation to detector fans representation. This reformatting of the data may be necessary to subsequently process the data in the most computationally efficient state.

Once the data has been preprocessed it is stored as reformatted raw data in a fast access memory. At this point further data processing is performed and may include detector spacing corrections which compensate for deviations in detector center-to-center distances, beam hardening corrections which correct for changes in the x-ray spectrum reaching different points in the object being scanned, and corrections due to differences in detector gains. Furthermore, a mathematical convolution (or filtering) of the data is performed using a high speed floating-point array processor which can perform mathematical operations on entire arrays of data in a relatively short period. This convolution process is necessary to avoid image reconstruction artifacts such as ringing and edge overshooting. Finally, the image is reconstructed using a hardware backprojector which performs two-dimensional reconstruction of the one-dimensional transmission profiles by backprojecting the detector fan readings and accumulating the contributions of all detectors fans to the resulting image matrix.

The main host computer handles the data flow, scheduling, and control of detector fan data from the fast access memory to the hardware backprojector. The resulting reconstructed image is then stored in an image memory which can then be easily displayed on a video monitor. Long term storage of the CT image is usually provided by archiving the digital images to magnetic tape or to optical disks.

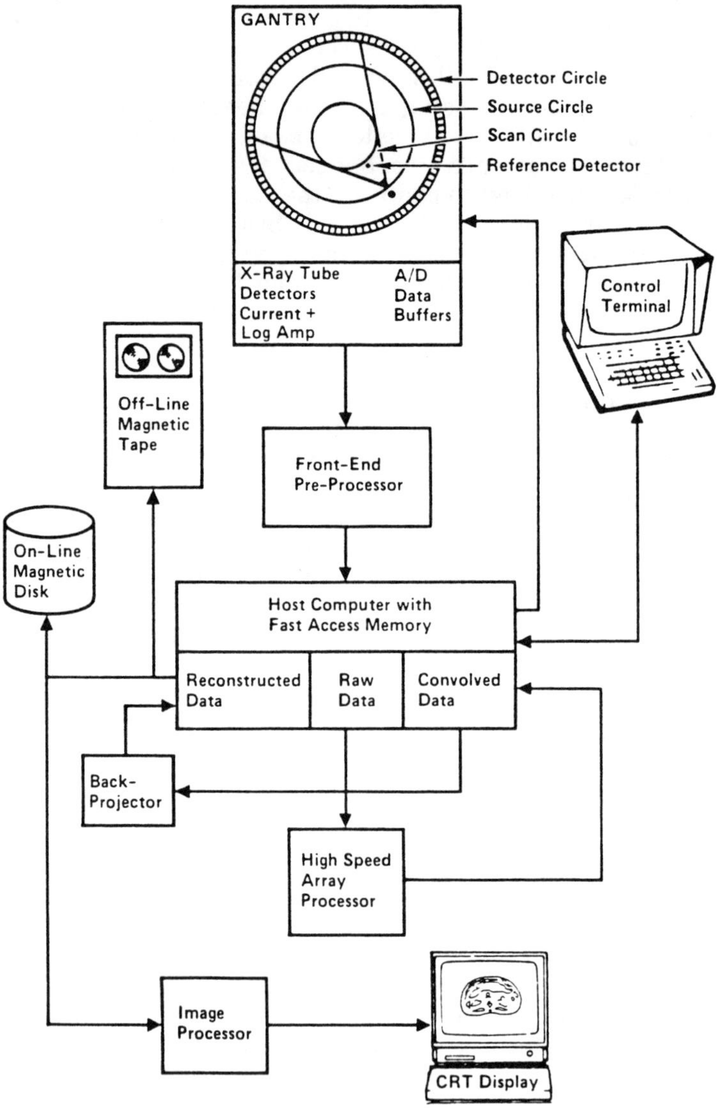

Figure 2 Diagram of the general hardware architecture of a CT scanner.
(from H.K. Huang, Elements of Digital Radiology, Prentice Hall, Englewood, N.J. 1987)

US: Figure 3 shows a general block diagram of a typical two dimensional ultrasound scanner. It is composed of a transducer, a transmitter circuit, a receiver circuit with time gain compensation (TGC), a digital scan converter (DSC), and a video display monitor.

The acoustical waves are generated by applying a high voltage pulse to a piezoelectric crystal, resulting in the creation of a longitudinal pressure sonic wave. The rate at which pulses are supplied by the transmitter circuit to the transducer is determined by a transmit clock, called the pulse repetition frequency (PRF). Typical PRF values range from 0.5 to 2.5 kHz. The frequency of the acoustic wave is determined by the thickness of the piezoelectric crystal and ranges from 1 to 15 MHz. The transducer can serve as both an acoustical transmitter as well as receiver since mechanical pressure waves interacting with the crystal will result in the creation of an electrical signal.

Received echo amplitude pulses, which eventually form an ultrasound image, are transferred into electrical signals by the transducer. A radio frequency receiver circuit then amplifies and demodulates the signal. The receiver circuit is a crucial element in an ultrasound scanner and must have a wide dynamic range (30 to 40 dB) in order to detect the wide range of reflected signals which are typically 1 to 2 volts at interfaces near the surface and microvolts at deeper structures. In addition the receiver must introduce little noise and have a wide amplification bandwidth.

The time gain compensator circuit allows the operator to amplify the echoed signal according to its depth of origin. This helps compensate for the higher attenuation of the signal seen from echos originating from deeper interfaces and results in a more uniform image. The operator is able to control the amount of gain at a particular depth in order to obtain the best possible image.

The output of the receiver is fed into the digital scan converter and used to determine the depth (Z dimension) at which the echo occurred. The depth at which the echo originated is calculated by determining the time that the echo takes to return to the transducer. The depth of the reflector can be obtained since time and depth are related and the depth is equal to one half the time interval from the transmission of the signal pulse to signal return times the velocity of sound in the traversal medium. The DSC performs A/D conversions of data, data preprocessing, pixel generation, image storage, data post-processing, and image display. The analog echo signals from the receiver circuit are digitized by an analog to digital converter in the DSC to typically 4 to 5 bits (16 to 32 gray levels). Fast A/D converters are normally used due to the wide bandwidth seen in most ultrasound echo signals and should have a sampling frequency of at least twice the highest frequency of interest in the image. Typical A/D sampling rates range from 10 to 20 MHz. The DSC image memory is a random access memory which typically is 512x512x4 or 8 bits for each memory plane. Echo strength data is addressed to the image memory according to its (X,Y,Z) positional data.

Preprocessing of the data may be performed in order to enhance the visual display of the data and to match the dynamic range of the subsequent hardware components. Echo signals are typically rescaled and often nonlinear (e.g., logarithmic) circuits are used to emphasize or de-emphasize certain echo amplitudes. Preprocessing the data will involve rescaling of the signal to compress the dynamic range and performing compensation of signal attenuation with respect to the depth of the originating echoed signal.

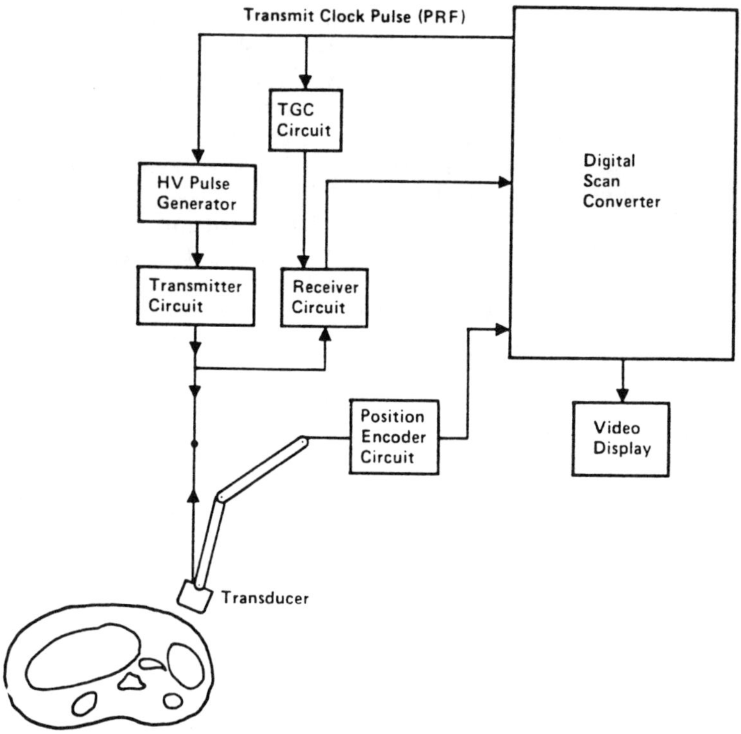

Figure 3 Diagram of the general hardware architecture of an ultrasound scanner *(from H.K. Huang, Elements of Digital Radiology, Prentice Hall, Englewood, N.J. 1987)*

MRI: A simplified block diagram of a typical MR imaging device is shown in figure 4, demonstrating several pieces of equipment necessary for the production, detection and display of the MRI signal. These include the magnet to produce the Ho field, RF equipment to produce the H_1 field (transmitter, amplifier and coil: transmitting mode) and detect the FID (coil: receiving mode, preamplifier, receiver and signal demodulator). X,Y and Z gradient power supplies and coils provide the magnetic field gradients necessary for encoding spatial position. Also necessary is the electronics and computer facility to orchestrate the whole imaging process (control interface with computer), digitize the image data (A/D converter), reconstruct (computer algorithms) and display the image (computer, disk storage, image processor and display system).

The creation of an MRI image requires obtaining data from many excitations with the magnetic field gradient in a variety of orientations. A

CT unit can sweep around part or all of the circle and obtain complete data for a two-dimensional image. The MRI imaging gradient must be swept in the third dimension as well. Also, unlike the CT system, the sweeping of the gradient is accomplished by increasing the electric current in some gradient coils and decreasing it in some others. With the large number of possibilities for the various parameters used in MRI, it is only practical to control the machine and its settings through a computer. Predefined setups are usually programmed for different procedures. The computer assures that the proper sequence of events takes place in order to gather the sophisticated information necessary to generate an image.

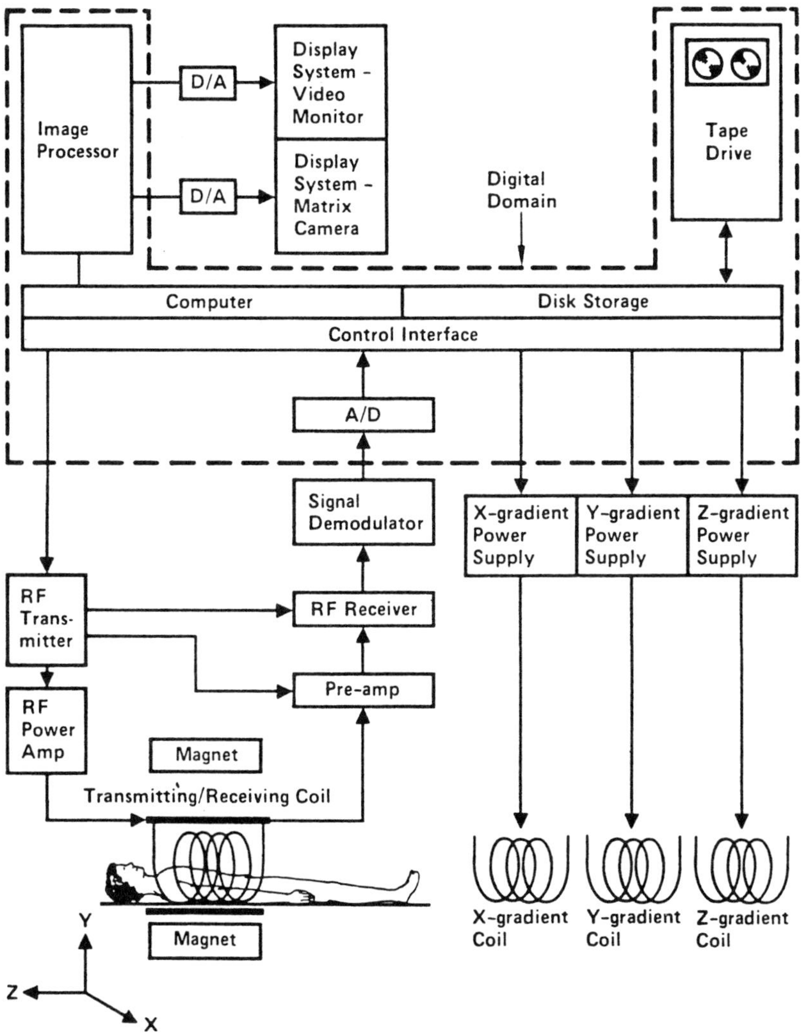

Figure 4 Diagram of the general hardware architecture of an MRI scanner. *(from H.K. Huang, Elements of Digital Radiology, Prentice Hall, Englewood, N.J. 1987)*

IMAGE PROCESSING AND ANALYSIS

The function of the computer for image manipulation and management in all three imaging modalities can be divided into four categories listed in the table shown below.

- **Image reconstruction**
- **Data management (traffic, storage)**
- **Picture reformatting & display**
- **Post-processing & analysis**

Image reconstruction: Most current systems have dedicated hardware for image acquisition and reconstruction. The host computer usually receives images that are already reconstructed in their final form. Both CT and MRI rely on fast array processors for reconstruction of images using filtered backprojection techniques. In MRI however a direct technique of 3D reconstruction based on Fourier transformation of the signal recorded has progressively replaced the conventional backprojection method.

In MRI much of the reconstruction of the images is essentially the same problem as found with CT, that is, there is a reconstruction of an image from projections. However, there are several differences in the MRI scheme, most of which call for even greater data manipulation. First, the MR data may refer to an entire three-dimensional volume rather than a single slice. The other major difference is that the projections themselves are usually only obtained as a result of breaking the MR signal into its various frequency components. The separation into frequency components is an example of Fourier analysis. Since Fourier reconstruction techniques are employed in the reconstruction process, this frequency analysis is advantageously built into the reconstruction process. That is, actual projections do not need to be obtained as explicit steps in the reconstruction.

A typical ultrasound unit is equipped with a dedicated microcomputer for data acquisition and image reconstruction. The main function of the computer is to apply geometric transformations to the data collected to display them in a two-dimensional image. Also, the raw data generated by the transducer are not suitable for display or interpretation. The computer is programmed to improve the quality of the data in a variety of ways before display. Echo intensity is corrected for attenuation at each depth by applying predefined, curve-generated correction factors for each depth. Averaging, smoothing and thresholding techniques may further improve the quality of the ultrasound images. Intensity data must then be placed in correct geometric location for a realistic two-dimensional display. Standard geometric and trigonometric computations are used together with interpolation techniques to determine the position of the echoes on the screen. The speed of dedicated processors allows the performance of these operations in real time.

Data storage: The management of data acquired and their storage are usually handled by a host computer. CT and MRI images are stored for short term on erasable magnetic disks, and long term archiving is usually done on magnetic tapes or optical disks. Ultrasound images are seldom stored in their digital form. In most radiology departments, permanent storage of ultrasound information is on film (hard copy of static images) or on video tape (dynamic images). With the recent development of multimodality picture archiving and communication systems (PACS) several manufacturers are starting to provide ultrasound images in a digital form readily available for archiving on magnetic or optical media.

The image sizes commonly used for the three different modalities are listed in the following table:

	Size	*Depth*
CT	512 x 512 pixels	16 bits
US	512 x 512 pixels	8 bits
MRI	256 x 256 pixels	12 bits

Based on an estimate of the average number of images acquired for each study in clinical routine, the corresponding storage capacity required is calculated in the following table:

	Number of Images	*Storage [Mbytes]*
CT	10 - 30	5 - 15
US	5 - 20	1 - 5
MRI	20 - 50	3 - 6

Picture reformatting & display: Once the digital data are stored in memory, several types of post-processing operations may be performed on it. They can be categorized according to changing the gray level display of the stored image, temporal smoothing of the data, or spatial operations.

Gray scale mean and windowing and nonlinear gray scale transformations are common. Most display systems provide a dynamic range of 256 gray levels and usually allow for manual adjustment of the intensity level and window. The data obtained from CT and MR are usually stored in 16 and 12 bits, respectively. Although the information obtained from MRI has a dynamic range that is usually smaller than 8 bits deep, CT images do require a larger dynamic range because the density information covers a much wider range. In addition to image intensity manipulation all display systems provide for image expansion (zooming) and panning through the screen, as well as real time display of dynamic images (movie).

PICTURE REFORMATTING
&
DISPLAY

- **Retrieval and display of images**
- **Gray scale adjustment (window and intensity)**
- **Zooming and panning**
- **Dynamic display**
- **Slice reorientation (Spatial interpolation)**

Additional features recently introduced in computer systems for display and processing of tomographic images usually include options for reformatting of the images in different orientations across the body volume. With sufficiently close transverse sections in CT, for example, it is possible to mathematically interpolate oblique slices. This type of processing probably falls more in the next category of post-processing tools. It is particularly useful for CT where only transverse sections can be obtained due to the orientation of the gantry. In MRI slices in any orientation can be directly acquired, and post interpolation of other orientations is less useful. In US images the slices are usually in limited number with variable orientations where resampling of the data is not feasible. Also US imaging allows for manual orientation of the tomographic plane in any direction during acquisition.

Post-processing & analysis: With the increasing speed and power of the computers more and more sophisticated tools for image analysis have become available for the different imaging modalities.

POST-PROCESSING
&
ANALYSIS

- **Measurement of sizes and volumes**
- **Measurement of image density**
- **Tissue characterization (texture analysis)**
- **Analysis of dynamic changes in the images**
- **3D reconstruction & rendering**
- **Image registration (2D & 3D)**

Using appropriate calibrations, most systems provide software capability for measuring object sizes and volumes. Density measurement is particularly used in CT (Hounsfield units) where some quantitative technique can measure directly the tissue density. Bone density can for example be measured by comparing the image density to precalibrated phantom vials. Density measurement is also emerging in MRI as a means for measuring tissue concentration of protons, and studies have shown that this parameter is reliable for tissue characterization and detection of some chemical or morphological alterations.

In US and MRI quantitative evaluation of the image texture seems to provide relevant information for tissue characterization related to morphological differences in the tissue structure. Differences in textures are also used for automatic outline of anatomical structures and image segmentation.

The three imaging modalities are capable of generating dynamic images (sequentially in time). It is possible, for example, to get dynamic images of a beating heart. From these dynamic images it is further possible to measure changes occurring over time like the extent of cardiac wall motion. Other dynamic parameters like the blood flow through the vessels can be measured in US using the doppler effect and in MRI using changes in signal intensity due to displacement of the blood volume from the imaging plane. Often these dynamic parameters are measured in every point of the image and the results are displayed using a predefined color scale. Such color coded images are also referred to as "parametric" images.

Three dimensional rendering of the tomographic data has become more popular these past few years. Most of the manufacturers currently provide with their systems the software for 3D reconstruction and display of the data. These reconstruction algorithms are very complex and often dedicated, high speed array processors are necessary to allow the calculations to be performed in an acceptable time frame.

CURRENT TRENDS

There is a common trend in medical imaging toward a standardization of the different tools that are applied for image processing and analysis. In the three modalities described here the following points represent the general trend:

- **Acquisition & reconstruction performed on dedicated hardware**: More and more systems rely on dedicated hardware for the acquisition and reconstruction of the images. The main computer usually only orchestrates the data flow and the display, storage and analysis of the images.

- **Common workstation for different imaging modalities**: With the advent of centralized picture archiving systems (PACS) most manufacturers tend to provide common workstations for different imaging modalities, and particularly for modalities that are very similar like CT, MRI and US.

- **Common image processing & analysis software**: Together with a common processing and display hardware comes a tendency to design software for image processing and analysis that can be applied to different image modalities.

- **3D rendering**: Three dimensional rendering is emerging as the future trend for displaying tomographic data.

- **Increased utilization of quantitative analysis techniques**: With the improvement in processing capabilities there is an increasing demand for more sophisticated quantitative analysis techniques of the images.

REFERENCES

1. T.B. Hunter, The Computer In Radiology (Aspen Publications,MD, 1986)

2. G.B. Greenfield, L.B. Hubbard, Computers In Radiology (Churchill Livingstone, N.Y., 1984)

3. H.K. Huang, Elements of Digital Radiology (Prentice Hall, Englewood, N.J. 1987)

4. H.U. Lemke, M.L. Rhodes, C.C. Jaffee, R. Felix, Computer Assisted Radiology, *Proceedings of the International Symposium CAR 87*, Springer Verlag, Berlin, Heidelberg, 1987.

5. R. Stuart MacKay, Medical Images and Displays (John Wiley & Sons, New York, 1984)

THREE-DIMENSIONAL IMAGING IN MEDICINE

Nicholas J. Mankovich, Ph.D.
Department of Radiological Sciences
University of California
Los Angeles, CA 90024

ABSTRACT

The past few years have seen a marked increase in the use of three-dimensional (3-D) imaging techniques. This tutorial introduces the methods behind the display of 3-D anatomical information. It assumes a familiarity with medical imaging but begins with a brief review of the most common imaging equipment used for 3-D image acquisition, including both Computed Tomography and Magnetic Resonance Imaging. Current 3-D medical practice transforms tomographic digital image studies into multi-planar reconstructions, 3-D volumetric images, 3-D surface images, or manufactured anatomical models. Multiplanar reconstruction interpolates and re-slices a patient image volume to produce novel radiographic views. Both volumetric and surface displays use lighting models to construct a displayable picture of the structures of interest. Volumetric rendering produces 3-D voxels by combining both opacity and surface shading information. Surface reconstruction explicitly identifies image objects and then subjects them to surface shading.

INTRODUCTION

All methods of radiological imaging attempt to impart an understanding of the condition of patient internal structures. Conventional radiographs use projections where all anatomical structures in the x-ray path have their images overlaid onto the film. By taking radiographs at various angles, the spatial relationship of these structures can be appreciated visually. In 3-D imaging, patient images must be acquired with care so that computers can meaningfully reconstruct the internal anatomy. Image acquisition devices must have precisely known geometry and the images must be optimized for the separation of organs or tissue types that we wish to view in three dimensions.

Once acquired, images are processed by a computer that filters the information so that only structures of interest remain and all noise, artifacts, and what we consider background is minimized or eliminated. The filtered image data can then be processed by a variety of algorithms designed to produce a display that contains meaningful information for the radiologist, surgeon, radiation therapist, or patient. Figure 1 shows the general procedure for the processing and ultimate display of 3-D information. This discussion concentrates on the basic methods of processing image data for 3-D display. The interested reader should consult Goldwasser et al.[1] for a review focussing on the hardware aspects of 3-D. Herman[2] presents an excellent review of the practical aspects of routine clinical 3-D.

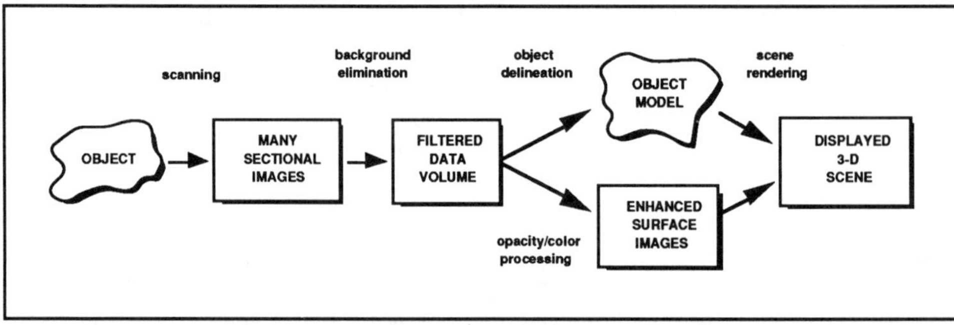

Figure 1. The basic processing associated with the production of 3-D displays including volumetric rendering (bottom) and surface reconstruction (top).

IMAGING MODALITIES

The most common methods for producing images suitable for medical 3-D processing are planar projection and tomographic imaging. Planar projection is used routinely in biplane imaging of the heart and in stereoscopic imaging of vascular structures. Because of the complexity of overlying anatomy, these methods concentrate on relatively high contrast objects, most notably vascular structures with injected contrast materials. By using careful patient positioning and the known geometry of the imaging equipment, the course of vessel trees can be reconstructed in three dimensions from two or more views. Unlike tomographic methods, the final 3-D information from planar projection methods tends to rely less on the visual use of the 3-D display and more on the measurement of vascular characteristics such as vessel size and blood volume.[3, 4, 5].

Tomographic imaging uses a variety of devices to produce images which resemble slices of patient anatomy. Because these represent the most broadly used modality for the reconstruction of 3-D structures, this discussion will focus on the principal tomographic modalities of computed tomography (CT) and magnetic resonance (MR) imaging. These approaches can be extended to other tomographic devices such as ultrasound[6, 7] and positron emission tomography[8, 9, 10] with appropriate consideration of the restrictions imposed by the fundamental properties of the modalities.

An introduction to the principles of CT and MR is beyond the scope of the current discussion, although Table 1 shows a comparison between the two in terms of characteristics important to 3-D imaging. Suffice it to say that CT and MR devices produce digital images with a known slice thickness across a region of patient anatomy with adjacent slices well aligned. These patient studies are characterized by imaging protocols and basic characteristics of the scanning devices including:

1. method of image production including device settings such as CT kVp, mAs, filtration and MR sequence timing (TE/TR, averaging, etc.).
2. slice thickness
3. slice spacing or interval (table motion or gradient encoding)

Characteristic	CT	MR
Matrix size	512x512	256x256
Density resolution	12 bits (4096 levels)	16 bits (actual 7 bits =128 levels)
Pixel dimension	0.5x0.5 mm typical	0.6 x 0.6 mm typical
Slice thickness	1.5—2.0 mm typical	2.0—3.0 mm typical
Slice spacing	2 mm typical (range 1—2 mm)	2 + mm typical
Voxel dimensions	0.5x0.5x2.0 mm	0.6x0.6x2.0 mm
#slices/3D study	30—70	20—128
Study protocol	Simple, choose location, technique, thickness, spacing	Complex: location, thickness spacing, parameters (TE,TR,...), and interleave
Patient risk	high radiation dose (4—6+ rad)	low, no ionizing radiation
Data storage	8—18 million voxels	1—8 million voxels
Advantage	separates bone from soft tissue	separates various soft tissues
Disadvantage	poor soft tissue separation	low signal to noise ratio

Table 1. Characteristics of Computed Tomography (CT) and Magnetic Resonance (MR) imaging as they relate to 3-D imaging.

4. field of view (hence pixel size)
5. reconstruction method including filtration, number of data points, etc.

Fundamental in all digital imaging is the picture element or *pixel*, representing the smallest unit of information within a single image. This concept has been extended to multi-image 3-D data sets where, in addition to the x,y dimension within the image, the pixel has a z dimension corresponding to the slice thickness. This 3-D pixel becomes a volume element or *voxel*.

Each voxel has an image intensity value that is proportional to some physical tissue characteristic. In CT images the voxel value represents the x-ray attenuation of tissue enclosed by the voxel. In MRI, the voxel intensity value represents a combination of the basic MR imaging parameters of T1, T2, and hydrogen density.

In order to be meaningful in the 3-D context, each voxel has a known location and size determined by its position within the reconstructed image and the image position relative to other images in the same patient study. Thus a voxel has a known x,y,z position and size. With the exception of MRI 3-D image acquisition, most tomographic images have a non-cubic voxel with a square dimension within the x,y image plane but a longer dimension along the z axis. In other words, a single tomographic slice is thicker than an image pixel.

With this basic framework for digital imaging laid down, we can now proceed to trace the different types of processing required to obtain a 3-D display while keeping in mind that all of the methods are being applied to a three-dimensional matrix of numbers, namely the 3-D image data set or image volume.

3-D RECONSTRUCTION METHODS

Many methods are used to transform voxel information into meaningful 3-D displays. In general, 3-D processing can be broken down into two basic methods known as volume imaging and surface reconstruction. The method chosen determines the amount of computer power and operator interaction required, the internal representation of 3-D data, and, to a lesser extent, the device used to display the result. Both methods rely on surface shading to produce a display that gives the impression of 3-D.

Volume imaging is characterized by the lack of any explicit delineation of image substructures. While the image volume is processed to produce a 3-D display that improves the viewer's perception of 3-D structure, at no time does the processing explicitly distinguish or locate internal structures. Thus, for example, while the bones of the spine are visible and discernible to the viewer, the computer programs never had to find the margins of the bone and process their surfaces independently from surrounding soft tissue. The programs did, however, process the images to enhance this separation, resulting in a display where the spine was clearly visible.

The simplest volumetric method is *multiplanar reconstruction (MPR)*, which requires a minimum of image processing to produce novel views into the original tomographic image volume[11, 12]. In practice, MPR first creates a projection image that the operator uses to designate a new slicing plane for the production of a reconstructed image. In this way, a data volume collected as axial images can be used to produce coronal, sagittal, or any other orientation of tomographic image.

The first step in MPR is the creation of a projection image. To minimize the effects of the lower z-axis resolution, the image volume is first interpolated to produce an isomorphic data volume (one with cubic pixels). The interpolated image volume is then used to form a projection image resembling a conventional projection radiograph. Figure 2 diagrams a CT data volume for a chest study being reprojected into a lateral chest radiograph. For each pixel on the display, a ray is cast back through the data volume and, in CT MPR, the voxel values are summed along this path in order to estimate the attenuation value at the display pixel of an x-ray beam following that same path. Once created and displayed on a system console or workstation, the operator uses a trackball or cursor to draw a line indicating the position of the desired re-slicing. This may be a simple planar cut as in Figure 3a or may be a curved surface following the shape of an organ as in Figure 3b, where the operator directs the MPR to follow the curvature of the spine. The computer then resamples the data, interpolating as necessary, and produces a set of parallel images that, in the spine example, will allow the radiologist to better follow the course of the spinal nerves.

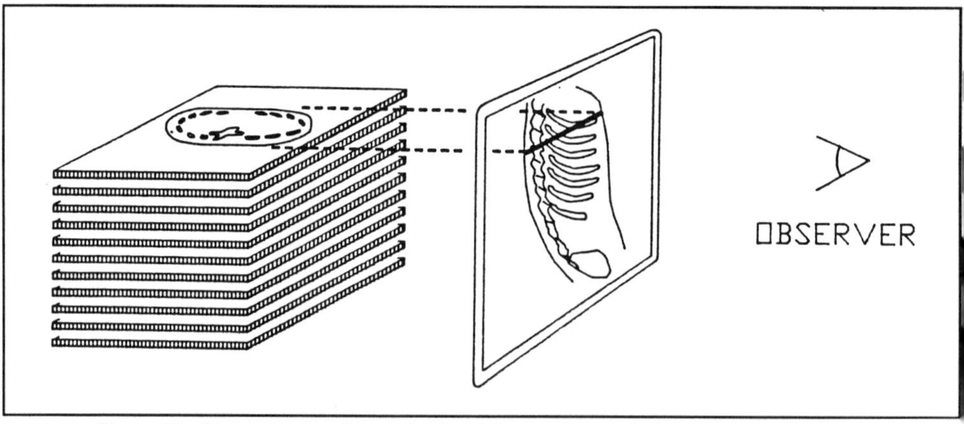

Figure 2. The process of reprojecting from an image volume onto a display plane to produce a projection image.

MPR retains the full intensity information present in the original tomographic image volume. Another volume process known as *volumetric rendering*[13] or *direct volume visualization*[14] performs the interpolation, projection, and resampling typical of MPR but applies additional image and display processing in order to produce 3-D images in which internal structures and their surfaces are rendered as opaque or partially transparent surfaces. Figure 4 summarizes the process of volumetric rendering. The brevity of this discussion is intended to give an overview of the process and the interested reader should refer to the more rigorous discussion in Levoy.[14]

Volumetric rendering starts with an interpolated data volume of cubic voxels. This isomorphic volume is processed along two paths and the resulting images are merged to form the final 3-D displayed image. The first path classifies voxels according to operator specified tissue value ranges. This classification is used to assign an opacity to each voxel. Unlike the surface generation algorithms discussed below, a particular voxel is allowed to be partially in tissue type A and partially in tissue type B resulting in an intermediate opacity. This is done by assigning voxel intensity ranges to each tissue type. Where a voxel intensity is such that it could contain a combination of tissue types, the classification table is given a percentage for tissue type A and a complementary (100-A%) value for type B. Although limited to two tissue types, his method of classification roughly corresponds to the partial volume effect present in tomographic image modalities where the sampled voxel actually contains contributions from two tissue types that happen to intersect within the voxel.

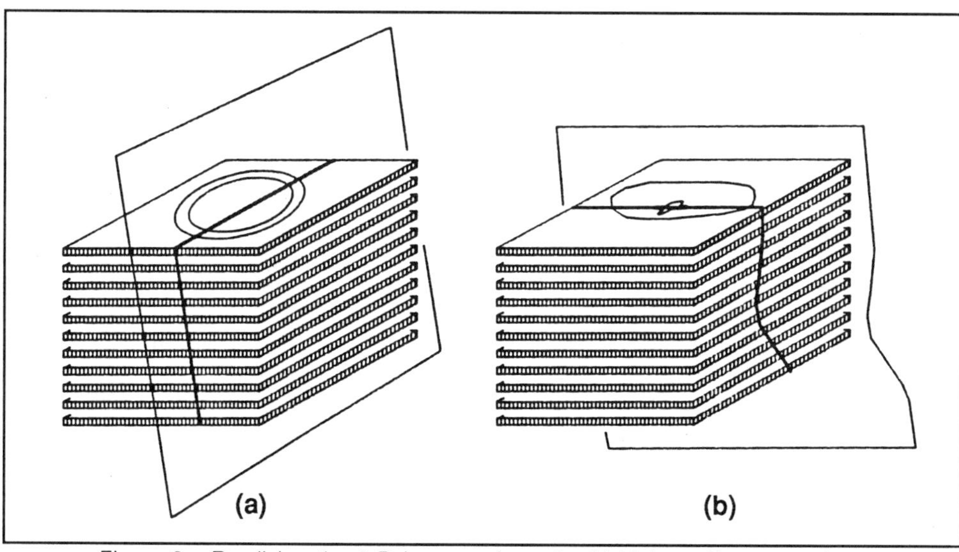

Figure 3. Re-slicing the 3-D image volume for Multiplanar Reconstruction (MPR) for (a) simple planar slicing and (b) curvilinear slicing.

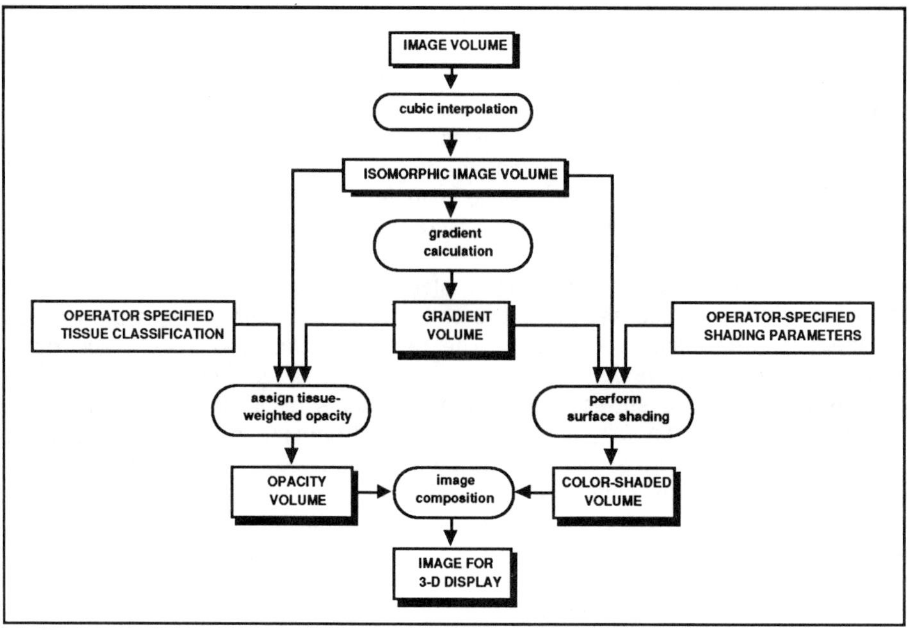

Figure 4. The processing of image information in the volumetric rendering method.

Opacity alone will not let us clearly visualize internal structures; to do this we must enhance the voxels on the surface of the structures, thus providing visual shape and location cues. The tissue opacities are weighted by the magnitude of the voxel gradient whose components are estimated by the nearest neighbor differences in the x,y, and z dimensions. In this manner, voxels near the surface of an internal structure are rendered most opaque, and those within the structure (as specified by their minimal gradient magnitude), transparent. This processing yields a new 3-D image data set called the opacity volume.

Independent of opacity processing, the isomorphic image volume is subjected to a shading process that results in the assignment of a color and intensity to each voxel. This is based on the classical surface shading methods reviewed below. Surface shading is usually employed when calculating the light reflected from a planar surface; it uses the relative orientation of the surface normal vector. With volumetric rendering, no such explicit surface is generated, although a local approximation to the surface normal is generated for each voxel by using the normalized gradient vector calculated, as above, using the nearest neighbor differences. Given a specified light source, voxels on the surface of the image facing the viewer are given an intensity related to their orientation to the light source, resulting in a new 3-D image data set called the color-shaded volume.

As the final step in volume rendering, the color-shaded volume is combined with the opacity volume by casting a ray from the observer's viewpoint back through the volumes while calculating the contribution of each on the final displayed 2-D projection of the 3-D scene. This composition process employs an underlying model of transparent gel superposition for the combination of color and opacity components. The result of this extensive volume processing is an image that conveys to the viewer the impression of looking through certain classes of tissue (e.g., soft tissue with low opacity) at other classes of tissue (e.g., bone with high opacity). Tissue interface appearance, and thus transparency, is controlled by modifying the contribution of the gradient operator in the classification and shading process.

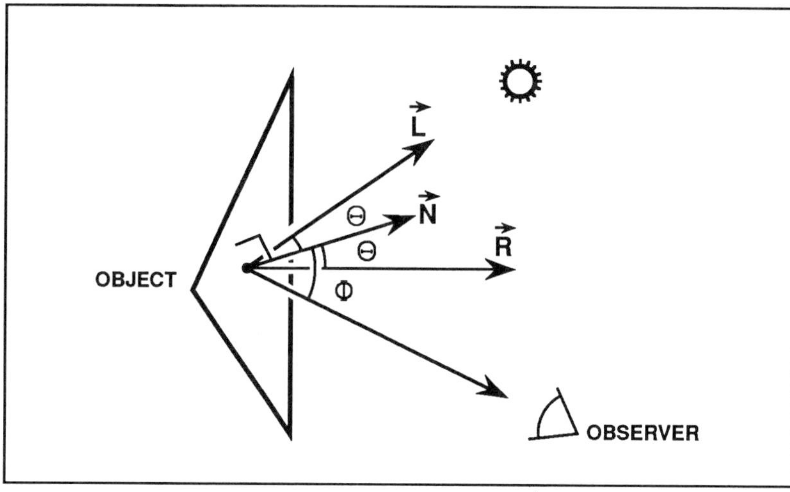

Figure 5. Basic elements of the simple lighting model including the light source vector, object surface patch, surface normal vector, and reflection and viewpoint vectors.

Surface shading techniques have been in computer graphics since the early 1970s. These techniques are used in both volumetric rendering and in surface reconstruction methods. While many elaborations on the basic scheme have been presented, all rely on a lighting model that describes how light interacts with mathematically defined surfaces to result in a displayable image. Figure 5 shows a simple representation of a lighting model given by Foley and Van Dam[15]. In its simplest form:

$$I_{displayed} = I_{diffuse} + I_{ambient} + I_{specular}$$

where any visible patch on a surface has a displayed intensity equal to the combination of reflection due to ambient light (non-directional), diffuse light, and specular light.

Given an object, a light source, and an observer, the diffuse light intensity reaching the eye of the observer can be calculated according to Lambert's cosine law stating that for dull matte surfaces, the intensity of light reaching the observer's eye is a function of the cosine of the angle between the normalized surface normal vector $\vec{N}$ and the light source vector $\vec{L}$ and can be written in linear fashion as

$$I_{diffuse} = I_l \, k_d \cos\theta$$

or in vector notation as

$$I_{diffuse} = I_l \, k_d \left(\vec{L} \cdot \vec{N}\right)$$

where I_l is the incident light intensity from the point source and k_d is the diffuse reflection coefficient constant.

In order to prevent areas in shadow from being rendered totally black, the ambient light term provides an overall light intensity, I_a, and an ambient coefficient constant, k_a, independent of the point source where

$$I_{ambient} = I_a k_a$$

The use of a specular light or highlight term modifies the observed intensity according to the angle between the reflection vector and the observer's viewing vector and is expressed as:

$$I_{specular} = I_l k_h \cos^n(\theta - \Phi)$$

where $\theta - \Phi$ is the angle between the viewer and the reflection vector. As the exponent n becomes large the object becomes a perfect reflector (mirrored) surface.

In its full form, the displayed light intensity is made up of color components accounting for color characteristics of source, object, and differential reflection and refraction. Many enhancements to the basic lighting model have been made to enhance realism and reduce computational complexity. In general, medical imaging models are simplified versions that attempt to minimize 3-D reconstruction time while providing the viewer with an understanding of the underlying structure. One of the most useful enhancements to this has been the introduction of non-linear edge effects (pseudo-Mach bands) that help to produce clear edge visibility in complex, overlapping objects.

Surface reconstruction methods have grown out of early work with computer graphics. Given a geometric description of an object in (x,y,z) space, we can use surface shading techniques to view it from any angle. The challenge inherent in medical 3-D imaging is the derivation of the geometric object description from a series of tomographic images. The process includes: (1) feature enhancement, (2) object location, (3) surface definition via small planar patches (tessellation), and finally (4) surface shading and image display.

There are many ways to locate objects in a 3-D image volume. The simplest algorithms process each image slice independently and then connect the located object boundaries into a surface. Sometimes the boundaries are outlined manually by an operator but this tends to be both time consuming and inaccurate and is usually discarded in favor of either automatic or computer-assisted methods.

Unlike volumetric rendering that uses overlapping tissue definitions, the simplest automatic boundary detection methods use a strict threshold to filter the image into a binary representation of the data. This binary representation contains non-zero values only where image intensities are in the specified range of the desired object. A simple example of this would be a CT image where all pixels with Hounsfield numbers greater than 170 would be set to white and all under 170 set to black. The resultant image would show all white cortical bone on a black background. Where there was compact bone of thickness less than the pixel size the Hounsfield number would be below 170 and hence the binary image would show a bony gap and produce pseudo-foramina. This inability to represent thin bone is one of the drawbacks of surface reconstruction but an experienced viewer knows where to expect these problems in the 3-D reconstructions.

Once the image has been filtered, the next step is to locate the object in Cartesian space. The general process of processing an image to obtain its component parts is known as *image segmentation*. Boundary detection algorithms trace the interface between object and background returning a list of (x,y) coordinates describing the perimeter. Alternatively, region growing algorithms start with a seed point within the object and then exhaustively search the surrounding pixels until all points within the object are found and the (x,y) location of all within-object pixels is recorded. Thus the boundary detection programs return a coordinate list describing the object surface, whereas the region growing algorithms have the ability to retain the entire solid structure of the object. Of course, while performing region growing, boundary information can be determined and retained. The advantage of the more data-intense region growing is that its resultant data structure retains full internal information about the binary object, thus allowing later slicing of the object to reveal its internal structure. 3-D data represented only by boundary points will reveal itself as an empty shell when sliced to reveal internal structure.

Once all images in the data volume have been filtered and processed to locate object interiors or boundaries, one more step is required in creating a true surface model. *Surface patching* or *tessellation* connects adjacent boundary loops by superimposing a polygonal planar patch between adjacent and corresponding boundary segments[16, 17]. The difficulty here is in determining adjacency and distributing the polygons in a way that produces a life-like surface. Objects with smooth borders and relatively little change in position across slices are easily handled by the existing algorithms. An example of this is the bone in a CT study of the upper portion of the skull. Irregular surfaces or those that change size rapidly from slice to slice are the most difficult. A worst case example of this is encountered when using conventional CT protocols to scan the basal skull region. Using the typical orbitomeatal line positioning, as the scan progresses superiorly from the neck, the images of the cervical spine and maxilla suddenly give way to the complex structure at the base of the skull. The connection of the boundary loops at these three levels requires some operator interaction and a knowledge of normal anatomy. Although usable, tessellation methods have concentrated on methods that exploit the known geometry of voxels to avoid working with the problem of boundary connection.

At the conclusion of surface tessellation, the computer contains a data structure describing the surface patches on one or more objects. The process described above treated the image as a matrix of (x,y) points at the center of each voxel and then constructed appropriately shaped planar pieces to fit between the points. In this manner the boundaries on adjacent slices are interpolated without having to produce the isomorphic data volume. Figure 6 shows this process applied to two adjacent skull boundaries.

Following the tessellation of the objects in a data volume, the entire data structure of surface patches is subjected to a hidden surface algorithm that leaves behind only those patches that are in the line of sight of the observer. In full-blown *ray-tracing algorithms*, a line is cast from the observer's eye through the display pixel back to objects in the scene[18] (Figure 7). This ray then encounters opaque or transparent surfaces and is further traceable back to the light source. By allowing the lighting model to include transparency and other surface attributes, the color and intensity of the displayed pixel can be calculated. Because of ray-tracing's heavy computational requirement, this full model is seldom employed in the day-to-day practice of 3-D medical imaging, where patient throughput is an important concern.

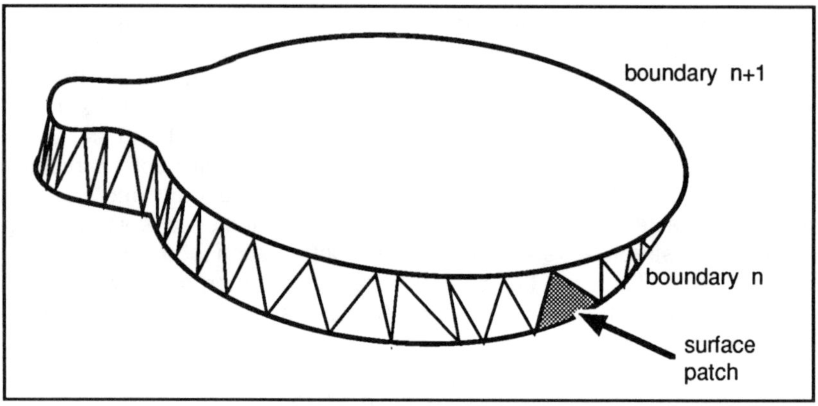

Figure 6. Two skull surface boundaries with the interpolated triangular surface tiles shown between boundaries.

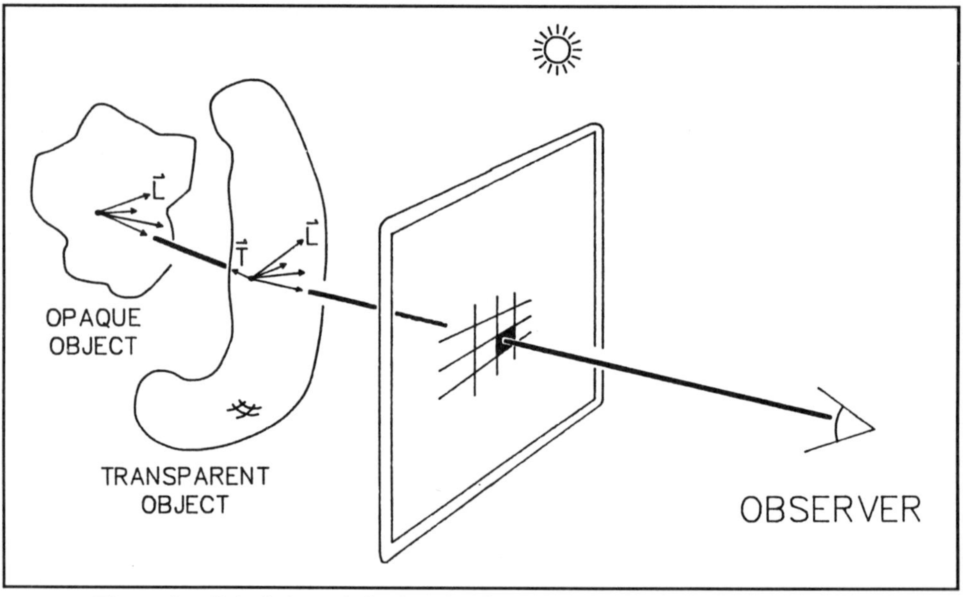

Figure 7. Calculation of displayed 3-D pixel value as determined by the application of ray tracing to the basic lighting model.

Another approach to the problem of tessellation uses the basic cubic nature of the interpolated isomorphic data volume to avoid the complexity of joining boundary loops. This process, sometimes called the *cuberille approach*, arises from the knowledge that each voxel has six faces, all oriented in either the xy, yz, or zx plane[19]. Unlike the previous patching method, the image data volume must be interpolated to a cubic voxel data volume. If region growing is applied to the interpolated data volume, then boundary processing can be applied to the object (x,y,z) list to produce list of adjacent voxels on a surface. By further processing the surface list, a list of exterior surface polygons (pixel sized squares) can be produced. These tiny squares fully describe the exterior (or interior) surface of the object. This is done without ever having to explicitly join boundary loops. Further, this segmentation process can proceed in three dimensions without recourse to slice-by-slice processing.

Many other methods have been developed that avoid the explicit patching step by relying on the voxel position information. These methods are intermediate between explicit joining patches and the single voxel face of the cuberille approach. One of the most widely used because it is part of a commercial 3-D package on a CT scanner is the method detailed by Cline et al.[20] that restricts the surface to a voxel 3-D neighborhood and reduces surface orientation to 15 possible surface orientations for tessellation.

Regardless of the method of obtaining the surface description, the result is a list of polygons that can be processed by standard surface shading methods. Some reflection on the size of a typical CT image study gives some indication of the computational intensity of this process. In a 3-D scan of the head with 68 2 mm image slices, each with 0.5x0.5 mm pixels, we found that after binary filtering 552,000 voxels remained of the original 17,826,000 image voxels. This yields an average of 8,100 bone voxels per slice. If five percent of these voxels are surface voxels then we might have 400 boundary points per slice. If we employ surface patching using strictly triangular segments then a symmetrical allotment of triangles will yield approximately two triangles per surface point. Thus the 67 interpolated sets of polygons will yield about 53,000 triangles. Even after hidden surface removal we still must perform shaded surface calculations on over 25,000 polygons. The application of the full shading equation to each of the polygons is a time consuming process, especially since it involves a trigonometric calculation. Carefully developed algorithms do much to reduce these calculations, but it remains a time-consuming process.

It is in the area of surface shading that the cuberille approach shows its superiority. Unlike the polygon methods, the square faces of the cubes have only six possible surface normal orientations. Thus, given a light source and the observer's viewpoint, only six surface intensity calculations need to be made to construct an intensity look-up table used when processing the surface voxel faces. This results in a great decrease in computational complexity and a speed-up in surface rendering. However, voxel imaging does require data set interpolation and its use of fixed-sized voxel faces does not allow the realism of specular reflection (highlights) without further increasing the computational burden.

DISCUSSION

There are many ways to go from sectional images to a 3-D display. Multiplanar reconstructions provide alternate ways of looking into the 3-D data volumes produced by tomographic imaging modalities. Volumetric approaches have the advantage of minimal object processing and therefore can rely on global image operations to produce a realistic 3-D rendering. But because they do not distinguish objects, they cannot be used for anything other than their visual application and thus there is no way to remove objects from the 3-D scene without further processing. Volumetric methods do provide a more realistic representation of tissue in voxels displaying the partial volume effect.

Surface reconstruction methods tend to be computation-intensive in their 3-D rendering because they rely mostly on neighborhood operations within the image such as border determination, boundary connection, and hidden surface removal. The advantage to surface reconstruction is the presence of an underlying object representation that, in a complex multi-object scene, allows objects to be moved, removed, or rendered differently from surrounding objects. This approach leads to increased interaction between the 3-D workstation operator and the 3-D medical scene and ultimately will allow the performance of simulated surgery. This will undoubtedly become a useful tool in the training of surgeons and in the planning of particular types of surgery.

Three-dimensional processing of tomographic images is now part of the standard image technology available to the surgeon or radiation therapist. Although present in only a few institutions and service centers, the 3-D display of patient data will grow as the technology is developed and the incremental cost to the patient is reduced[2]. The technology used to produce the 3-D displays will be extended to allow the presentation of information from more than one modality[9]. Impressive results are already being shown in the combination of functional images (PET) and anatomical images (MRI). The therapist is already using these techniques to superimpose radiation fields on patient anatomy to produce beam's-eye views of treatment plans.

The display of 3-D information is expanding with the introduction of new display technologies in holography and the varifocal mirror, as well as improved image interaction devices. The field of anatomic model construction is growing, allowing the 3-D information from tomographic images to be turned into manufactured models of patient anatomy. This has been primarily used for the construction of hip or cranial prostheses but as costs reduce, there is no reason that this cannot be yet another mechanism for reporting the results of a patient scan. Surgeons can use the models to plan and perhaps practice surgery.

REFERENCES

1. Goldwasser, S. M., R. A. Reynolds, D. A. Talton and E. S. Walsh. 1988. Techniques for the rapid display and manipulation of 3-D biomedical data. Comp. Med. Imag. and Graphics. 12(1):1-24.
2. Herman, G. T. 1988. Three-dimensional imaging on a CT or MR scanner. Journal of Computer Assisted Tomography. 12(3):450-458.
3. Fessler, J. A. and A. Macovski. 1989. Three-dimensional reconstruction of vessels with stenoses and aneurysms from dual biplane angiograms. Medical Imaging III: Image Processing. SPIE 1092:22-32.
4. Smets, C., D. Vandermeulen, P. Suetens and A. J. Oosterlinck. 1989. Knowledge-based system for the 3-D reconstruction and representation of the cerebral blood vessels from a pair of stereoscopic angiograms. Medical Imaging III: Image Processing. SPIE 1092:130-138.
5. Kitamura, K., J. Tobis and J. Sklansky. 1988. Estimating the 3-D skeletons and transverse areas of coronary arteries from biplane angiograms. IEEE Transactions on Medical Imaging. 7:173-187.
6. Vandermeulen, D., C. Smets, P. Suetens, G. Marchal, J. Gybels and A. J. Oosterlinck.

1989. Knowledge-based 3-D segmentation of blood vessels on a spatial sequence of MRI and ultrasound images. Medical Imaging III: Image Processing. SPIE 1092:142-152.

7. Greenleaf, J. F. 1983. Computerized tomography with ultrasound. Proceedings of the IEEE. 71(3):330-337.
8. Henri, C., L. Collins, T. M. Peters, A. C. Evans and S. Marrett. 1989. Three-dimensional interactive display of medical images for stereotactic neurosurgery planning. Medical Imaging III: Image Processing. SPIE 1092:67-74.
9. Peters, T. M., J. A. Clark, A. Olivier, E. Marchand, G. Mawko, M. Dieumegarde, L. V. Muresan and R. Ethier. 1987. Integrated stereotactic imaging with CT, MR imaging and digital subtraction angiography. Radiology. 161:821-826.
10. Levin, D. N., C. A. Pelizzari, G. T. Y. Chen, C.-T. Chen and M. Cooper. 1988. Retrospective geometric correlation of MR, CT, and PET images. Radiology. 169:817-823.
11. Fishman, E. K., D. Magid, B. R. Mandelbaum, W. W. S. Jr., P. Weiss, R. Hadfield, B. Mudge, S. E. Kopits, A. F. B. Jr and S. S. Siegelman. 1986. Multiplanar (MPR) imaging of the hip. Radiographics. 6(1):7-53.
12. Johnson, G. A. and M. Korobkin. 1982. Image techniques for multiplanar computer tomography. Radiology. 144:829-834.
13. Drebin, R., L. Carpenter and P. Hanrahan. 1988. Volume rendering. Computer Graphics. 22(4):65-74.
14. Levoy, M. 1988. Direct visualization of surfaces from computed tomography data. Medical Imaging II. SPIE 914:828-839.
15. Foley, J. D. and A. van Dam. 1983 "Fundamentals of interactive computer graphics." Addison-Wesley. Reading, Massachusetts.
16. Batnitzky, S., H. Price, K. Lee, L. Cook, P. Cook, S. Fritz, S. Dwyer and C. Watts. 1982. Three-dimensional computer reconstructions of brain lesions from surface contours provided by computed tomography: A prospectus. Neurosurgery. 11(1):73-84.
17. Christiansen, H. N. and T. W. Sederberg. 1978. Conversion of complex contour line definitions into polygonal element mosaics. Computer Graphics. 12:187-192.
18. Kuchkuda, R. 1988. "An introduction to ray tracing." Theoretical Foundations of Computer Graphics and CAD. Earnshaw ed. Springer-Verlag. Berlin Heidelberg.
19. Artzy, E., G. Frieder and G. T. Herman. 1981. The theory, design, implementation and evaluation of a three-dimensional surface detection algorithm. Computer Graphics and Image Processing. 15:1-24.
20. Cline, H. E., W. E. Lorensen, S. Ludke, C. R. Crawford and B. Teeter. 1988. Two algorithms for the three-dimensional reconstruction of tomograms. Medical Physics. 15:320-327.

MICROPROCESSORS IN ANGER SCINTILLATION CAMERAS

Jonathan M. Links, Ph.D.
The Johns Hopkins Medical Institutions
Baltimore, Maryland 21205-2179

The Anger scintillation camera (named after its inventor, Hal Anger of Donner Laboratory) is the most widely used imaging device in nuclear medicine. The camera consists of a collimator, a sodium iodide scintillation crystal, an array of photomultiplier tubes, and a unique positioning logic circuit. The collimator is essentially a lead slab with a series of holes in it, which discriminates against x-rays or gamma-rays traveling in undesired directions (i.e., those not traveling along the long axes of the holes are absorbed in the lead). The sodium iodide scintillator gives off a flash of visible light when an x-ray or gamma-ray interacts with crystal molecules and deposits energy (hereafter referred to as an "event"). This light flash is detected by the array of photomultiplier tubes, and the event's X,Y position determined by the positioning logic, generally utilizing an implementation of a centroid or center-of-mass approach.

Microprocessors play a major role in Anger scintillation cameras in three ways: (l) on-board, on-line correction of intrinsic distortions; (2) control of data acquisition; and (3) full system integration, including data processing.

DISTORTION CORRECTION

Three major performance parameters of Anger cameras are spatial resolution, field uniformity, and linearity. Spatial resolution is defined as the ability of a camera to distinguish between separate sources of radioactivity. It is generally measured with point sources of radioactivity, and is typically specified as either the minimum distance by which two sources must be separated to be distinguished as separate in an image, or as the full-width-at-half-maximum (FWHM) of a profile through a single point source. Field uniformity is defined as the ability of a camera to depict a uniform source as uniform, and is generally measured by presenting the camera with a uniform field of radioactivity (often referred to as a "flood"), and quantifying the variation across the image. Linearity is defined as the ability of a camera to depict a straight line as straight, and is generally measured by quantifying the deviations in the image of a straight line source.

It was originally believed that non-uniformities were primarily the result of variations in detection sensitivity across the face of the sodium iodide crystal, and that limitations in spatial resolution and linearity were due to analog non-linearities in the positioning logic electronics. It is now known that all three parameters are mainly affected by "spatial distortion," defined as the incorrect assignment of an event's X,Y location. An approach to correction for this spatial distortion has been implemented in microprocessor form in virtually all current Anger cameras. In this approach, a phantom study is performed to define the mispositioning of events. This phantom typically consists of an orthogonal array of point sources. A correction matrix is generated, which calibrates the repositioning of points so that their position in the image corresponds to that in the actual phantom. During acquisition of an actual patient study, this correction is applied on-line, on an event-by-event basis as individual x-rays or gamma-rays are detected.

In addition to this correction for spatial distortion, other on-line corrections are usually performed. Primary among these is correction for any residual non-uniformity which remains after spatial distortion correction. This is accomplished through the use of a reference image of a uniform "flood" source, identical to that used to assess uniformity. Another correction which is usually implemented is that for variation in the energy response across the face of the camera. This is accomplished through the use of a reference table generated from measurements of the so-called Z or energy signal of the camera as a finely-focussed point source is moved across the crystal.

These approaches generally require that the signals be put in the digital domain in the camera head electronics. There is no agreement as to when digitization should take place in the electronics chain, nor what "digital resolution" (i.e., matrix size) is required (e.g., 64 x 64 versus 256 x 256). One side benefit of digitization of the pulses is the opportunity for digital pulse shaping, which improves count rate performance.

ACQUISITION CONTROL

Anger scintillation cameras were originally completely analog in nature. All images were produced by the exposure of a piece of film viewing a cathode ray tube display. The shutter of a photographic camera was left open during the entire acquisition period, and an image was built event-by-event on the film. Many nuclear medicine studies are still acquired in this fashion. Indeed, even studies of some dynamic processes, such as the first pass of a bolus of radioactivity through the circulation, or its arrival at certain organs, can be acquired with technologist-operated advance of the photographic film. However, there are certain studies whose time course dictates the use of a faster, more accurate approach, and microprocessors play a major role.

Two major classes of studies which require microprocessor control are rapid dynamic processes, and repetitive dynamic processes. An example of a rapid dynamic process is the first transit of a bolus through the circulation, particularly if phases of the cardiac cycle are to be resolved. This requires approximately twenty or more frames per second, a frame rate which is clearly beyond the capabilities of manually advanced film.

An example of a repetitive dynamic process is the acquisition of a "gated blood pool" study, in which a series of heart beats (cardiac cycles) are superimposed under microprocessor control to produce a single, representative cardiac cycle. In this approach, the number of frames spanning the cardiac cycle is first specified. The computer measures the average heart rate, sets up the specified number of frames in memory, and computes the acquisition time per frame (as the average duration of a beat divided by the number of frames). Acquisition itself is triggered by a physiologic signal (the R wave from the ECG). When acquisition begins, data are dumped to the first frame for the specified time, then to the second frame, and so on. If the patient's heart rate is completely regular, at the end of dumping data to the last frame, another R wave occurs, and data are dumped back in the first frame, on top of the data already there. If another heart beat starts prematurely, the final frames do not receive data for that beat, as any occurence of an R wave resets acquisition to the first frame. At the end of the entire acquisition period, the computer memory holds the summed data for a number of beats, as one set of images spanning the cardiac cycle.

Single photon emission computed tomography (SPECT) is now playing a major role in nuclear medicine. This type of study is generally performed with specially adapted rotating Anger cameras. In this approach, the camera rotates around the long axis of the patient, acquiring a number of images (called projections), which form the basis for reconstruction. Microprocessor control is particularly important during acquisition of so-called non-circular orbits, in which the camera head follows either an ellipse or the actual contour of the patient as it rotates. A computer is also required for image reconstruction, of course. Indeed, SPECT has driven the market to concentrate on the use of fully integrated camera/computer systems. These integrated systems simplify the installation, set-up, and use of SPECT systems, because there is generally better "communication" between the camera and the computer in an integrated system from one manufacturer.

One final approach which spans distortion correction and acquisition control is that developed by Siemens. Dubbed WAM, for weighted acquisition module, this acquisition technique seeks to reduce the deleterious effects of scattered photons on image formation by weighting the contribution of each event to the final image. In conventional acquisition, events are either accepted or rejected by the pulse height window, which is typically placed symmetrically around the photopeak of the pulse height spectrum in an attempt to discriminate against scattered events. In WAM, all events contribute toward the formation of the image, but in a weighted fashion. The weights are empirically determined, and are radionuclide and collimator specific. The weights are a function of the actual energy deposited in a given interaction (i.e., to distinguish between photoelectric and Compton scatter events).

FULL INTEGRATION

Many nuclear medicine studies are routinely processed by computer. Indeed, an argument can be made that nuclear medicine was the first discipline in radiology to become fully computerized. Digital analysis generally requires digital acquisition, as film-based digitization is not widely available in nuclear medicine. Thus, there is strong functional and economic motivation to integrate an Anger camera and a computer into a single system, in which the microprocessor, or a group of microprocessors, performs the tasks of distortion correction, acquisition control, and processing. Indeed, it is difficult these days to purchase an analog Anger camera without a computer!

REFERENCES

Rao, D. V., Chandra, R., Graham, M. C. (eds.): Physics of Nuclear Medicine. AIP, New York, 1984

Rollo, F. D. (ed.): Nuclear Medicine Physics, Instrumentation, and Agents. C. V. Mosby Co., Saint Louis, 1977

Sorenson, J. A., Phelps, M. E.: Physics in Nuclear Medicine. Grune & Stratton, Inc., Orlando, 1987

CARDIOLOGY APPLICATIONS

L. Stephen Graham, Ph.D.
V. A. Medical Center, Sepulveda, CA 91343
and
University of California School of Medicine, Los Angeles, CA 90024

ABSTRACT

The application of computers in cardiovascular nuclear medicine has increased markedly in the past five years. The most common procedure is the gated blood pool study. It is used to provide qualitative estimates of chamber size, cardiac function, and detect the presence of aneurysms. Gated blood pool studies are also used to make quantitative measurements of global and regional ejection fraction and to quantify valvular regurgitation. First pass studies are often used to measure the ejection fraction of the right ventricle, to measure cardiac output, and to detect the presence of left-right and right-left shunts. Bone scanning agents have been found to be useful in detecting acute infarcts in the myocardium. The myocardium itself can be imaged with Tl-201, a potassium analog in tracer amounts. Although planar imaging techniques are often used, single photon emission computed tomography (SPECT) provides Tl-201 images of increased sensitivity and specificity. Positron emission tomography (PET) with N-13 ammonia, F-18 deoxyglucose, and C-11 amino and fatty acids is now being used to provide high resolution images of myocardial blood flow and metabolism.

INTRODUCTION

At the present time, cardiovascular nuclear medicine procedures account for as much as 50% of all studies in the nuclear medicine departments of some large medical centers. Although this percentage is smaller in many community hospitals, virtually all perform some type of nuclear medicine study to qualitatively and quantitatively measure cardiac function. The basic equipment that is needed is a scintillation camera, a nuclear medicine computer, and an ECG gate. To expand the scope of cardiovascular studies, treadmills, exercise tables, and bicycle ergometers must also be available. Regardless of institutional size, cardiovascular nuclear medicine has been found to be a cost-effective method for studying patients with known or suspected organic heart disease.

GATED BLOOD POOL STUDIES

Gated blood pool studies are used to provide a qualitative estimate of chamber size, cardiac function and to detect the presence of aneurysms. In addition, they can yield quantitative information concerning global and regional ejection fraction, ejection velocity, and patterns of ventricular emptying. Red blood cells labeled with Tc-99m are used after IV administration of a non-radioactive tin-containing bone scan agent followed in 20-30 minutes by the radiopharmaceutical Tc-99m pertechnetate. Data collection into a computer is performed by physiological gating using an ECG. Prior to initiating data collection, a short acquisition is used to establish the average time of the patient's cardiac cycle. During the imaging sequence, the appearance of an R wave in the ECG is used to initiate collection of an image in a portion of the computer's memory. After a period corresponding to 1/16 (the most commonly used fraction) of the average beat length, the data acquisition is moved to another portion of the computer's memory. This process continues each 1/16th of the cardiac cycle until a total of 16 separate images are formed. When the next R wave occurs this process is repeated using the same 16 images in the same sequence. A larger number of images (e.g., 32) is often used if the study is performed while the patient is exercising[1]. Imaging is completed when 200 to 500 cardiac cycles have been collected. Figure 1 presents a set of images for the left anterior oblique (LAO) projection. A complete study usually consists of images obtained from three different orientations: left anterior oblique, anterior (ANT), and left lateral (LAT). A non-gated SPECT study can be used to provide the additional information that is needed to calculate left ventricular volume[2].

Because cardiac cycle length is not constant, most systems have software for discarding "bad" beats. Some software provides only partial bad beat rejection. A cardiac cycle more than 10% (a typical value) different than the average causes data acquisition to cease until the next beat that falls within ± 10% of the average cardiac cycle occurs. This process continues until the required number of cycles has been collected. Most state-of-the-art systems have complete arrhythmia rejection. Each beat is buffered in memory so that only "good" beats are used to form the data set. Both of these techniques have the characteristic that the computed ejection fraction is higher than the patient's "effective" ejection fraction.

Prior to the analysis of a gated blood pool study, it is often appropriate to apply temporal and/or spatial smoothing, particularly in the case of exercise studies. Temporal smoothing often takes the form of a 1:2:1 smooth in the time domain. Spatial smoothing is usually a 1:2:1 two-dimensional spatial smooth but may utilize an appropriately designed Fourier filter. Application of two or three dimensional smoothing facilitates the determination of blood pool edges because noise is reduced.

Both semiautomatic and fully automatic software has been written for finding edges of the right and left ventricles. Most are based on a combination of second-derivative and thresholding techniques. Band-pass, edge enhancing two dimensional frequency domain filters have also been used[3]. Once regions-of-interest (ROIs) have been set on the end diastolic and end systolic frames, the ejection fraction is calculated by the equation:

$$EF = \frac{\text{End Diastolic Counts} - \text{End Systolic Counts}}{\text{End Diastolic Counts}} \qquad (1)$$

Totally automatic algorithms have been developed for processing gated blood pool studies that only require input of the patient's name[4]. After subtracting the minimum in the entire data set for each pixel, a 1:2:1 filter is applied in space and time to increase the signal/noise ratio for edge detection routines. The four stroke volume images with the highest center-of-mass (CM) are summed to create a composite of atrial end diastole. The four stroke volume images with the lowest CM are summed to create a composite of ventricular end diastole. The atrial composite is then subtracted from the ventricular composite to minimize the problem of overlapping structures. Scanning horizontal profiles are used to locate local minima between the right and left ventricle. After application of interpolative background subtraction, a nine point Laplacian operator is used to define the left ventricle edges. A background region is also defined automatically. From these data the ejection fraction is computed and the ventricular volume curve is displayed. A heart rate histogram is also shown as a part of the report. The histogram indicates the quality of the gating and the repeatability of single heart beats. If the histogram is wide or two peaks are shown, averaging will effect both the visual and quantitative analysis.

Other quantitative measurements of cardiac function can be made. These are: maximum emptying and filling rates[5], first third ejection fraction, regional ejection fractions, stroke volume images, and paradox images. The maximum emptying rate is an approximate representation of ventricular contraction rate and gives some indication of contractility. The maximum filling rate indicates myocardial compliance. Some authors have suggested that the first third ejection fraction is a sensitive indicator of cardiac dysfunction[6-8].

Stroke volume images are created by subtracting the frame that represents end systole from the end diastole frame (Figure 1). The resulting image presents spatial information on the relative regional change in volume. A paradox image is obtained by subtracting the end diastole frame from the end systole frame (Figure 1). This image is helpful in showing dyskinetic motion that is present in an aneurysm.

Normally atrial contraction can be observed since the atria are approximately 180° out of phase.

The pattern of ventricular emptying can be seen by an analytical technique known as phase imaging[9-12]. Phase images are created by fitting each pixel in the gated blood pool data set to a cosine function. Although other harmonics can be used, most implementations of phase imaging utilize only the first[13]. Normally, all pixels within the left ventricle will have phase values that are similar to one another. All pixels in the atria will behave similarly, but collectively they will be 180 out of phase relative to the ventricular set. By careful study, which is sometimes assisted by color coding, the small differences in phase within the ventricles can be seen. In the normal patient the pattern of emptying begins in the interventricular septum and spreads out through the right and left ventricles. Although the interpretation of these images is not simple, irregular patterns of contraction that cannot be readily seen on LAO gated studies are visualized.

Recently another method of evaluating gated blood pool and other dynamic studies of the heart was introduced by Cavailloles et al[14]. The algorithms for physiological factor analysis were originally developed by Bazin and Di Paola[15] and Di Paola et al[16]. Abnormalities are recognized by the appearance of a kinetic feature that is known not to be present in nonpathological cases. In a gated study consisting of p frames of size n x n pixels, the data can be considered to be a set of n-squared vectors of length p. In mathematical form[17]:

$$X_{ij} = [x_{ij}(1), x_{ij}(2), x_{ij}(3), \ldots x_{ij}(p)] \tag{2}$$

This method assumes all factors in the set can be described by a linear combination of a limited number of physiological factors. These factors represent the physiological kinetics that produced the observed count-time profile. In mathematical form,

$$X_{ij} = v_{ij}V + a_{ij}A + b_{ij}B + N_{ij} \tag{3}$$

where V represents the kinetics of the ventricles, A represents the kinetics of the atria, B represents the kinetics of background, and N is the random noise at ij. v_{ij}, a_{ij}, b_{ij}, and N_{ij} are the relative contributions of the ventricles, atria, background, and random noise. If there is a contribution at any location of abnormal kinetics, i.e., neither ventricular, atrial, nor background, the vector becomes

$$X_{ij} = v_{ij}V + a_{ij}A + b_{ij}B + p_{ij}P + N_{ij} \tag{4}$$

where p_{ij} is the contribution of a pathological kinetic factor P. An example of the use of this technique can be found in reference 18.

INFARCT IMAGING

Several years ago it was observed that patients with acute myocardial infarcts showed uptake of Tc-99m labeled pyrophosphates in the infarct[19]. Subsequently it was found that the peak sensitivity occurred two to five days post infarction. Triangulation of the infarct is attained by collecting LAO, ANT, and LAT views. Uptake of the radiopharmaceutical in the infarct is visually graded and assigned a value of zero through four. Zero and one represent normal levels of activity and are usually associated with the blood pool and superimposed radionuclide. A grade of two is questionable. Its significance may depend on whether the uptake is localized or diffuse. Grades of three and four are generally diagnostic for myocardial infarction (Figure 2). SPECT studies have markedly improved interpretation of infarct imaging by eliminating nonspecific blood-pool activity that produces false positives[20].

FIRST PASS STUDIES

First pass studies involve the collection of data during the passage of a small bolus of activity as it passes through the cardiac chambers and lungs. Several different radiopharmaceuticals are used but the label is normally Tc-99m. Data can be collected in "list" or "frame" mode[21]. For list mode acquisition, the X,Y coordinates of individual events are "listed" in memory along with periodic time marks of 1 to 10 milliseconds. Images of the desired matrix size are created retrospectively and can have any time duration down to the length of the timing marks. In frame (sometimes called histogram) mode, images of prespecified size are collected directly in memory. After the first image has been formed the second image is stored at a different location in memory while the first is being transferred to disk. That portion of memory is then set to zero and is reused for the third image, and so on. Of course this method does not allow any change in matrix size; frame duration can only be modified by adding frames together. First pass studies simplify the measurement of right heart ejection fraction by minimizing interference from left heart tracer activity[22].

Cardiac output can also be measured with the first pass method[23]. Twenty millicuries of Tc-99m are injected as a small bolus and images are collected at a rate of one/second. After correction of the data for coincidence losses in the camera computer system[24,25], a region-of-interest (ROI) is drawn around the left ventricle. Another is drawn just outside the ventricle to measure background. These ROIs are used to generate time-activity curves and the background, normalized for the number of pixels in the ROI, is subtracted to produce a net curve. The net curve is fitted with a gamma variate for determination of the counts under the first pass envelope[26]. The average number of counts per time interval at equilibrium is also calculated from the net curve. Together with the patient's blood volume, these data are used to calculate the cardiac output by the equation:

$$CO = \frac{C_{equilibrium}}{\int^{\infty} C(t)\ dt} \times \text{Blood Volume} \qquad (5)$$

The denominator of equation (5) is the total number of counts under the first pass curve. Blood volume can be measured directly using the principle of isotopic dilution[27] or with a knowledge of the patient's height and weight, can be estimated from a nomogram[28]. Nomograms are less accurate but if the patient is serving as his/her own control for serial studies they may be satisfactory. An alternative method for measuring cardiac output involves calculating the chamber volumes by the method described in reference 2 and using the equation

$$CO = (\text{ED Volume} - \text{ES Volume}) \times \text{Heart Rate} \qquad (6)$$

Because of the non-invasive nature of the method, first pass studies have been found to be useful in measuring left-right and right-left shunts, especially in children[29]. A sharp bolus is injected in a peripheral vein and data are collected at a rate of 2.5 frames per second for a time of 60 seconds. Before processing the data it is important to check the "quality" of the bolus. A diffuse bolus will produce significant errors in calculation of the magnitude of the shunt although techniques have been described for removing this effect by deconvolution. Deconvolution is not needed if the full-width-at-half-maximum (FWHM) is less than 2.5 seconds.

For quantitation of the shunt, a ROI is set over the peripheral lung fields and a time-activity curve is generated. A gamma variate is fitted to the first pass envelop (A) and is subtracted from the raw data to get a difference curve. The difference curve is also fitted with a gamma variate (B). The magnitude of the shunt, the Qp (pulmonary):Qs (systemic) ratio, is calculated by the equation

$$Qp:Qs = A/(A - B) \qquad (7)$$

It has been demonstrated that this method provides accurate determinations of Qp:Qs within a range between 1.2 and 3.0.

Right-left shunting can be evaluated with macroaggregates of albumin (MAA). In the normal patient MAA lodge in the capillaries of the lungs so only the lungs are visualized. If a shunt is present, some of the MAA bypass the lungs and lodge in other capillary beds such as those of the brain and kidneys. Some can also be found in the general circulation. Quantitation consists of setting one ROI over the lungs (Qp) and another over the entire body, including lungs (Qs). First pass methods can also be used to evaluate right-left shunting because of temporal relationships. After injection of the bolus, activity appears in the left side of the heart shortly after it enters the right side.

MYOCARDIAL PERFUSION IMAGING WITH TL-201

Tl-201 is an analog of potassium and is commonly used for imaging the myocardium. Its primary use is to examine patients who are at risk for coronary artery disease (CAD). Tl-201 is also used to follow known CAD patients on medical therapy or who have been treated with bypass surgery, angioplasty, or thrombolysis. Tl-201 imaging is the only noninvasive technique in routine clinical use that directly evaluates blood flow to the myocardium[30].

In planar imaging three views are used for triangulation. The radiopharmaceutical is injected during peak exercise (stress) and the patient is imaged immediately. Approximately four hours later the patient is imaged again (resting study). Regions of reduced activity that are present during stress but which fill in when the exercise study is performed represent viable myocardium which is ischemic on exercise. By contrast, "cold" spots at exercise that are still "cold" at rest generally represent myocardial tissue that is not viable or which has been replaced by scar tissue. Due to the poor statistics associated with Tl-201 imaging, computer processing of the images is an important consideration. The first step is normally background subtraction. Because the superimposed background is not uniform, interpolative background subtraction is used[31]. A rectangular box is drawn that encompasses the myocardium. To calculate the background for a given point inside the box, counts in the 4 pixels on the box along the corresponding X and Y axes are interpolated. This process is repeated until a value has been calculated for all points inside the box. A representative study is presented in Figure 3.

A number of years ago it was found that sensitivity and specificity of Tl-201 studies were significantly improved by computer analysis in the form of circumferential profiles [32, 33]. Software has been written that radially searches from the center of the myocardial image outward to find the pixel with the maximum counts. This process is repeated at six degree intervals until the entire myocardium has been covered. These data are displayed graphically as a plot of maximum counts vs angle for both exercise and rest images (Figure 3). Some programs also contain a circumferential profile that represents a normal patient without CAD. Comparison of the two curves is aided by calculation of the percent washout curve. It is calculated using the equation:

$$\% \text{ Washout} = 100 \times (\text{Stress} - \text{Delayed})/\text{Stress} \qquad (8)$$

Sensitivity and specificity of Tl-201 studies have been increased even more by the use of SPECT. Emission tomographic studies are performed immediately after exercise and at rest. After reconstruction of the data, transverse sections are mathematically transformed into sections that are parallel to the long and short axis of the heart[34]. A set of images perpendicular to the long axis is presented in Figure 4. Maximum count circumferential profiles of the long and short axis tomograms are used to define the three dimensional distribution. These data can be mapped into a two dimensional polar distribution. In the "bullseye" format activity at the apex is displayed in the center of the image. Sections progressively positioned toward the base of the heart are displayed as rings of increasing radius[35].

NEW IMAGING AGENTS

A number of new investigational agents have recently become available for imaging the myocardium[36]. One class, the isonitriles (e.g., methoxyisobutyl isonitrile, MIBI), shows special promise because they use Tc-99m as a label. The higher energy, relative to Tl-201, gives better system spatial resolution and improved statistics because of decreased attenuation and scattering. However, there is one disadvantage. The radiopharmaceutical does not redistribute so stress and resting studies must be done on separate days from separate injections.

In-111 labeled monoclonal antibodies specific for cardiac myosin are expected to be introduced in 1988[30]. This agent was reported to be highly sensitive and specific for the diagnosis of myocardial infarction in the first 24 hours following an infarct[37]. This would provide more rapid diagnosis than is possible with Tc-99m pyrophosphates. Labeled antibodies, coupled with the use of SPECT, should play an important role in more rapid and precise determination of the location and extent of an infarct.

I-123 labeled fatty acids have been used for imaging the myocardium because they measure aerobic metabolism. However, the high background activity and the problems associated with imaging I-123 may be the reason for the infrequent use of labeled fatty acids[38,39].

A number of pharmaceuticals labeled with positron emitting radionuclides are now being used. These include F-18 deoxyglucose (FDG), N-13 ammonia, and C-11 fatty acids[36]. N-13 ammonia is used to measure blood flow to the myocardium while FDG measures glucose metabolism. As indicated above, fatty acids can be used to measure aerobic metabolism, the normal method of metabolism. In the presence of ischemia or anoxia the heart switches to anaerobic glucose metabolism. Amino acids labeled with C-11 have been utilized to measure protein synthesis[40].

SUMMARY AND CONCLUSIONS

The scintillation camera interfaced to a computer can be used to qualitatively and quantitatively measure cardiac function with varying degrees of accuracy. In some cases (gated blood pool, first pass, and SPECT studies) the computer is an absolute necessity; in others (infarct imaging) it is useful but not an absolute requirement. To extract diagnostic information certain operations are common to virtually all applications: the ability to set regions-of-interest and to generate time-activity curves, the ability to manipulate the curves and to describe the data mathematically, and the ability to operate on the data to increase SNR[41].

It is important to remember that it is often difficult to compare results from different systems because of variations in software, not hardware. Even for the same computer system, changes in software can produce marked changes in the results of a study. This is most clearly stated by Yuille:

> "Computers and computer processing techniques are of fundamental importance to the performance and success of nuclear cardiology. The understanding and application of these techniques are not always straightforward, and subtle differences in software performance can have a major impact on the success of a technique at a given institution. Thus, it is extremely important for physicians overseeing and interpreting these studies to evaluate their results continuously and objectively."[16].

REFERENCES

1. S. L. Bacharach, M. V. Green, J. S. Borer, et al, J. Nucl. Med. 20, 189 (1979).
2. G. R. Caputo, M. M. Graham, K. D. Brust, et al, Am. J. Card. 56, 781 (1985).
3. K. S. Shanmugan, IEEE Trans On Pattern Analysis and Machine Intelligence. Vol. PAMI-1, No. 1, (January, 1979).
4. J. Areeda, E. Garcia, K. Van Train, et al, Computers in Cardiology, IEEE Proceedings, 109-112 (1982).
5. J. A. Bianco, D. G. Makey, W. K. Laskey et al, J. Nucl. Med. 20, 1 (1979).
6. L. L. Johnson, K. Ellis, D. Schmidt et al, Circ. 52, 378 (1975).

7. A. Battler, R. Slutsky, J. Karliner, et al, Amer. J. Cardiol. 45, 197 (1980).
8. R. Slutsky, J. S. Karliner, A. Battler, et al, Circ. 61, 1083 (1980).
9. J. W. Verba, I. Bornstein, N. P. Alazraki, et al, J. Nucl. Med. 20, 625, 1979. Abstract.
10. W. E. Adam, A. Tarkowska, F. Bitter, et al, Cardiovas. Radiol. 2, 161, 1979.
11. J. M. Links, K. H. Douglass, and H. N. Wagner, J. Nucl. Med. 21, 978 (1980).
12. J. M. Links, J. S. Raichlen, H. N. Wagner, et al, J. Nucl Med. 26, 27 (1985).
13. J. Machac, S. F. Horowitz, D. Broder, et al, J. Nucl. Med. 25, 1294 (1984).
14. F. Cavailloles, J. P. Bazin, and R. Di Paola, J. Nucl. Med. 25, 1067 (1974).
15. J. P. Bazin and R. Di Paola, in C. Raynaud (ed.), Nuclear Medicine (Pergamon Press, Paris, 1982), p. 35.
16. R. Di Paola, J. P. Bazin, F. Aubry, et al, IEEE Trans. Nucl. Sci., NS-29, 1310 (1982).
17. M. Goris, Sem. Nucl. Med. 17, 18 (1987).
18. L. Philippe, I. Mena, J. Darcourt, et al, J. Nucl. Med. 29, 159 (1988).
19. F. J. Bonte, R. W. Parkey, K. D. Graham, et al, Radiol. 110, 473 (1974).
20. N. Tamaki, T. Mukai, Y. Ishii, et al, Eur. J. Nucl. Med. 9, 106 (1984).
21. D. E. Lieberman, Computer Methods: The Fundamentals of Digital Nuclear Medicine (The C. V. Mosby Company, St. Louis, 1977).
22. M. L. Nusynowitz, A. R. Benedetto, R. A. Walsh, et al, J. Nucl. Med. 28, 950 (1987).
23. J. M. Links, L. C. Becker, J. G. Shindledecker, et al, Circ. 65, 82 (1982).
24. R. Adams, G. J. Hine, and C. D. Zimmerman, J. Nucl. Med. 19, 538 (1978).
25. R. Adams and I. Mena, Med. Phy. 15, 415 (1988).
26. M. D. Harpen and M. L. Lecklitner, Med. Phy. 11, 690 (1984).
27. J. C. Sisson, Total Blood Volume, in J. W. Keyes (ed), Manual of Nuclear Medicine Procedures, 3rd edition (CRC Press, West Palm Beach, 1978), pp. 155-156.
28. S. B. Nadler, J. U. Hidalgo, and T. Bloch. Surg. 51, 224 (1962).
29. S. Treves, Sem. Nucl. Med. 10, 16 (1980).
30. D. L. Yuille, Appl. Rad., 39 (June 1988).
31. M. L. Goris, S. G. Daspit, P. McLaughlin, et al, J. Nucl Med. 17, 744 (1976).
32. J. Maddahi, E. V. Garcia, D. S. Berman, et al, Circ. 64, 924 (1981).
33. E. Garcia, J. Maddahi, D. Berman, et al, J. Nucl. Med. 22, 309 (1981).

34. J. A. Borrello, N. H. Clinthorne, W. L. Rogers, et al, J. Nucl. Med. 22, 471 (1981).
35. E. V. Garcia, K. Van Train, J. Maddahi, J. Nucl. Med. 26, 17 (1985).
36. H. Schelbert, Sem. Nucl. Med. 17, 145 (1987).
37. B. A. Khaw, T. Yasuda, H. K. Gold, et al, J. Nucl. Med. 28, 1671 (1987).
38. N. D. Poe, G. D. Robinson, L. S. Graham, et al, J. Nucl. Med. 17, 1077 (1976).
39. L. E. Feinendegen, K. Vyska, C. Freundlieb, et al, Eur. J. Nucl. Med. 6, 191 (1981).
40. J. R. Barrio, F. J. Baumgartner, E. Henze, et al, J. Nucl. Med. 24, 937 (1983).
41. W. E. Barnes and E. E. Gose, Sem. Nucl. Med. 17, 58 (1987).

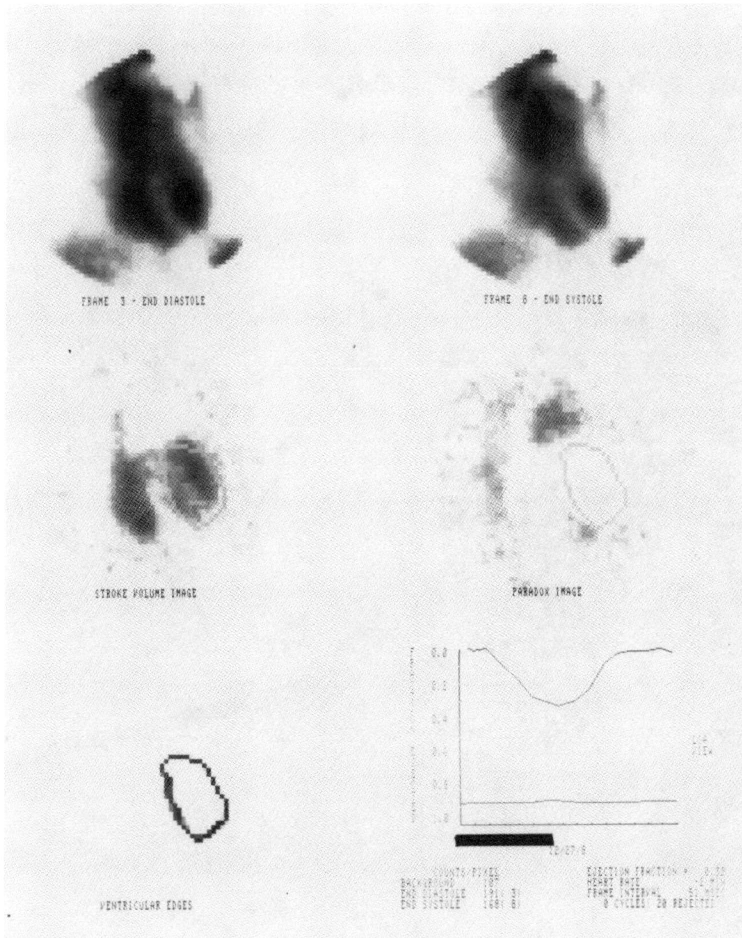

FIGURE 1. Image set for a Tc-99m red blood cell gated blood pool study. Patient was an 81 year old male with multiple myeloma, chronic obstructive pulmonary disease, congestive heart failure, and persistent bilateral pleural effusions. Top Row: Left anterior oblique (LAO) view at end diastole and end systole. Center Row: Stroke volume and paradox images. Bottom Row: Superimposed images (16) of the ventricular edges and volume curve. Images show marked dyskinesis of the right ventricular apex and akinesis of the left ventricular apex with an ejection fraction of 32%.

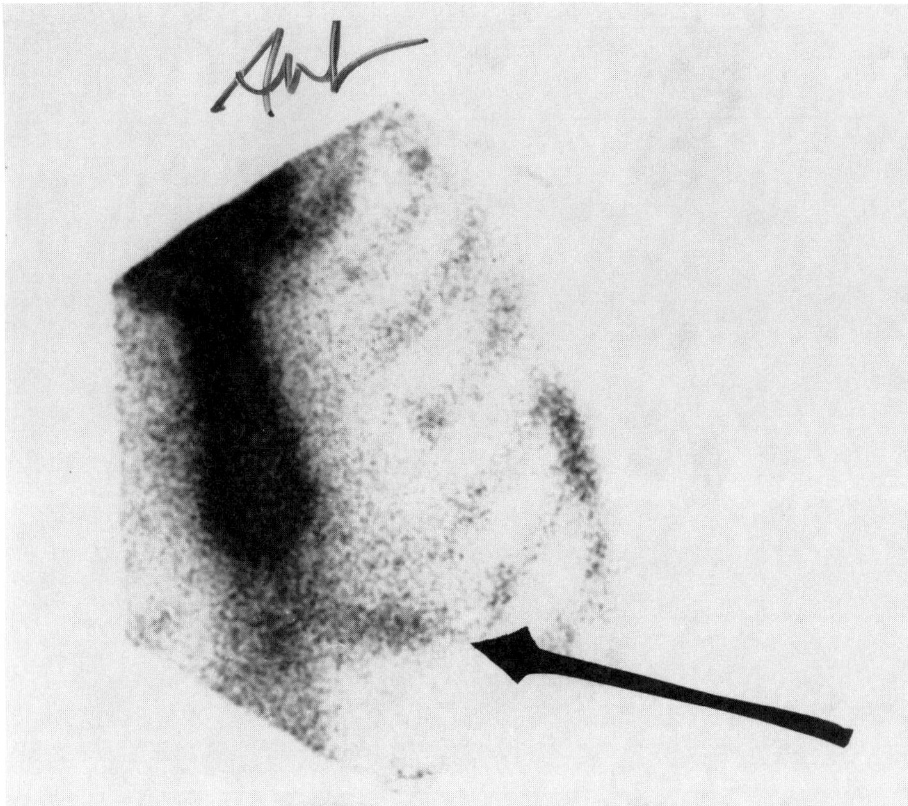

FIGURE 2. Image from a Tc-99m pyrophosphate study. Patient was a 50 year old male with a first episode of sharp sub-sternal chest pain and ECG changes consistent with acute myocardial infarct (MI). A left bundle branch block prevented localization of the MI. Anterior view shows a band of activity graded 3+ ("hotter" than rib) consistent with acute inferior MI.

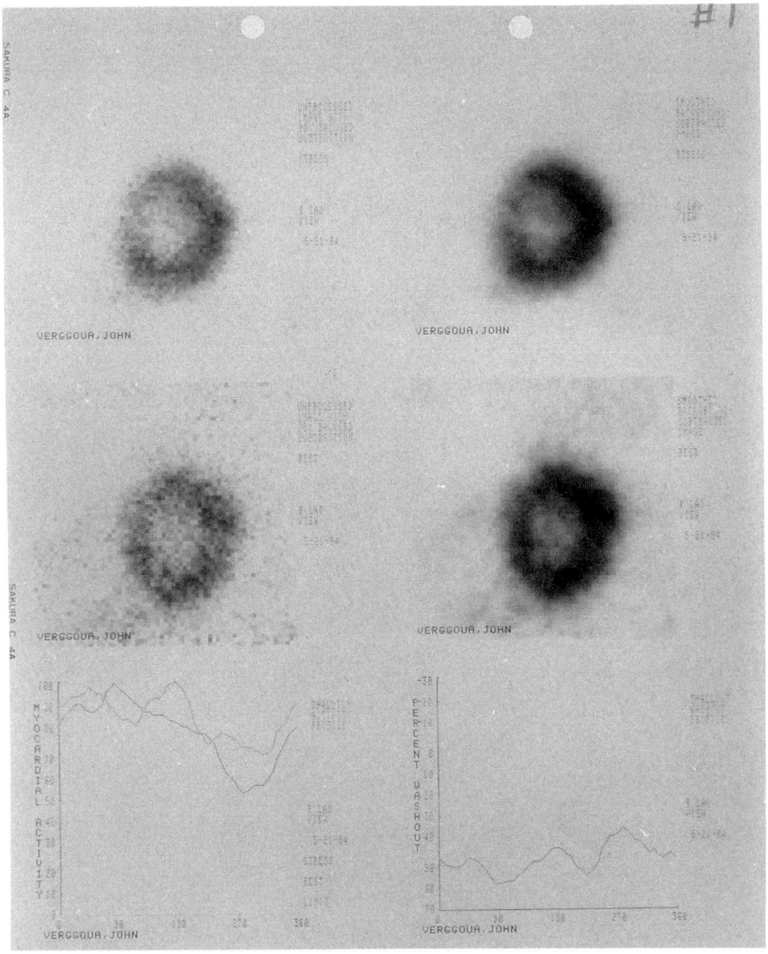

FIGURE 3. Stress and planar LAO images of the myocardium obtained with Tl-201. Patient was a 65 year old male with history of myocardial infarction in 1974. Top Row: Unprocessed (20% threshold) and processed (interpolative background subtraction) stress images. Middle Row: Unprocessed and processed resting images. Bottom Row: Circumferential profiles for stress and resting images and washout curve. Normal washout values are 50%.

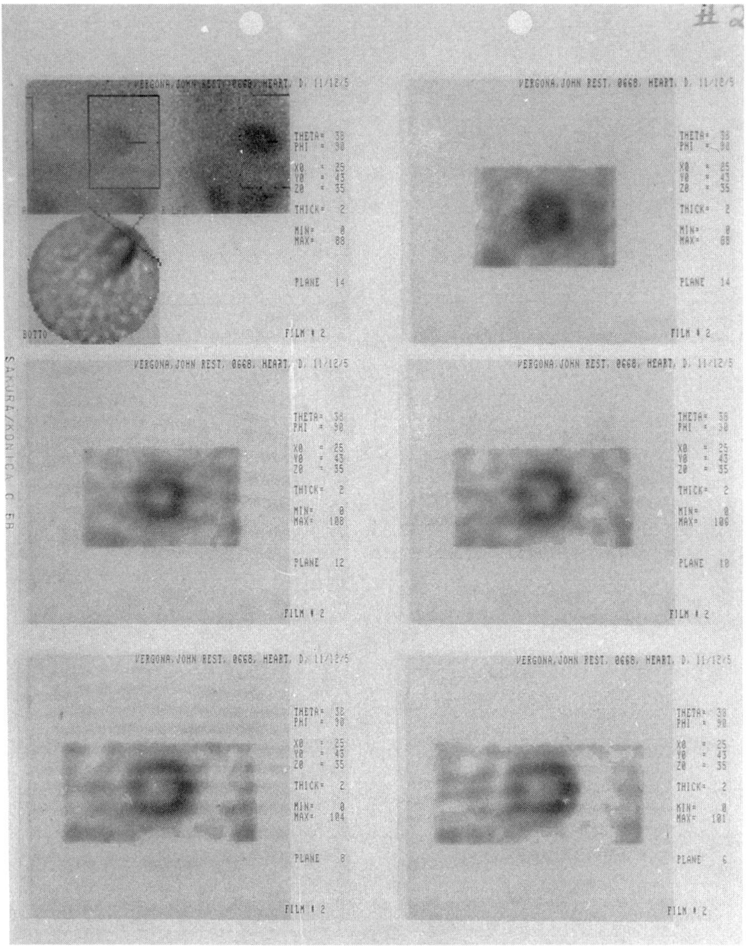

FIGURE 4. Reconstructed images from a Tl-201 SPECT study on th
patient described in Figure 3. Images have been transformed s
they are transverse to the long axis of the heart. Study
demonstrates minimally reversible myocardial perfusion defects
the septum.

OTHER NUCLEAR MEDICINE APPLICATIONS

L. Stephen Graham, Ph.D
V.A. Medical Center, Sepulveda, CA 91343
and
UCLA School of Medicine, Los Angeles, CA 90024

ABSTRACT

Renal function is evaluated with a number of different radiopharmaceuticals: Tc-99m DTPA, Tc-99m DSMA, I-131 and I-123 iodohippurate. Analysis may include visual inspection of static and dynamic images but primarily involves evaluation of time-activity curves generated from regions-of-interest set on the kidneys, bladder, and background. Functional parameters such as glomerular filtration rate (GFR), effective renal plasma flow (ERPF), and tubular transit time can be calculated from these data. Increasing use is now being made of studies to measure gastric function. Gastric emptying time can be calculated from a sequential series of images taken after oral administration of labeled foods and liquids. Single photon emission computed tomography (SPECT) studies of the heart, liver, skeleton, brain, and lungs are now routine in many large medical centers. Computers are used for acquisition and gantry control as well as reconstruction, image processing, and preparation of hard copy output. Positron emission tomography (PET) studies have found a wide variety of clinical applications in addition to being an invaluable tool in the study of basic physiology. Computers provide virtually the only means of quantitating field uniformity, spatial resolution, contrast resolution, noise, and temporal resolution of scintillation cameras. Recently there has been an increase in the use of computers to store and display data from instrument performance measurements as a function of time.

RENAL STUDIES

Studies of renal perfusion and function can be made with a number of different radiopharmaceuticals. Tc-99m DTPA and DSMA are commonly used to assess blood flow to the kidneys and to determine the functional status of the glomeruli as demonstrated by extraction of these agents from plasma. Static images of either of these agents provide good pictures of renal anatomy. Glomerular filtration rate (GFR) can be calculated by many methods including comparing the counts in each kidney between the second and third minute (corrected for depth) to the total number of counts injected[1]. This negates the need for collection of blood or urine samples.

One of the earliest dynamic studies performed with radionuclides utilized I-131 iodohippurate[2]. This agent is used to measure renal tubular function. Two NaI(Tl) probes with the appropriate electronics were interfaced to strip chart recorders and the observed counts recorded over a period of 30 minutes after the injection of the radiopharmaceutical. The basic principle of this study has not changed. However, the data are now collected with a scintillation camera interfaced to a computer and regions-of-interest (ROIs) are used to separate activity in the kidneys and bladder from activity in other structures. From time-activity curves a number of quantitative measurements can be made. These include time-to-maximum, full-width-at-half-maximum (FWHM), and time-to-fifty-percent clearance, etc. Figure 1 shows a typical set of curves. Effective renal plasma flow (ERPF) can be quantitated by comparing the counts in each kidney between the second and third minute (corrected for depth) to the total number of counts injected[3-5]. A nomogram is used to convert the calculated ratio to ERPF.

Some physicians find that a measurement of tubular transit time is useful. The calculation of mean transit time requires deconvolution of the time-activity curve from the kidney with an input function, namely the time-activity curve obtained from an ROI set over the aorta[6,7]. Because of the poor statistics in those data, it is normally necessary to filter the data before deconvolution[8]. If this step is not performed, the calculated function will show marked oscillations and contain negative values that have no physiological meaning. The selection of a filter is critical because minimal filtering will produce an unstable measurement while excessive filtering will modify the characteristics of the true function.

GASTRIC EMPTYING STUDIES

The use of radiopharmaceuticals to physiologically quantitate gastric function is relatively new[9,10]. Both radiolabeled solid foods (e.g., chicken liver) and liquids (e.g., water-DTPA) are used. When the emptying of solid food and water is measured simultaneously, two different labels are used. Usually the radionuclides are Tc-99m and In-111. The data set consists of a series of images obtained with the stomach in the center of the scintillation camera's field of view and measured over a period of two to three hours. Images are acquired every five minutes for the first 30 minutes, then every fifteen minutes for the remainder of the time. ROIs are set over the stomach with care taken to avoid including activity that has moved into the small intestine. Counts from each ROI are plotted as a function of time, corrected for physical decay, and the time for clearance of 50% of the activity calculated from a linear least squares fit of the data. Caution must be exercised in selecting the data for the curve fit because some patients exhibit an initial delay in the emptying pattern. This is the so-called "lag" phase[11].

Although the method described above answers many of the questions that are asked by the referring physician, it is often of value in basic physiology studies to have more accurate data. The simplest approach is to use the geometrical mean of anterior and posterior views. Accuracy can be improved by even further by making corrections for downscatter in dual isotope studies, for the anterior shift as the labeled material moves from the upper (fundus) to the lower part (antrum) of the stomach, and for the loss of counts that are in the stomach but which fall outside the ROI[12,13]. It is also possible to apply depth dependent convolution techniques to produce an accurate representation of the scatter from the high energy photons that falls in the low energy window[14]. Subtraction of the downscatter image from the image obtained with the low energy window yields an image devoid of scatter from the high energy component. Gastroenterologists have found this image helpful in following the shift of food particles of two different sizes within the stomach during digestion, i.e., from the fundus to the antrum.

EMISSION TOMOGRAPHY

Single photon emission computed tomography (SPECT) is being used with increasing frequency in nuclear medicine. The resurgence of brain imaging with I-123 iodoamphetamine (IMP) is due to the work of Kuhl et al[15] and Holman et al[16]. Kuhl et al and Cohen et al[17] demonstrated that the images obtained with IMP were essentially the same as those obtained with F-18 deoxyglucose, an agent well known to measure glucose metabolism in the brain[18]. The advantage of SPECT is the ability to use the imaging instrumentation that is readily available for routine clinical studies in many hospitals while positron emission tomography (PET) is limited to a few research centers.

Despite the promise of IMP, SPECT imaging with this agent is complicated by the relatively poor uptake of activity in the brain, approximately 7% at 20 minutes, and the limited amount of activity that can be administered because of the absorbed dose to the patient and cost considerations[17]. To minimize the problem of poor statistics it is useful to filter the images before reconstruction. Two dimensional Butterworth[19], Gaussian[20], and Metz[21-23] filters can be used. Preprocessing is then followed by reconstruction with a ramp filter.

The Metz filter is a form of resolution recovery. The mathematical form presented by King is:

$$M(f) = MTF(f)^{-1} \times \{1 - [1 - MTF(f)^2]^X\} \tag{1}$$

where X controls the extent to which the inverse filter (first term) is followed before the low-pass portion (second term) dominates. King et al optimized X by minimizing the normalized mean-squared error in images of a phantom[21]. In addition, the filter can be adjusted until it is approximately one when the blurred object power spectrum equals the 50% (or any value selected by the user) confidence limit for fluctuations in the noise power spectrum[24]. In essence this means the filter is adjusted to have a value of 1.0 when there is a 50% chance it would otherwise be amplifying frequency terms where the noise dominates the object. A comparison of the 2D Butterworth and the Metz filters on the same patient is shown in Figure 2.

Preliminary clinical tests are currently being performed with a radiopharmaceutical labeled with Tc-99m that can be used for imaging the brain[25]. Tc-99m HMPAO will allow the use of low energy collimators which provide high spatial resolution. However, despite the injection of much larger activity levels (typically 20 mCi) considerable noise is still present in the images and preprocessing with smoothing filters may still be necessary. In addition, blood background levels are often high.

Glucose metabolism in the brain is measured with F-18 deoxyglucose (FDG) for a wide variety of clinical conditions[18]. At UCLA FDG is used for the clinical study of patients with epilepsy. During the ictal state (seizure in progress) regions that are involved show high levels of glucose metabolism. In contrast, hypometabolism is seen in the same regions during the interictal state. FDG studies are used to identify these regions before neurosurgery.

FDG is also used to measure the extent of strokes and to study patients with brain tumors and Alzheimer's disease. Although the reconstructed images alone provide information about glucose metabolism, mathematical models are commonly used to calculate actual glucose metabolism on a mg per 100 cc/min basis. Also, PET studies with FDG have been used extensively to study basic physiology of the brain. Changes in glucose metabolism following visual, auditory, and language stimulation are a few of the areas that have been studied. Many C-11 labeled receptor binding agents are now being used to study neurotransmitters and opiate receptors, e.g., C-11 methyl spiperone[26].

SPECT studies are regularly used to image the liver, skeletal structures, and the lungs[27]. Emission studies of the liver involve relatively standard acquisition and processing but interpretation is difficult because of the variable and irregular shape of the liver. ECT of the skeleton generally requires the use of finer matrices (e.g. 128 x 128), more views (128), and some type of preprocessing to reduce noise. The lung is more difficult to image satisfactorily because of variable attenuation. Most of the currently available systems provide software that assumes constant attenuation.

QUALITY CONTROL

Probably the most underutilized application of computers in nuclear medicine is the ability to provide quantitative information about the performance of scintillation cameras. Some software is now provided by a few vendors but generally such software must be written by the user.

Field uniformity of scintillation cameras can be measured by a number of techniques, each having certain advantages and disadvantages. The calculation of integral and differential uniformity according to the NEMA protocol requires that each pixel in the 64 x 64 matrix have at least 4000 counts[28]. In a 38 cm diameter circular field camera 14 million counts are required to meet this condition. Fewer counts can be used, such as would be the case for routine quality control, but the calculated values will be correspondingly higher and cannot be compared to specifications published by the vendor. Field uniformity can also be calculated as the number of pixels within, or above and below, the mean $\pm$ a specified percent[29]. Figure 3 presents an example of images obtained by this method. Thresholding techniques can then be used to zero all pixels outside the specified limits and show the user where the pixels outside the limits are located. The percent of pixels inside the specified limits can be used as a measure of uniformity.

In my experience, the parameter that is most sensitive to changes in field uniformity is the "Uniformity Index" described by Cox and Diffey[30]. In essence it is the total variance minus the variance that is to be expected from Poisson statistics.

Stochastic and non-stochastic effects on field uniformity can be studied by the use of power spectra[31]. While this method is probably not useful for quality control, it may prove valuable for studying the differences that exist between cameras produced by different vendors and may prove useful in acceptance testing. Whichever method is selected for quantitating field uniformity, the calculated parameters should be displayed in graphic form so that variations in camera performance as a function of time can be monitored. At the present time only a few vendors make this option available as a part of the standard software package.

Although spatial resolution is an important parameter, it is usually evaluated by estimating the FWHM from the smallest bars that can be resolved in an image. The rule-of-thumb is that the FWHM is 1.8 times the width of the smallest bar that can be resolved[32]. Exact measurements of intrinsic spatial resolution can be made with a NEMA slit pattern; extrinsic values can be obtained using a line source such as a catheter filled with Tc-99m[33]. At the present time, only two vendors provide software for calculating the FWHM as a part of their standard software. However, the data can be obtained from any camera/computer system using keyboard "profile" or "slice" commands. The data can be listed on the screen or printed out, and the FWHM and FWTM calculated by hand. Alternatively, the user must write a program that will calculate these values and the modulation transfer function (MTF)[34].

Fourier analysis of bar pattern images is still another possible approach to quantitation of spatial resolution. Profiles across the bars in each quadrant of a four-quadrant test pattern can be Fourier transformed and the magnitude of the dominant frequency measured. This parameter is quite sensitive to small changes in camera spatial resolution.

Quality control of SPECT systems is discussed in a number of publications[35-37]. A phantom can be used to provide quantitative information about spatial resolution, contrast resolution, integral uniformity, and root-mean-square noise[38,39]. The "System Spatial Resolution Phantom" (Nuclear Associates) or any line source can be used to measure reconstructed spatial resolution. This is best done with a 128 x 128 matrix, 128 views, at least 50K counts per view, and reconstructed with a ramp filter[40]. Attenuation correction is not necessary. Critical evaluation of the results involves comparison of the FWHM (averaged in the X and Y direction in the reconstructed transverse section) with the FWHM of the same line measured at the same collimator-line distance as the average radius-of-rotation. The FWHM obtained from the SPECT study should be within ± 10% of the value obtained from the planar measurement.

Contrast resolution, RMS noise, and integral uniformity can be measured from data collected with a Data Spectrum (Jaszczak) phantom. The raw data should be collected with a 64 x 64 matrix, 64 views, 200K counts per view, and a 15 cm radius of rotation. Reconstruction should be done with a Hann filter using a 1 Nyquist cutoff. Integral uniformity can be calculated from a transverse section that has no structures. Contrast resolution can be calculated from the transverse section that contains the "cold" spheres. By maintaining records of the calculated values, changes in system performance can be monitored.

SUMMARY AND CONCLUSIONS

Software for clinical applications requires careful testing and must be thoroughly documented. Updates of software provided by vendors must also be checked against previously used software. This can often be done by maintaining a set of clinical studies to serve as benchmarks for the testing of modified or new software. It is also imperative that the physicians review results with a critical eye.

Greater use should be made of computers for the analysis of field uniformity and spatial resolution in scintillation cameras. Maintenance of a data base can be useful in monitoring instrument performance as a function of time. This methodology can, and should, be expanded to include tomography because of its complex nature.

REFERENCES

1. G. Gates, Clin. Nuc. Med. 8, 400 (1983).
2. G. V. Taplin, O. M. Meredith, H. Kade, et al, J. Lab & Clin. Med. 48, 886 (1956).
3. I. M. Thompson, F. G. Boineau, B. B. Evans, et al, J. Urol. 129, 461 (1983).
4. A. Chachati, A. Meyers, J. P. Godon, et al, J. Nucl. Med. 28, 829 (1987).
5. E. J. Fine, M. Axelrod, J. Gorkin, et al, J. Nucl. Med. 28, 1393 (1987).
6. J. Reeve and J. C. W. Crawley, Clin. Sci. Mol. Med. 47, 317 (1974).
7. R. W. Kenny, D. M. Ackery, and J. S. Fleming, Br. J. Radiol. 48, 481 (1975)
8. R. Puchal, J. Pavia, A. Gonzalez, et al, Phys. Med. Biol. 33, 831 (1988).
9. J. H. Meyer, I. L. GacGregor, R. Gueller, et al, Gastro. 61, 675 (1976).
10. R. A. Wright, D. Thompson, and I. Syed, J. Nucl. Med. 22, 772 (1981).
11. J. A. Siegel, J. L. Perri, R. A. Vitti, et al, J. Nucl. Med. 27, 903 (1986). Abstract.
12. G. VanDeventer, J. Thomson, L. S. Graham, et al, J. Nucl. Med. 24, 187 (1983).
13. J. H. Meyer, G. VanDeventer, L. S. Graham, et al, J. Nucl. Med. 24, 197 (1983).
14. L. S. Graham, V. Porter-Fink, J. H. Meyer, et al, Proc of First International Symposium on Applications of Computers in Nuclear Medicine, Oct 19-21, 1988. In Press. Abstract.
15. D. E. Kuhl, J. R. Barrio, S. C. Huang, et al, J. Nucl. Med. 23, 196 (1982).
16. B. L. Holman, R. G. L. Lee, T. C. Hill, et al, Arch. Neurol. 41, 1060 (1984).
17. M. B. Cohen, L. S. Graham, L. S. Yamada, Int. J. Radiat. Appl. Instrum. 37, 749 (1986).
18. M. E. Phelps, J. C. Mazziotta, and S. C. Huang, J. Cerebral Blood Flood Metab. 2, 113 (1982).
19. D. R. Gilland, B. M. W. Tsui, W. H. McCartney, et al, J. Nucl. Med. 29, 643 (1988).
20. M. T. Madsen and C. H. Park, J. Nucl. Med. 26, 395 (1985).
21. M. A. King, R. B. Schwinger, P. W. Doherty, et al, J. Nucl. Med. 25, 1234 (1984).
22. M. A. King, R. B. Schwinger, B. C. Penny, Med. Phy. 13,

139 (1986).
23. M. A. King, R. B. Schwinger, B. C. Penny, J. Nucl. Med. 29, 1327 (1986).
24. M. A. King, S. J. Glick, B. C. Penny, et al, J. Nucl. Med. 28, 1192 (1987).
25. R. A. Holmes, S. B. Chaplin, K. G. Royston, et al, Nuc. Med. Communications 6, 443 (1985).
26. H. N. Wagner, H. D. Burns, R. F. Dannals, et al, Science 221, 1264 (1983).
27. Sem. Nucl. Med. 17 (July 1987).
28. Performation Measurements of Scintillation Camera, National Electrical Manufacturers Association (NEMA) Standards Publication NU 1-1986 (NEMA, Washington, D.C., 1986).
29. J. W. Keyes, G. R. Gazella, and D. R. Strange, J. Nucl. Med. 13, 525 (1972).
30. N. J. Cox and B. L. Diffey, Brit. J. Radiol. 47, 820 (1974).
31. L. W. Grossman, M. P. Anderson, R. J. Jennings, et al, Phy. Med. Biol. 31, 941 (1986).
32. H. D. Royal, P. H. Brown, and B. C. Claunch, J. Nucl. Med. 20, 977 (1979).
33. R. K. Tyson and S. R. Amtey, Med. Phy. 5, 480 (1978).
34. J. R. Price, L. C. Wu, and H. Dawalibi, Eur. J. Nucl. Med. 8, 262 (1983).
35. B. Y. Croft, Single Photon Emission Computed Tomography (Yearbook Medical Publishers, Chicago, 1986).
36. R. J. English and S. E. Brown, SPECT Single-Photon Emission Computed Tomography: A Primer (The Society of Nuclear Medicine, New York, 1986), 148 pp.
37. R. F. Mould (ed), Quality Control of Nuclear Medicine Instrumentation, CRS-38. (The Hospital Physicists' Association, London, 1983), 169 pp.
38. L. S. Graham, Acceptance Testing of SPECT Systems, CEL (The Society of Nuclear Medicine, N. Y., 1988). Audio-visual.
39. Quality Control of Nuclear Medicine Instrumentation, TECDOC-317, Draft of Chapter 8, "Single Photon Emission Computerised Tomographic (SPECT) Systems Using Rotating Scintillation Cameras" (IAEA, Vienna, 1986).
40. Rotating Scintillation Camera SPECT Acceptance Tesing and Quality Control, AAPM Report No. 22 (American Institute of Physics, N. Y., 1987).

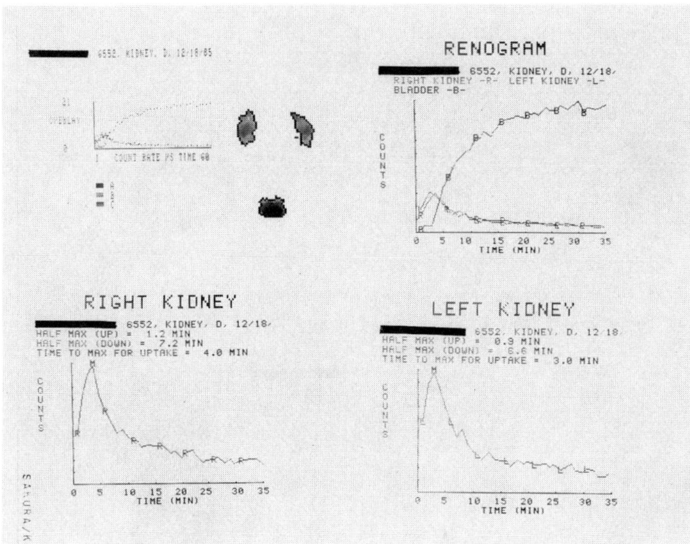

FIGURE 1. Data set for I-131 orthoiodohippurate renal study. Top Row: ROIs for kidneys and bladder and time-activity curves. Bottom Row: Individual kidney curves. Patient has hypertension and elevated serum creatinine. Conclusion: Normal renal function.

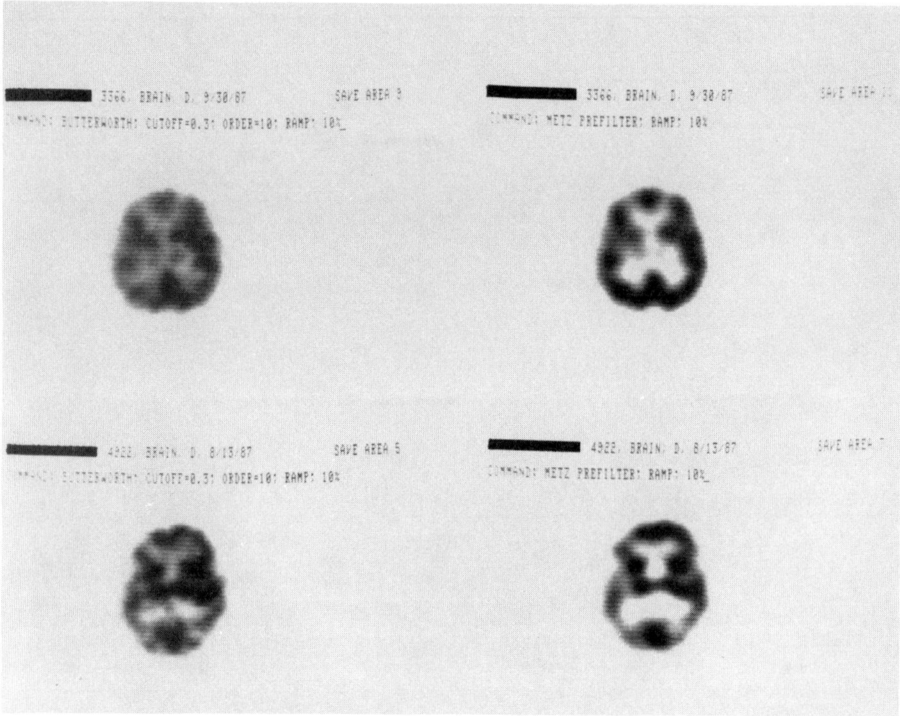

FIGURE 2. I-123 IMP SPECT brain images. The top row presents a transverse section from a normal patient. The bottom row present a transverse section from a patient with Alzheimer's disease as shown by decreased activity in the parietal cortex. Images in left column were preprocessed with a 2D Butterworth filter using a cutoff of 0.3/pixel and an order of 10. Images in the right column were preprocessed with an image dependent Metz filter using a confidence limit of 50%. Reconstruction was performed with a Ramp filter for both patients. A 10% lower threshold was applied to all images.

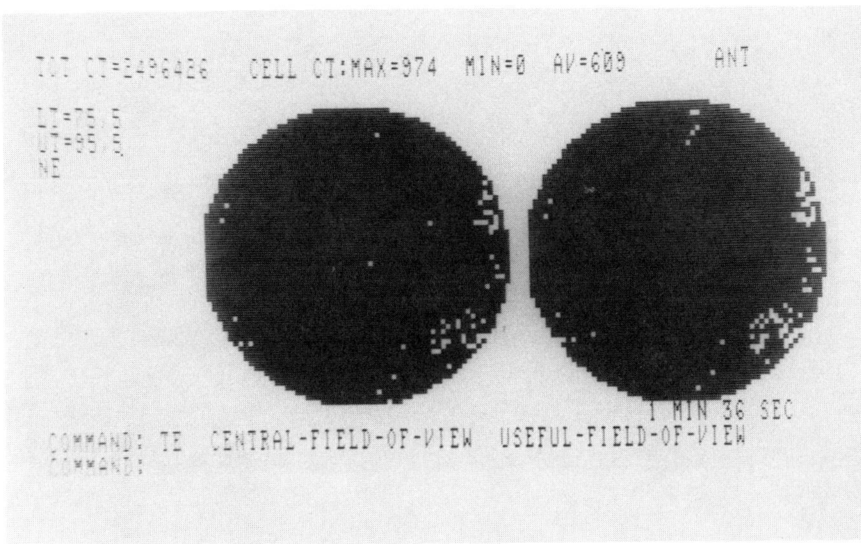

FIGURE 3. Computer measurement of extrinsic flood field uniformity. In the central-field-of-view (CFOV) 99.5% of the pixels fall within +/- 5% of the mean. In the useful-field-of-view (UFOV) 98.5% fall within +/-5% of the mean. The location of the pixels outside the limit are shown in white.

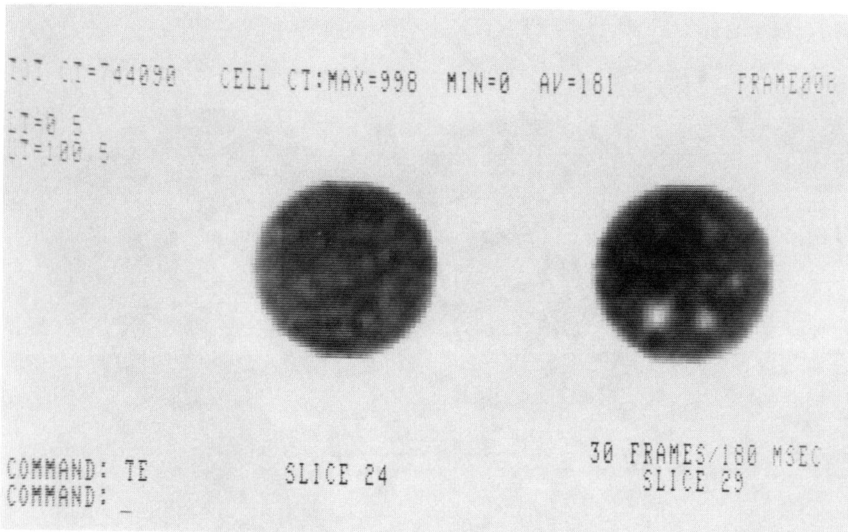

FIGURE 4. Reconstructed images of a Data Spectrum ("Jaszczak") phantom. Slice 24 is from a uniform section. Slice 29 is from a section that shows the cold spheres most clearly. ROIs can be used to quantitate uniformity, noise, and contrast.

III. APPLICATIONS IN RADIATION THERAPY

DEVELOPMENTS IN RADIATION THERAPY TREATMENT PLANNING

B.A. Fraass, Ph.D. and D.L. McShan, Ph.D.
Department of Radiation Oncology
University of Michigan Medical Center
Ann Arbor, Michigan 48109

INTRODUCTION

The use of computers to assist in the planning of radiation therapy treatments has become a standard of care throughout the U.S. and other countries. Since the first attempts published in 1955 (42), radiation therapy treatment planning (RTTP) has become increasingly sophisticated and complex as both computer hardware and computer software have become more capable. Several recent articles illustrate the progress which has been made over the last 30 years (23,29,32); a review is not repeated here.

In this paper, two aspects of recent work in treatment planning are presented. The first is a discussion of the expansion which has taken place in what the term "treatment planning" means. This phrase has typically been taken to mean determining radiation beam positions with respect to the patient's anatomy (often with only manual contour information available), and then performing dose calculations. In recent years, the term treatment planning has been expanded to include preparation, planning, optimization, evaluation and verification of the treatment plan with which the patient is treated. The second part of this paper is the characterization of the basic kinds of RTTP systems, i.e., two-dimensional planning systems three-dimensional systems.

A GENERALIZED DEFINITION OF TREATMENT PLANNING

Radiation therapy treatment planning (RTTP) has consisted of a number of standard steps for many years. First, one determined the patient outline using a mechanical method to trace the patient's external contour (plaster casts, solder wire). Internal structures of interest were entered onto that contour from geometrical measurements made from radiographs, and both sets of contours were entered into the computer using a digitizer tablet of some sort. The next step was to determine the beam technique to be used; the field size and isocenter (or SSD entrance point) were then entered and displayed. Finally, the dose distribution was calculated on that single slice, and the resulting isodose lines displayed.

In the last few years, treatment planning has become a much more elaborate and complex task. In order to maintain a full three-dimensional description of the patient, radiation beam, dose distribution, and to verify the accuracy of the implementation of the treatment plan, the following steps are now routinely part of the treatment planning process in some departments (23,

29, 41) :

- o Localization and immobilization. In order to treat a patient with any amount of reproducibility, the patient must be fully immobilized throughout all localization, imaging, and treatment procedures. This initial procedure (typically performed on the simulator) establishes the basic treatment position that will be used throughout all procedures, and uses various immobilization techniques to make it possible to re-establish that position for each procedure. At this same procedure, a basic coordinate system that will be used throughout the planning and treatment process is established. Typically this is done with a set of orthogonal radiographs which document a basic origin to the treatment coordinate system. This origin is defined also by placing marks on the patient that can be realigned to the alignment lasers which define the isocenter of the simulator or treatment machines. Immobilization is the subject of much work at the present time (6, 17,39).
- o CT scanning. Numerous papers have addressed the use of CT data for treatment planning (14,22). Once the patient has been marked and immobilized at the localization procedure, he/she can be set up on the CT table in the same way. Immobilization devices and patient position must be configured so that they may be used in the CT scanner without space, position, or artifact problems. The patient should be scanned in the treatment position, with a flat table top, and with markers to define the original coordinate system center that was determined in the original localization simulation (22). A large number of slices is often useful: between 30 and 60 slices are commonly obtained. In areas in which the anatomy is changing rapidly, such as the head and neck, the separation between slices may need to be as small as 1.5 mm (for the orbit, for example). In the abdomen and thorax, a large number of slices may be required to cover the area of interest, even if CT slice thicknesses as large as 1 cm are used.
- o Other imaging (MRI, PET, SPECT, Ultrasound, Angiograms, etc). In many clinical situations, the imaging information from more than one study is used to make diagnoses and define target volumes. Magnetic Resonance Imaging (MRI) is most commonly used, but PET, SPECT, and other studies are also of interest to the treatment planner. Although the methodology for the quantitative use of this information is in its infancy (see, for example, ref 10,15,37,40), these studies are clearly part of the planning process even when used only in a qualitative way.
- o Image correlation. If the description of the patient's anatomy is to be based on more than one imaging study (for example, CT and MRI), then the various imaging studies must be correlated geometrically with each other. This is a major research area at the present time, since people in all fields of imaging need the same kinds of correlation tools (see, for example, refs. 10,18,31). This is not new to RTTP. People have used CT and simulator films and manual contours for a long time; the difference now is that the quantitative tools which can actually perform correlation, and can measure and display the lack of consistency in patient imaging studies and treatments are now being incorporated in planning systems and in clinical

use.

- o Establishment of the 3-D description of the patient's anatomy. Design of target volumes and normal structure volumes must now be done in a three-dimensional way, with much more information at the physician's fingertips than was available in the past. The 3-D descriptions and display methods which are available (25, 33) make obvious the inconsistencies which inevitably occur when 2-D techniques (drawing contours on single slices) are used for volume definitions.
- o Geometrical plan optimization. This part of the planning process is well-known, as it involves orienting the radiation beams with respect to the target anatomy. This process has become much more complex, however, as the transition from a single 2-D slice to a full 3-dimensional display of both anatomy and beam orientation has become possible. In particular, the use of the Beam's Eye View (BEV) display (9,11,27) has added a great deal of information to the geometric part of the planning process.
- o Dosimetric plan optimization and evaluation. This part of treatment planning (i.e., dose calculations) has always been important. Dose calculation algorithms for photons (3) and electrons (16) have been reviewed, as have some methods for plan evaluation and optimization (13,19,24,25,29). In modern treatment planning, however, the calculation, optimization, and evaluation phases of planning are often integrated, conceptually if not physically. Much research work is directed toward the solution of the "inverse problem," in which the calculation, optimization, and plan evaluation are all included in the algorithm which creates the plan by starting with the desired solution(1,2).
- o Plan verification and treatment verification. Techniques for verification of the treatment plan are now an important part of the planning process, since the accuracy with which complex plans are implemented depends on the ability to document and verify that the plan can actually be applied as desired. Planning and documentation of the dose received by the patient are major functions of a treatment planning system. Verification of how the patient is actually treated is thus a part of the planning process, since any changes, errors, or uncertainties in the treatment of the patient should be folded into the planning for the radiation fields which follow the initial treatment. The importance of this subject is highlighted by the large number of investigators working in the area of verification (5,28) and real-time portal verification imaging (20,21,30,35).
- o Correlation of follow-up findings with plan dosimetry. One of the ways in which physicians learn how to treat patients more effectively is to correlate the results of previous treatments with the patient outcomes. As we gain more knowledge of the 3-D anatomy and the 3-D volumetric dose distribution, there is a possibility to improve the correlation of the results of patient treatments with the knowledge of how each individual patient was actually treated.

WHAT MAKES TREATMENT PLANNING THREE-DIMENSIONAL ?

Over the past several years, there has been much discussion of two-dimensional ("2-D") and three-dimensional ("3-D") treatment planning. As a way of illustrating the differences between the different capabilities of the above-named kinds of treatment planning, a classification scheme is proposed which lumps treatment planning into three basic categories. In addition to the 2-D and 3-D systems mentioned above, a "2.5-D" system is described which encompasses many of the "state-of-the-art" treatment planning systems which are available today.

1. BASIC 2-D RTTP SYSTEM

The components of what is defined here as a basic 2-D treatment planning system are listed in Table 1. The only patient anatomy information which is included in the planning is one or more contours which are digitized into the computer using a sonic or e/m digitizer system. External contours are typically obtained using solder wire, plaster, or a pin contour device while the patient is simulated. Internal anatomy may be outlined by using orthogonal films or other information to determine (usually by hand) the location of the structure(s) of interest with respect to the external contour. If internal inhomogeneities are determined, their density is assigned in a bulk fashion.

TABLE 1. BASIC 2-D RTTP SYSTEM

Anatomy:	1-3 manual contours entered with digitizer.
Beams:	2-D definition only. Beam in axial plane.
Dose Calcs:	1- or 2-D model only. Little scatter, blocks, or density correction sophistication.
Evaluation:	Isodose lines on contours.

The next step in the planning process typically involves defining the beams to be used to treat the patient. In a 2-D system, the beam description is typically a representation of the 2 beam edges which are visible on the axial contour through the central axis of the beam. Collimator angle and block edges will typically not be shown, and blocks are handled by changing the field size (and position) to mimic the edges of the shaped radiation field on the one particular plane which is investigated. The beam may only be moved in the axial plane, so collimator angle, table angle, and often the isocenter location in the "z" direction cannot be changed from the normal default positions. See Fig. 1.

Dose calculation models in a basic 2-D RTTP system are 1- or 2-dimensional in scope. The effects of shielding blocks, inhomogeneities, and body shape are taken into account only in a simple (typically 1-dimensional) way. The models rarely take into account any aspect of the third dimension ("z"). Although multiple contours sometimes may be used, the dose calculation to the "off-axis" planes treats them basically as if they were located at the central plane, and shows only the differences in the dose which are due to differences in the body contour. Dose display and evaluation are performed using isodose lines displayed one contour at a time (graphically or using a pen plotter).

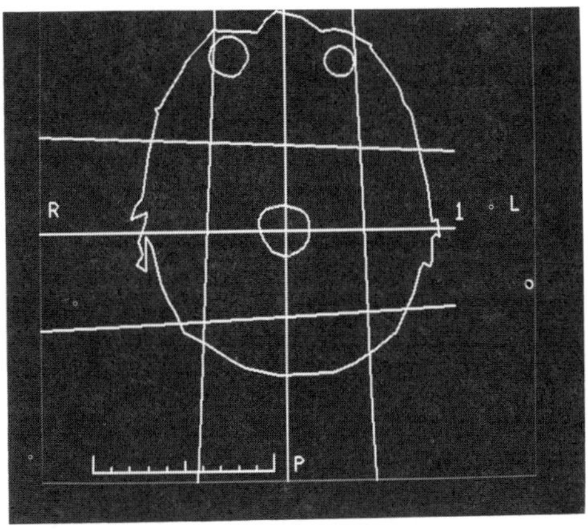

Fig. 1. 2-D Treatment Planning. Contours and beam outlines are shown for a simple 2-D field arrangement in a brain tumor case.

2. CT-BASED 2.5-D RTTP SYSTEM

Many commercial treatment planning systems now are more likely to be categorized in the CT-based 2.5-D category than in the more basic 2-D category. In this kind of system (summarized in Table 2), the description of the patient anatomy is in one way or another based on the anatomical information afforded by a set of CT scans. In many systems, the CT information can be read from a magnetic tape into the planning system, the CT images can be displayed on the RTTP system display monitor, and various methods can be used to outline (contour) the external contour of the patient as well as internal structures. In addition, the computer system then has access to the CT densities and can use them as part of the dose calculation algorithms. A somewhat less quantitative way to use the CT data is to optically display the CT images (one at a time) on a digitizer, and to then trace in the CT-based contours. In a true 2.5-D system, the system knows the relative location of all the CT slices in 3 dimensions.

The amount of beam definition information which is known to a 2.5-D system is much larger than that for a 2-D system. Beams can typically have rotated collimators, and block shapes are handled in a more complete way (than in a 2-D system). Since the anatomical information is presented in CT slices which have a known location in the direction orthogonal to the slice ("z" direction), then the projection of blocks and collimator edges is often handled in a more correct fashion than in a 2-D system (see Figure 2). Full divergent 3-D projection of all beam geometry for all 2-D slices is, however, still not usually accomplished.

TABLE 2. CT-BASED 2.5-D RTTP SYSTEM

Anatomy:	Contours defined using CT data, CT density information available.
Beams:	> 2-D definition. Blocks, collimator angle, other perturbations are addressed. Beam in axial plane.
Dose Calcs:	1- to 2.5-D calc models. 2.5-D ==> some aspects of geometry and/or dose calcs are addressed in the "z" direction. Sophisticated density corrections are possible.
Evaluation:	Isodose lines overlaid on CT images.

Dose calculations in a 2.5-D system make a general attempt to handle some of the 3-D aspects of the dose calculation problem. There are many levels of sophistication with which this is done, however. For example, many 2.5-D systems will incorporate a "z"-dependence into the dose profile calculation so that off-axis effects are taken into account, but a full 3-D divergent calculation of the geometry for the beam is often not done. Note that with this definition, much of the very early work in treatment planning was at least 2.5-D (or more) in scope (4,38,43). Handling of blocked field calculations can be done in a more complete way than the simple one-dimensional approach described above, since the beam description will contain most of the information needed to describe the block shapes and their basic effects of the dose for the plane(s) being calculated. The availability of the CT density information allows fairly sophisticated inhomogeneity correction methods to be implemented in the 2.5-D system (for example, ETAR (36)) . Dose calculations can be performed on multiple 2-D slices through the patient if multiple CT slices are entered into the system. Evaluation of the dose calculations is still based on isodose lines displayed on single slices (typically overlaid on a gray scale CT image). Dose-Area Histograms can sometimes be used as a first attempt to perform quantitative analysis of the dose distribution (see ref 7 for an example of a 2.5-D analysis method).

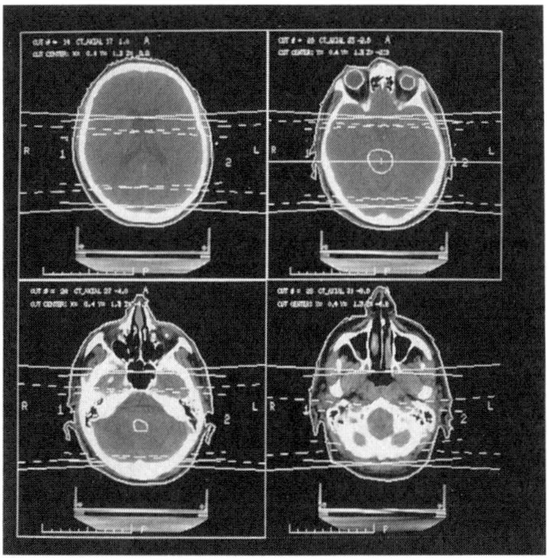

Fig.2. Illustration of 2.5-D Treatment Planning. This 4-panel display shows the lateral opposed beams (and block edges: dotted lines) used to treat the same brain tumor as Figure 1, but with the plan viewed on a number of different CT slices.

3. 3-D RTTP SYSTEM

3-D treatment planning has been the subject of much discussion over the last several years. Very few, if in fact any, existing treatment planning systems (commercial or otherwise) presently meet all of the requirements for "completely 3-D" treatment planning which are discussed below, although there are a few systems which satisfy most of the requirements. The basic requirements for 3-D RTTP are summarized in Table 3.

TABLE 3. 3-D RTTP SYSTEM

Anatomy:	3-D structure definitions. 3-D electron density description.
Beams:	3-D beam/block description. Blocks, collimator and table angle handled. Beam's Eye View and 3-D perspective views.
Dose Calcs:	3-D calc models. General approach to 3-D geometry. Full 3-D scatter included.
Evaluation:	3-D graphical and analytical tools.

Definition of the patient anatomy is typically based on a series of CT scans in 3-D treatment planning (between 20 and 100 slices are used clinically in our department). The description of the anatomy is not, however, simply a series of 2-D slices and contours. In a 3-D system, both the abstracted anatomical information (analogous with contours) and the density information (CT densities) must be fully 3-dimensional objects. In the case of most 3-D systems, contours of various parts of the anatomy are drawn on multiple 2-D slices (CT images) and then meshed into a full 3-D surface description (26, 33). This means, for example, that at the top and bottom of each structure, the system must be able to determine how to complete the 3-D description of the structure which is to be constructed from the multiple contours. Additionally, 3-D display of surfaces created from multiple planar contours illustrates quite quickly that the consistency of contour definition from slice to slice is not perfect (see Fig. 3), and tools are needed to "sculpt" the 3-D structures so that the final surface accurately represents the best understanding of the person who identified the structure.

3-D use of the imaging information is also important. If a number of CT slices are available, then they define a volume of CT data which should be able to be displayed in a number of different ways. In particular, the use of CT data reformatted into sagittal, coronal, and oblique planes is extremely valuable to the treatment planner. A second use for such data is the creation of a 3-D density grid, typically obtained from the CT densities. This 3-D density grid is needed for sophisticated 3-D dose calculation algorithms which attempt to take into account the actual density variation through the patient.

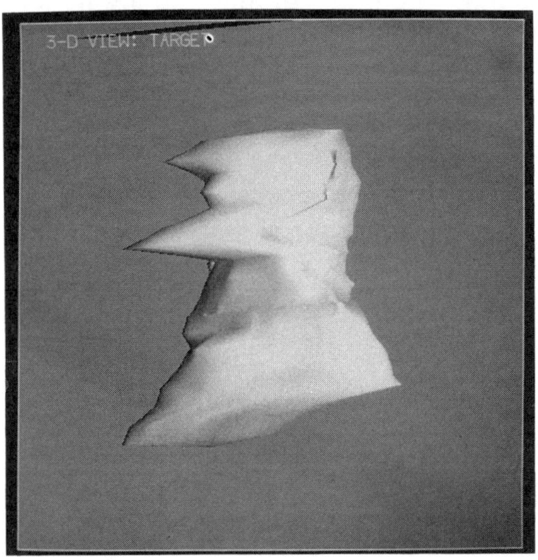

Fig. 3. Solid surface display of a mediastinal target volume outlined from serial CT scans. This surface was created from the original CT contours and has not been sculpted to remove jags in the surface which are due to inconsistencies in the contour drawing from one slice to the next.

One of the results of the quantitative use of the imaging information which is used to determine the anatomical description of the patient is that all the information available is "needed". This means that other information such as MRI scans, simulator films, and other studies must be quantitatively used inside the planning system. This forces the inclusion of the capability of integrating multiple sets of imaging information inside the system, for example, the concurrent use of CT and MRI (10). These various datasets must be geometrically correlated, and information defined on one dataset must be transferable to another.

In a fully 3-D system, beams and radiation sources must be defined completely in three dimensions. This means that all motions and adjustments available on a therapy machine must also be supported inside the planning system. In order to make it possible to set all of the parameters three-dimensionally, then a number of different display capabilities are needed. These include:

- o Fully-correct projection of beams, blocks, contours, etc., onto any 2-D slice information, including oblique reformatted CT images.
- o 3-D projection views of the anatomy and beams (see Fig. 4).
- o Beam's Eye View (BEV). This is a view of the anatomy displayed in a projective geometry as seen from the source of the radiation (the target of the linear accelerator).

Fig. 4. 3-D perspective view of a 3 field (including vertex field) plan used to treat the same patient shown in Figs. 1 and 2. Sagittal CT, coronal MR and axial MR images are shown.

It should be possible to change any machine, collimator, block shape or other parameter and see the correct relationships between machine, patient, and beam on all available display modes. In addition, the BEV capability leads directly to the use of the planning system for the design of field shapes (see ref 8 in this volume). If that block shape information is to be useful, then it must be possible to output it so that the block can be fabricated.

Fully three-dimensional dose calculations are the subject of a great deal of research work at the present time. The discussion here will be limited to the basic problems which must be addressed in such a calculation. While considering the list that follows, it is important to remember that increasing sophistication of the dose calculation algorithm is usually at odds with one of the basic tenets of "treatment planning", as opposed to "dose calculation". If one is "treatment planning", then the interactivity (speed) of the dose calculations is extremely important, because the amount of planning and plan optimization that can be performed is dependent on the time it takes to perform an iteration through the dose calculation/evaluation process. The decisions made by each set of RTTP system designers can reasonably be significantly different, as they trade off interactivity and dose calculation speed versus the accuracy of the dose calculation algorithm in the various situations it is expected to be used in. Fully three-dimensional dose calculations include:

1. Calculation on a 3-D grid of points.

2. Use of a 3-D description of the patient (i.e., use of 3-D surfaces and 3-D density grids as opposed to multiple 2-D slices).
3. Calculation of 3-D divergence effects in all geometrical calculations.
4. 3-D beam description contained in the calculational model.
5. Model accounts for 3-D patient shape (scatter as well as primary attenuation).
6. 3-D scatter effects from blocks (and transmission through blocks) are accounted for.
7. 3-D effects of inhomogeneities are accounted for.
8. Electron equilibrium and interface effects are included.

Dose evaluation capabilities must be fairly extensive in a 3-D system. Dose distributions should be displayed as three-dimensional isodose surfaces rather than isodose lines; and these 3-D dose surfaces should be able to be displayed with respect to both the 3-D surface anatomy as well as the 3-D CT image data. Isodose lines should be displayed on any 2-D plane (including oblique reformatted images). Dose volume histogram (DVH) analysis (12) is an extremely useful complement to the dose display features, since it summarizes the relationship of the dose to the defined structures used by the treatment planner. This DVH data can be used by several models which attempt to determine the probability of normal tissue complications and/or tumor control (19,24,34,44). In addition, it is very important that there be somewhat sophisticated plan comparison tools, since the complexity and amount of information available with 3-D plans are too large to allow simple side-by-side comparisons alone.

CONCLUSIONS

The definition of the term "treatment planning" has evolved a great deal over the last several years. The introduction of 3-dimensionality and quantitative use of different imaging modalities has resulted in a generalization of the term's definition. Now, definition of treatment position and immobilization, use of CT, MRI, and other imaging modalities, dose distribution display and analytical evaluation techniques, and even treatment verification and patient follow-up studies are contained within the scope of the term "treatment planning."

In connection with this evolution, a great deal of progress in the development and use of sophisticated treatment planning software has been made. In the latter part of this paper, an attempt is made to delineate some of the semantic definitions which cloud the area of "three dimensional treatment planning" work. 2-D, 2.5-D, and 3-D treatment planning are thus distinguished by various differences in the definition of patient anatomy, beams, dose calculations, and plan evaluation tools.

REFERENCES

1. NH Barth: An Inverse Problem in Radiation Therapy. Med Phys 14: 452, 1987 [Abstract].

2. A Brahme: Optimization of stationary and moving beam radiation therapy techniques. Radiotherapy and Oncology 12: 129-140, 1988.
3. JR Cunningham: Computer Algorithms for Photon Beams, in Radiation Oncology Physics - 1986, AAPM Medical Physics Monograph 15, AIP, New York,1987, pp. 512-531.
4. JR Cunningham, PN Shrivastava, JM Wilkinson: Program IRREG - Calculation of dose from Irregularly Shaped Radiation Beams, Computer Progr. Biomed 2: 192,1972.
5. PB Dunscombe, K Fox: The precision of determining compliance with prescribed fields from conventional portal films. Phys Med Biol 33 (Sup. 1):86, 1988 [Abstract].
6. A Dutreix: Keynote address: Prescription, precision, and decision in treatment planning. Int J Radiat Oncol Biol Phys 13:1291-1296, 1987.
7. BA Fraass, AS Lichter, DL McShan, BR Yanke, RF Diaz, KS Yeakel, J van de Geijn: The influence of lung density corrections on treatment planning for primary breast cancer. Int J Rad Onc Biol Phys 19:179-190, 1988.
8. BA Fraass, DL McShan, KJ Weeks: Computerized Field Shaping. this volume.
9. BA Fraass, DL McShan, KJ Weeks: 3-D Treatment Planning: III. Complete Beam's-Eye-View Planning Capabilities, The Use of Computers in Radiation Therapy, ed. IAD Bruinvis et al, Elsevier Science Publishers B.V., (North-Holland), 1987, p 193-196.
10. BA Fraass, DL McShan, RF Diaz, RK Ten Haken, A Aisen, S Gebarski, G Glazer, AS Lichter: Integration of magnetic resonance imaging into radiation therapy treatment planning. Int J Rad Onc Phys Biol 13:1897-1908, 1987.
11. M Goitein, M Abrams: Multi-dimensional Treatment Planning: II. Beam's-Eye-View, Back Projection, and Projection Through CT Sections. Int J Radiat Oncol Biol Phys 9:789-797, 1983.
12. M Goitein, T Miller: Planning Proton Therapy of the Eye, Med Phys 10:275 -283,1983.
13. M Goitein, M Urie, JE Munzenrider, R Gentry, et al: Evaluation of treatment planning for particle beam radiotherapy. Int J Radiat Oncol Biol Phys. (In Press).
14. M Goitein, J Wittenberg, M Mendiondo, J Doucetter, C Friedberg, J Ferrucci, L Gunderson, R Linggood, WU Shipley,and HV Fineberg: The Value of CT Scanning in Radiation Therapy Treatment Planning: A Prospective Study. Int. J. Rad. Onc. Biol. Phys. 5:1787-1798, 1979.
15. RM Henkelman, PY Poon, MJ Bronskill: Is Magnetic Resonance Imaging Useful for Radiation Therapy Planning. in Proceedings of the Eighth International Conference on the Use of Computers in Radiation Therapy. IEEE Computer Society. Toronto, Canada. 181-185,1984.
16. KR Hogstrom: Evaluation of Electron Pencil Beam Dose Calculations, in Radiation Oncology Physics - 1986, AAPM Medical Physics Monograph 15, AIP, New York,1987, pp. 532-561.
17. M Hunt, G Kutcher, C Burman, D Fass, L Harrison, S Leibel, Z Fuks: The

Effect of Positional Uncertainties on the Treatment of Primary Nasopharynx Cancer. Med Phys 16: 456, 1989 [Abstract].

18. ML Kessler, S Pitluck, GT Chen: Techniques and applications of image correlation in radiotherapy treatment planning. Front Radiat Ther Oncol 21:25-32, 1987 .

19. GJ Kutcher, C Burman, L Brewster, M Goitein, R Mohan: Histogram Reduction Method for calculating complication probabilities for 3D treatment planning evaluations. (in press) .

20. WC Lam, M Partomwah, DJ Lee, MD Wharam, KS Lam: On-Line Measurement of Field Placement Errors in External Beam Radiotherapy. Br J Radiol 60: 361-367, 1987.

21. J Leong: A Digital Image Processing System for High Energy X-ray Portal Images. Phys Med Biol 29: 1527-1535, 1986.

22. AS Lichter, BA Fraass, J van de Geijn, HA Fredrickson, E Glatstein: An Overview of Clinical Requirements and Clinical Utility of Computer Tomography (CT) - Based Radiotherapy Treatment Planning. Computer Tomography in Radiation Therapy, C.C. Ling, C.C. Rogers and R.S. Morton, editors. Raven Press, New York. pp. 1-21, 1983.

23. AS Lichter: Clinical practice of modern radiation therapy treatment planning. A Categorical Course in Radiation Therapy Treatment Planning. Editors, Paliwal BR, Griem ML, Radiological Society of North America, Chicago, pp. 7-12, 1986.

24. JT Lyman, AB Wolbarst: Optimization of radiation therapy III: A method of assessing complication probabilities from dose volume histograms. Int J Radiat Oncol Biol Phys 13:103-109, 1987.

25. DL McShan: Treatment Plan Evaluation and Optimization, in Syllabus: A Categorical Course in Radiation Therapy Treatment Planning, ed. B.R. Paliwal and M.L. Griem, RSNA Chicago,1986, pp. 33-39.

26. DL McShan, BA Fraass: 3-D treatment planning: II. Integration of gray scale images and solid surface graphics. The Use of Computers in Radiation Therapy, ed., IAD Bruinvis, et al. Elsevier Science Publishers B.V., (North-Holland), pp. 41-44, 1987.

27. DL McShan, DR Hauman, LE Reinstien, AS Glicksman: An Interactive Three-Dimensional Radiation Treatment Planning System. Br J Radiol 15(Suppl.):144-146,1979.

28. H Meertens, M van Herk, J Bijhold: First clinical experience with an ionization chamber device for portal imaging. Phys Med Biol 33 (Sup. 1):45, 1988 [Abstract].

29. R Mohan, LJ Brewster, LJ Chui, CS Kutcher, JS Laughlin, Z Fuks: A Comprehensive Three-Dimensional Radiation Treatment Planning System. Int J Radiat Oncol Biol Phys 15: 481-495,1988.

30. P Munro, JA Rawlinson: A Digital Fluoroscopic Imaging Device for Radiotherapy Localization. Phys Med Biol 33(Sup.1):45, 1988 [Abstract].

31. CA Pelizzari, GTY Chen: Registration of multiple diagnostic imaging scans using surface fitting. The Use of Computers in Radiation Therapy. I.A.D. Bruinvis et al. (Editors) Elsevier Science Publishers B.V. (North-Holland), pp 437-440, 1987.

32. JA Purdy, JW Wong, WB Harms: Treatment Planning Computers (Present and Future), in Radiation Oncology Physics - 1986, AAPM Medical Physics Monograph 15, AIP, New York,1987, pp. 495-511.
33. J Rosenman, GW Sherouse, H Fuchs, SM Pizer, AL Skinner, C Mosher, K Novins, JE Tepper: Three-dimensional Display Techniques in Radiation Therapy Treatment Planning, Int J Rad Oncol Biol Phys 16:263-269 (1989).
34. TE Schultheiss, CG Orton, and RA Peck: Models in Radiotherapy: Volume Effects. Med. Phys. 10:410-415 (1983).
35. S Shalev, T Lee, S Cosby, K Leszczynsk: On-line Digital Portal Imaging. Med Phys 14: 706, 1987 [Abstract].
36. MR Sontag, JR Cunningham: The equivalent tissue-air ratio method for making absorbed dose calculations in a heterogeneous medium. Radiology 129:787-794, 1978.
37. MR Sontag, JM Galvin, L Axel, P Bloch: The Use of NMR Images for Radiation Therapy Treatment Planning. Proceedings of the Eighth International Conference on the Use of Computers in Radiation Therapy. IEEE Computer Society. Toronto, Canada. pp.168-172, 1984.
38. TD Sterling, H Perry, L Katz: Automation of radiation treatment planning IV. Br J Radiol 37:544-550, 1964.
39. HD Suit, D Phil, J Becht, J Leong, M Stracher, WC Wood, L Verhey, M Goitein: Potential for improvement in radiation therapy. Int J Radiat Oncol Biol Phys 14: 777-786, 1988 .
40. RK Ten Haken, BR Yanke, A Aisen, BA Fraass, A Gerhardsson, AS Lichter, DL McShan, C Perez-Tamayo, AF Thornton: Design of MRI Scan Protocols for Use in 3-D, CT-Based Treatment Planning. Submitted to Int J Rad Oncol Biol Phys.
41. RK Ten Haken, C Perez-Tamayo, RJ Tesser, DL McShan, BA Fraass, AS Lichter: Boost Treatment of the Prostate Using Shaped, Fixed Fields. Int J Rad Oncol Bio Phys 16:193-200, 1989.
42. KC Tsien: The Application of Automatic Computing Machines to Radiation Treatment Planning. Brit J Radiol 28: 432-439, 1955.
43. J van de Geijn: The Computation of Two and Three Dimensional Dose Distributions in Cobalt-60 Teletherapy. Br. J. Radiol. 38: 369-377, 1965.
44. AB Wolbarst: Optimization of radiation therapy III: The critical voxel model. Int J Radiat Oncol Biol Phys, 10: 741-745, 1984.

CT-MRI TREATMENT PLANNING INTERFACE

C. W. Cheng, Ph.D.
Joint Center for Radiation Therapy, Department of Radiation Therapy, Harvard Medical School, Boston, MA 02115

ABSTRACT

Computed tomography (CT) has been used routinely in radiotherapy treatment planning (RTP) for some years. Patient positioning can be verified accurately using the technique of digitally reconstructed radiograph (DRR). Because of the excellent soft tissue contrast and the ability to image in arbitrarily oriented planes, magnetic resonance imaging (MRI) has been thought to be a valuable imaging tool for radiotherapy, in much the same way as CT. The use of MRI, however, is not as straightforward nor as well established as CT. The general consensus is that MRI can be an excellent complementary imaging tool to CT in RTP but is not advisable to use alone. In this manuscript, the DRR technique used in CT based treatment planning, the integration of MRI in RTP and the capabilities required for a full 3D treatment planning system will be reviewed.

INTRODUCTION

Accurate localization of the target volume and determination of its spatial relationship to adjacent normal tissues are essential parts of the dose optimization problem in RTP. Although the conventional method of localizing target volume and critical structures on simulation films using clinical information, surgical clips, etc., may be adequate in some cases, there are occasions in which simulation films alone do not contain enough information to allow the clinician to delineate the target volume or certain critical structures. CT has revolutionalized RTP over the past decade. It provides detailed anatomical information in relation to the target volume. One of the most important aspects of CT is its capability of showing the patient's anatomy in transverse cross-sections. The clinicians and treatment planners can choose the appropriate beam configuration to encompass the target volume and minimize the dose to normal tissues, with more accuracy and confidence. More recently, MRI has shown itself to be superior to CT in imaging of certain anatomical sites, particularly the central nervous system. The ability of MRI to produce images in arbitrarily oriented planes offers yet another option to clinicians to visualize tumor regions in a true three dimensional sense which is particularly useful in delineating the tumor extent.

The objective of this manuscript is to review the required capabilities of a treatment planning system to integrate CT and MRI in RTP. In the next section, the limitations of two-dimensional treatment planning will be briefly discussed. The impact and the use of CT and MRI in

RTP will then be considered, followed by a discussion on correlating CT and MR images in RTP. Finally, the feasibility, the current status and the future development of using CT and MRI in RTP will be discussed.

LIMITATIONS OF TWO-DIMENSIONAL TREATMENT PLANNING

When both the patient anatomy and the target volume change little throughout the treatment field, two dimensional (2D) dose calculation may be satisfactory. However, very often either the patient contour and/or the target volume vary in size and position. It is then necessary to carry out calculations in multiple planes to ensure coverage. Furthermore when a medical decision has to be made as to whether or not a region of concern should be treated because of the possibility of a complication, dose distributions calculated throughout the region are important information which a clinician can utilize. All these point to the necessity of three-dimensional (3D) treatment planning. By 3D planning, one usually relates it to 3D dose calculation. However, it also includes the acquiring of multiple contours, and not just the external but also target volume and critical structures in correct spatial relationship to the external contours.

CT BASED TREATMENT PLANNING

The conventional method of localizing target volume and critical structures is often done radiographically. Specifically, the clinician outlines the target volume and critical structures on simulation films based on the information acquired from the patient's diagnostic workup, the locations of the surgical clips (if available), the enhanced contrast of certain organs (e.g., IVP used for delineation of kidneys) and the anatomy (e.g., the position of the spinal cord). This appears to be adequate in some cases. However, there are occasions in which plane radiographs alone offer little help in the localization of tumor or critical organs. An example is the localization of kidneys without the use of IVP because the patient is allergic to the iodinated contrast agent. Other structures, such as pancreas and spleen, are difficult to delineate on a plane radiograph. Under such circumstances, CT is usually the preferred technique for solving the localization problem.

The impact of CT in RTP has been studied by a number of people[2-10]. Their results are in broad agreement with each other in that, depending on the site of the disease, approximately 50% of the patients studied had changes in their treatment plans as the results of CT information. CT is used in treatment planning for delineation of target volumes and critical structures as well as for tissue heterogeneity correction in the dose calculation. It is also used to obtain patient contours and design treatment portals.

There are two types of CT scans. A diagnostic scan, with the patient in the conventional diagnostic position on a curved CT couch and with suspended respiration for each scan, is not useful for treatment planning purpose. A CT scan for radiotherapy treatment planning must be carried out with (1) a flat table insert placed on top of the CT couch, and (2) the patient set up in the anticipated treatment position. For example in our institution, a lung patient would be scanned with the head in a foam cast, with both arms above the head and a triangular sponge under the knees.

The transfer of CT information to the actual treatment situation is the most critical step in these applications. Different methods have been employed to correlate CT and treatment information.[1,7,11-17] These techniques use skin marks for patient setup and identical laser systems in CT and treatment/simulator rooms for coordinate transfer. Due to movement of subcutaneous tissue, the positions of skin marks relative to internal structures may change. They are therefore not reliable reference points for patient positioning. To ensure that the patient's position is reproduced accurately on the simulator couch, one should compare the anatomy in both positions. This can be done using DRR from CT scans which can be carried out to an accuracy of 0.5 cm or better[18].

Briefly, the CT scans, including a scout view and the axial scans, are transferred to a treatment planning computer using magnetic tape and displayed on a graphic terminal. The clinician, after reviewing the scans, uses a trackball (for example) to trace out the target volume on as many slices as are necessary. Anatomical structures such as kidneys, spinal cord and rectums are also drawn, if necessary. DRR from the CT scans then can be made in the same longitudinal plane as anticipated for the simulation films[4]. It should be pointed out that the DRRs are reconstructed with the same beam divergence as that of the simulation films. This is a very important aspect of the DRR technique. Without the correct divergence, a direct comparison cannot be made between the simulation films and the corresponding DRR. The DRR can be generated at any gantry angle. Usually an anteroposterior (AP) and a lateral view are produced for setup verifica- tions. To facilitate comparison between simulation films and DRR, the central axis (C.A.) of a DRR is chosen so that it falls on a readily identifiable anatomical structure such as the edge of a vertebral column in a thorax scan. The contours drawn by the clinician on each slice can also be projected on the DRR as line contours. The DRR's are then transferred to the CT computer via magnetic tape. Hardcopy in the form of 8cm x 8cm radiographic film is obtained using a multiformat camera connected to the CT unit. These miniaturized images are then magnified photographically to a convenient size for use in simulation, usually with the same magnification factor as the simulation films so that a direct comparison can be made. These hard copies derived from CT then define the patient position, the target and all relevant anatomy to be used as reference during simulation.

During the simulation session, the patient is first placed on the simulator couch the same way as on the CT table. Skin marks on the patient are aligned with the simulator room lasers as closely as possible (Fig. 1a). A 10cm x 10cm field size is used for these preliminary setup procedures. By operating the simulator in the fluoroscopic mode and by adjusting the couch position, the center of the x-ray field can be placed on the same anatomical landmarks as the isocenter of the DRR (Fig. 1b). AP and lateral films are then taken to verify that the patient position is indeed correct. In case a discrepancy is observed between the DRR and the actual simulation films, the position of the patient must be adjusted and setup films taken again. Once the patient position is verified by matching the DRR and simulation films, the simulation of the treatment can proceed as usual. Digital readouts in the simulation room are used to obtain the exact translations of the couch. The treatment isocenter position in the CT coordinate system can therefore be determined (Fig. 1c).

The technique of DRR is illustrated with a lung case. For the planning of lung cancer, the CT spans the entire lung volume. For the lung case discussed here the gross disease region was scanned with 0.5 cm spacing and the superior and inferior margins were scanned with 1.0 cm spacing. As shown in Fig. 2a, the isocenter on the DRR is at the bottom of T5 and on the right edge of the vertebral column in the AP projection. The superior edge of the 10cm x 10cm field was near the bottom of T3. Using this information the patient can be set up easily in the x- and y-direction from the DRR in the AP view. In Fig. 2b, the isocenter was on the anterior aspect of the vertebral column, and the posterior edge of the 10cm x 10cm field was on the spinal process. Although the lateral fluoroscopic image of the patient had rather poor contrast, only the height of the couch needed to be adjusted to set up the patient in the z-direction. This was done by rotating the gantry 90° and comparing the field edges on the lateral projection with the corresponding field edges on the fluoroscope. The setup films (Fig. 3) verified that the patient position was set up correctly. Since oblique portals were used in the treatment, DRRs at these oblique angles with target volume, vertebral column and cord superimposed were made after the simulation (Fig. 4a). Shielding blocks were then designed based on these projections (Fig. 4b).

MRI IN RADIOTHERAPY

The superior soft tissue contrast of MRI and its capability of producing images in arbitrarily oriented planes enable the clinician to define the target volume and critical structures with great accuracy. This is particularly true for treatment of brain tumors. On the other hand, there are practical problems associated with the use of MRI in RTP. (1) The nonuniformity of the magnetic field gives rise to geometric distortion near the periphery of the patient. (2) Patient motion artifacts are not uncommon due to the long scan time. (3) Electron density information is not available from MRI scans for inhomogeneity correction in the dose calculation. (4) Bone is poorly visualized in MRI scans, thus rendering both localization and verification difficult based on bony landmarks.

The use of MRI in radiotherapy treatment planning has been studied qualitatively by a number of people[19-21]. Shuman et al[22] have reported the successful use of MRI in treatment planning, and more recently Schad et al[23] have employed MRI, CT and PET in the treatment planning of brain tumors. The general consensus is that MRI can be used complementary to CT in RTP but should not be used alone because of the disadvantages mentioned above.

As in CT based treatment planning, the patient is set up in the treatment position on a flat table insert with the same immobilization device as in other procedures, if possible. However, unlike the DRR technique used in CT in which skin marks are secondary reference marks, the placement of skin markers is critical in MRI and catheters with strong MR signals should be used. This is because the skin marks are used in addition to anatomical structures to correlate CT and MR images. The catheters used in MRI should be placed at the same positions as those used in CT and additional markers are usually required for the non-axial images. Fig. 5 illustrates the differences in catheter placements in CT and MRI. Image transfer to the treatment planning computer is also via magnetic tape. The clinician again will draw in the target volume and critical structures on the MR images. The similarities between CT based RTP and MRI in RTP, however, stop here.

The capability of MRI to produce images in axial, coronal and sagittal planes demands additional manipulation of the images in RTP application. For example, a coronal scan may be obtained 40 minutes after the initial axial scan and can easily be geometrically inconsistent with the axial images as it is difficult for a patient to remain immobile through such a long period of time. In MRI, different pulse sequences are used to image the same tissue in different image planes. These images may then have different signal characterisics. To use the MR images in RTP, the first step is to generate a geometrically correct data set with consistent gray scale assignment. The method has been reported in great detail in Reference 24 by of Fraass et al. Briefly, the axial, sagittal and coronal images are aligned so that they follow the correct geometrical relationships to each other. Specifically, the contours obtained from one set of image data, e.g., axial scans, are overlayed on top of the image data from another set, e.g., sagittal scans; translation and/or rotation of a slice or a set of slices with respect to the other set can be carried out interactively to correct for the inconsistencies between the two data sets. Similar operations can be performed to check and force the coronal scans to be consistent with the other two image sets. This procedure is to ensure that the anatomical information obtained from the different image planes is properly aligned and correct. An example of the overlay capability of the planning system can be found in Reference 24. Gray scale reassignment is then carried out to generate a consistent MR image set with the appropriate gray scale. Having generated a self-consistent set of image data, the next step is to correlate the MRI data set with the corresponding CT images.

IMAGE CORRELATION

The objective in image correlation is to identify the MR images with corresponding CT scans so that the additional information available from MRI can be transferred to the analogous CT slices.

Image correlation can be significantly simplified if a careful planning of the sequence of procedures is made prior to MRI and CT scans. This includes the choice of appropriate immobilization device and the determination of the positions of the skin marks. If skin marks are placed so that one can identify which axial MR image corresponds to which CT image, then it is relatively straightforward to overlay the contours from the MR image set on top of those from CT and use translation and/or rotation to interactively align the two sets of images. An example of comparing CT and MR images for a brain tumor is shown in Fig. 6. The contours of the target volume and normal structures which can be easily seen on the MR image are transferred to the analogous CT slice. Alternatively, when internal structures are easily identified in both CT and MR image sets, it is most effective to match the two sets of contours by a combination of operations such as image overlay, contour rescaling and least-squares fitting to minimize the deviations from all the matching points or 'tie points' as described in Reference 24 (Fig. 7). Because of the potential geometric distortion in MR images, CT-derived anatomy should be used as the geometrically accurate template in the least-squares fitting procedure.

When it is not possible to correspond an MR and a CT image slice by slice, or when the imaging planes of the two studies are not parallel, e.g., the scans may be from different institutions, 3D image correlation techniques such as point to point fitting and surface fitting[25,26] can be used. Once this is done, the imaging information obtained from the MRI can be transferred to the CT scan. For example, by resectioning the surface from MRI scans along the planes defined by CT using the transformation parameters obtained in the surface fitting method, structures outlined in the MR scans can be transferred to the corresponding CT slices. DRR can be reconstructed from the CT scans, with accurately defined tumor extent and normal structures obtained from both CT and MRI. The treatment planning procedures can then be carried out as illustrated above.

DISCUSSION

We have reviewed the techniques in using CT and MRI in RTP. The technique of using CT in RTP is well established. CT is used in RTP both for localization and inhomogeneity correction. On the other hand, it is not advisable to use MRI alone in RTP because of the limitations inherent in MR images. However its superior soft tissue contrast compared to CT images, and its capability of producing images in arbitrarily oriented planes, suggest that MRI is an excellent technique complement to CT in the localization problem. Appropriate immobilization and careful skin marking are prerequisites for meaningful integration of both CT and MRI data in RTP. The planning system must be able to accept both CT and MR images. Capabilities of producing DRR from CT scans as well as comparing and manipulating CT and MR images and performing the surface fitting procedure for image correlation are also required for the planning system. While we have not addressed the problem in 3D dose calculation, a planning system capable of carrying out all the image acquisition, manipulation and correlation described above should necessarily incorporate in it the capability of 3D beam positioning and dose calculation in arbitrarily oriented planes with CT based inhomogeneity correction. A fully 3D treatment planning system can then be realized.

CT is now routinely used in RTP in most clinics. However, unless DRRs from CT scans are used to verify patient positioning, skin marks alone are not sufficient to guarantee that the patient's CT position is reproduced on the simulator couch. At our institution, the DRR technique is used routinely in the CT based RTP for different treatment sites. Our experience indicates that anatomy is the only reliable reference for patient positioning. The DRR technique described above can be streamlined further if a dedicated multi-format camera is available and hardwired to the planning computer. Hard copies of the DRRs can be produced without the extra step of transferring the DRRs to the CT computer via magnetic tape.

Unlike CT, MRI is still not widely used in the clinics for treatment planning purposes, probably because of its cost. Also the disadvantages associated with the use of MRI in RTP may be part of the reasons. Image distortion due to the magnetic field anisotropy can be quantified and characterized with phantom measurements[29], and can be corrected with special algorithms.[24,27,28] However, severe distortion has been observed for certain MR images, such as those in the thorax and in the pelvic region. A full 3D correction algorithm is required to restore the image quality efficiently. Although in diagnostic MRI scans, a 0.25 cm slice thickness is possible, in therapeutic applications 0.5 cm or 1.0 cm thickness is used instead, to shorten the scan time. Even then, the imaging time, in the multi-slice format, is of the order of 2-3 minutes for 0.5 cm slice thickness. This is long compared to CT scans. Since improvements in signal-to-noise ratio (SNR) could lead to equivalent image quality with a shorter imaging time, the use of magnetic field strengths of 0.3 Tesla or higher appears to be a solution to reduce the imaging time.[30] Finally, the failure of MRI to image bone is a big handicap in its application in the localization problem for some specific sites, such as the thorax and the pelvis. Very often bony structures provide the only reliable landmarks for verification of patient setup, e.g., between a simulation film and a DRR or a portal film. The absence of bony structure in MR images greatly reduces their value and requires extra steps in the image correlation process, particularly in transferring MRI data from sagittal or coronal planes to the CT scans. Future development should include more efficient algorithms in the image correlation process.

ACKNOWLEDGEMENT

The author would like to thank Dr. Fraass for providing the MRI figures and Drs. Langer, Siddon and Svensson for reviewing the manuscript.

Figures 1, 2 ,3 and 4 are reprinted with permission from Int. J. Rad. Oncol. Biol. Phys. vol. 13, Cheng el al. "A coordinate transfer of anatomical information from CT to treatment simulation", 1987, Pergamon Press plc.

Figures 5, 6 an 7 are reprinted with permission from Int. J. Rad. Oncol. Biol. Phys. vol. 13, Fraass et al. "Integration of magnetic resonance imaging into radiation therapy treatment planning: I. Technical considerations", 1987, Pergamon Press plc.

REFERENCES

1. M. Goitein, Med. Phys. 4, 580, (1982).

2. H.J. Dobbs, R.P. Parkers, Clin. RAdiol. 35, 433, (1984).

3. B. Emami, A. Melo, B.L. Carter, J.E. Munzenrider, A.J. Piro, Am. J. Roentgenol. 131, 63, (1978).

4. M. Goitein, M. Abrams, M. Rowell, Pollari, D., J. Wiles, Int. J. Rad. Oncol. Biol. Phys. 9, 789, (1983).

5. M. Goitein, J. Wittenberg, M. Mendiondo, J. Doucette, C. Friedberg, J. Ferrucci, L. Gunderson, R. Linggood, W.U. Shipley, H.V. Finberg, Int. J. Rad. Oncol. Biol. Phys. 5, 1781, (1979).

6. P. Hobday, N.J. Hodson, J. Husband, R.P. Parker, J.S. MacDonald, Radiolog. 133, 477, (1979).

7. J.E. Munzenrider, M. Pilepich, J.B. Rene-Ferrero, I. Tchakarova, B.L. Carter, Cancer 40, 170, (1977)

8. D.P. Ragan, C.A. Perez, Am. J. Roentgenol. 131, 75, (1978).

9. R.I. Rothwell, D.V. Ash, W.G. Jones, Clin. Radiol. 34, 103, (1983).

10. R.J. Stanley, S.S. Sagel, R.G. Levitt, Am. J. Roentgenol. 127, 53, (1976).

11. A. Breit, H. Gfittner, A. Atzinger, Cancer 48, 1341, (1981).

12. M.T. Gillin, W.D. Foley, R.W. Kline, L.E. Kun, Int. J. Rad. Oncol. Biol. Phys. 7, 1603, (1981).

13. B.R. Griffin, W.P. Shuman, K.H. Luk, D. Tong, Int. J. Rad. Oncol. Biol. Phys. 10, 555, (1984).

14. G.L. Jelden, E.S. Chernak, A. Rodriguez-Antunez, J.R. Haaga, P.S. Lavik, R.S. Dhaliwal, Am. J. Roentgenol. 127, 179, (1976).

15. M. Tatcher, S. Elgrabli, A. Kuten, U. Kleinhaus, Radiology 136, 236, (1980).

16. R.A. Zimmerman, L.T. Bilanluk, G. Grundy, P. Littmen, Radiology 119, 230, (1976).

17. D.R. Haynor, A.W. Borning, B.A. Griffin, J.P. Jacky, I.J. Kalet, W.P. Shuman, Radiology 158, 537, (1986).

18. C.W. Cheng, L.M. Chin, P.K. Kijewski, Int. J. Rad. Oncol. Biol. Phys. 13, 1559, (1987).

19. M.R. Sontag, J.M. Galvin, L. Axel, P. Bloch, Proc. Eighth Int. Conf. on the Use of Computers in Rad. Therapy, Toronto, Canada 168, (1984).

20. C.W. Coffey, H.C. Hines, P.C. Wang, S.L. Smith, Proc. Eighth Int. Conf. on the Use of Computers in Rad. Therapy, Toronto, Canada 173, (1984).

21. R.M. Henkelman, P.Y. Poon, M.J. Bronskill, Proc. Eighth Int. Conf. on the Use of Computers in Rad. Therapy, Toronto, Canada 181, (1984).

22. W.P. Shuman, B.R. Griffin, D.R. Haynor, J.S. Johnson, D.C. Jones, L.D. Cromwell, A.A. Moss, Radiology 156, 143, (1985).

23. L.R. Schad, R. Boesecke, W. Schlegel, G.H. Hartmann, V. Sturm, L.G. Struss, W.J. Lorentz, J. Comp. Ass. Tomog. 11, 948, (1987).

24. B.A. Frass, D.L. McShan, R.F. Diaz, R.K. Ten Haken, A. Aisen, S. Gebarski, G. Glazer, A.S. Lichter, Int. J. Rad. Oncol. Biol. Phys. 13, 1897, (1987).

25. C.A. Pelizzari, G.T.Y. Chen, Proc. Ninth Int. Conf. on the Use of Computers in Rad. Therapy, Scheveningen, the Netherlands, 437, (1987).

26. M. Kessler, Proc. Ninth Int. Conf. on the Use of Computers in Rad. Therapy, Scheveningen, The Netherlands, 441, (1987).

27. C.M. Lai, J. Phys. E. Sci. Instrum. 15, 1093, (1982).
28. K. Sekihara, M.Kuroda, H. Kohno, Phys. Med. Biol. 29, 15, (1984).

29. M.M. Covell, D.O. Hearshen, P.L.Carson, T.P. Chenevert, P.Shreve, A.M. Aisen, F.L. Bookstein, B.W. Murphy, W. Martel, Med. Phys. 13, 815, (1986).

30. L.E. Crooks, M. Arakawa, J. Hoenninger, B. McCarten, J. Watts, L. Kaufman, Radiology 151, 127, (1984).

FIGURE CAPTIONS

Fig. 1 Coordinate transfer from CT to simulator and vice versa. R is the origin of the reference coordinate system. P is the position of the projection center and I is the position of the isocenter.

Fig. 2 AP and lateral projections reconstructed from CT of the lung case.

Fig. 3 AP and lateral setup films of the lung case.

Fig. 4 An oblique view of the target volume and other critical structures reconstructed from CT (top). Design of the shielding blocks on the planning film based on the oblique projection (bottom).

Fig. 5 Schematic illustration of the differences in catheter placement.

Fig. 6 Comparison of the CT and MR imaging for brain tumors. Target and normal structures outlined on the MR image (top) are transferred to the analogous CT slice (bottom).

Fig. 7 2-D image unwarping procedure, see Reference 24 for details.

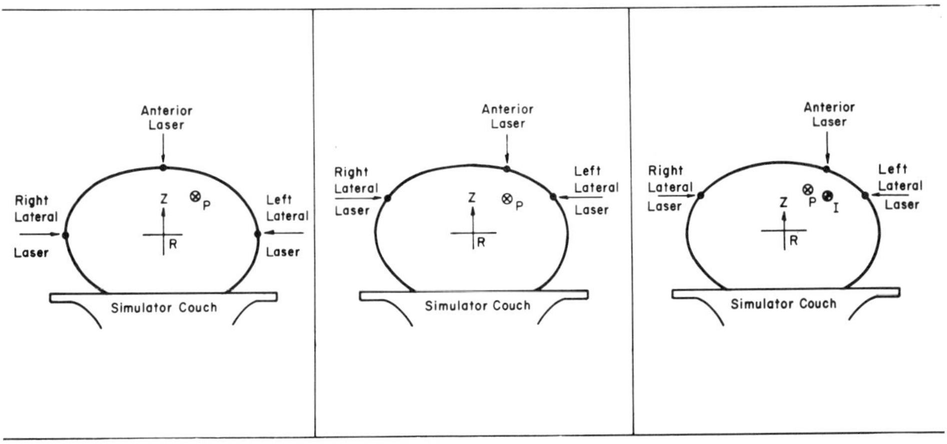

Fig. 1 Coordinate transfer from CT to simulator and vice versa. R is the origin of the reference coordinate system. P is the position of the projection center, and I is the position of the isocenter.

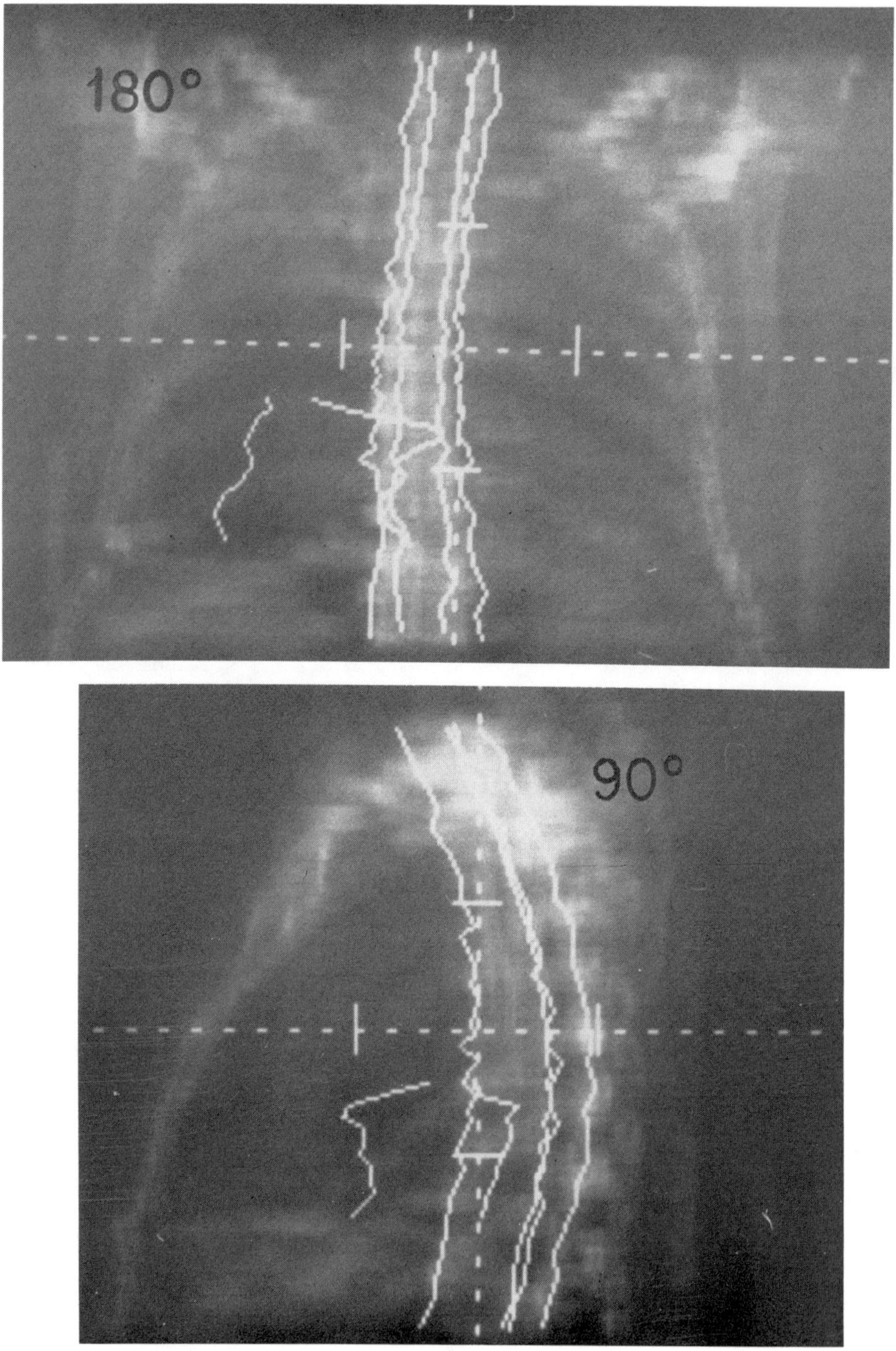

Fig. 2 AP and lateral projections reconstructed from CT of the lung case.

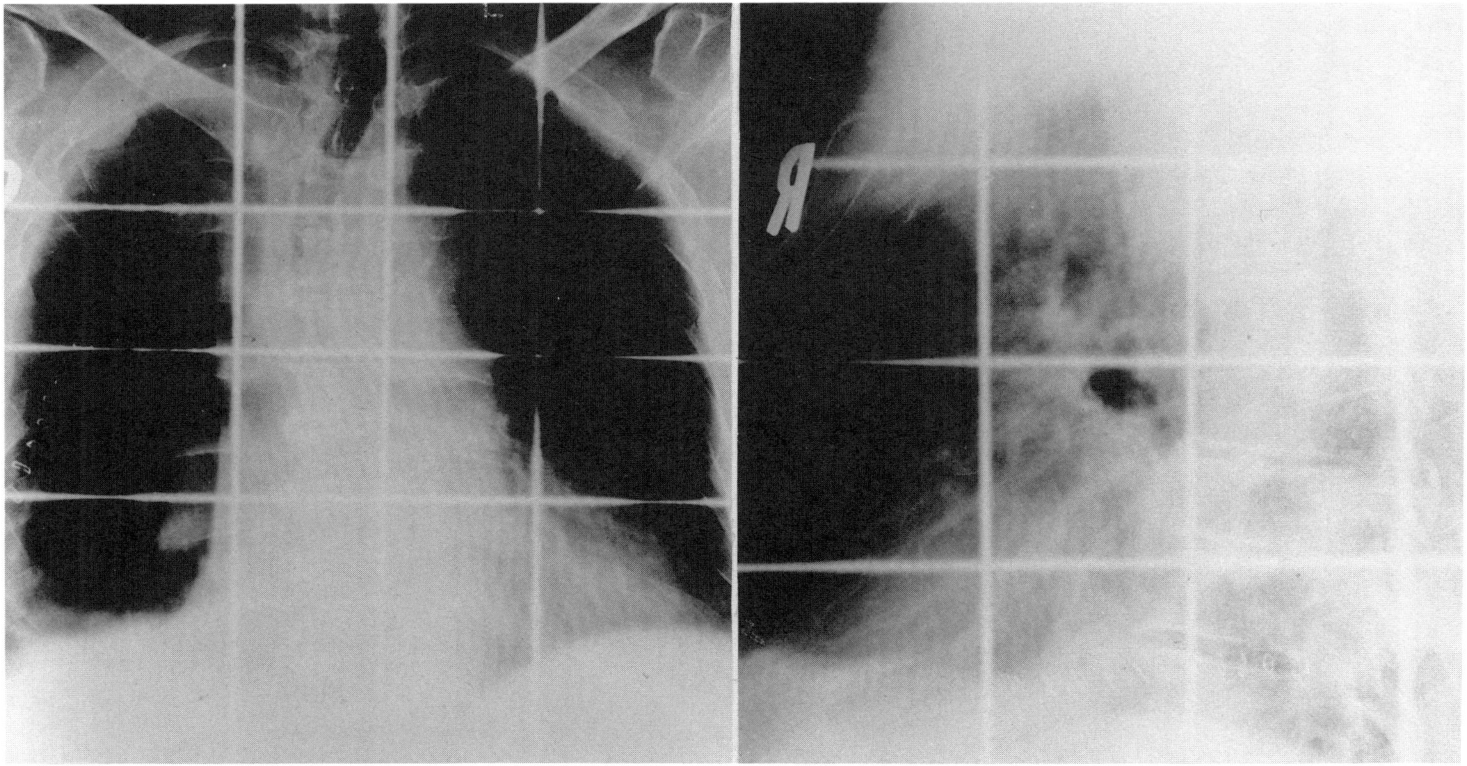

Fig. 3 AP and lateral setup films of the lung case.

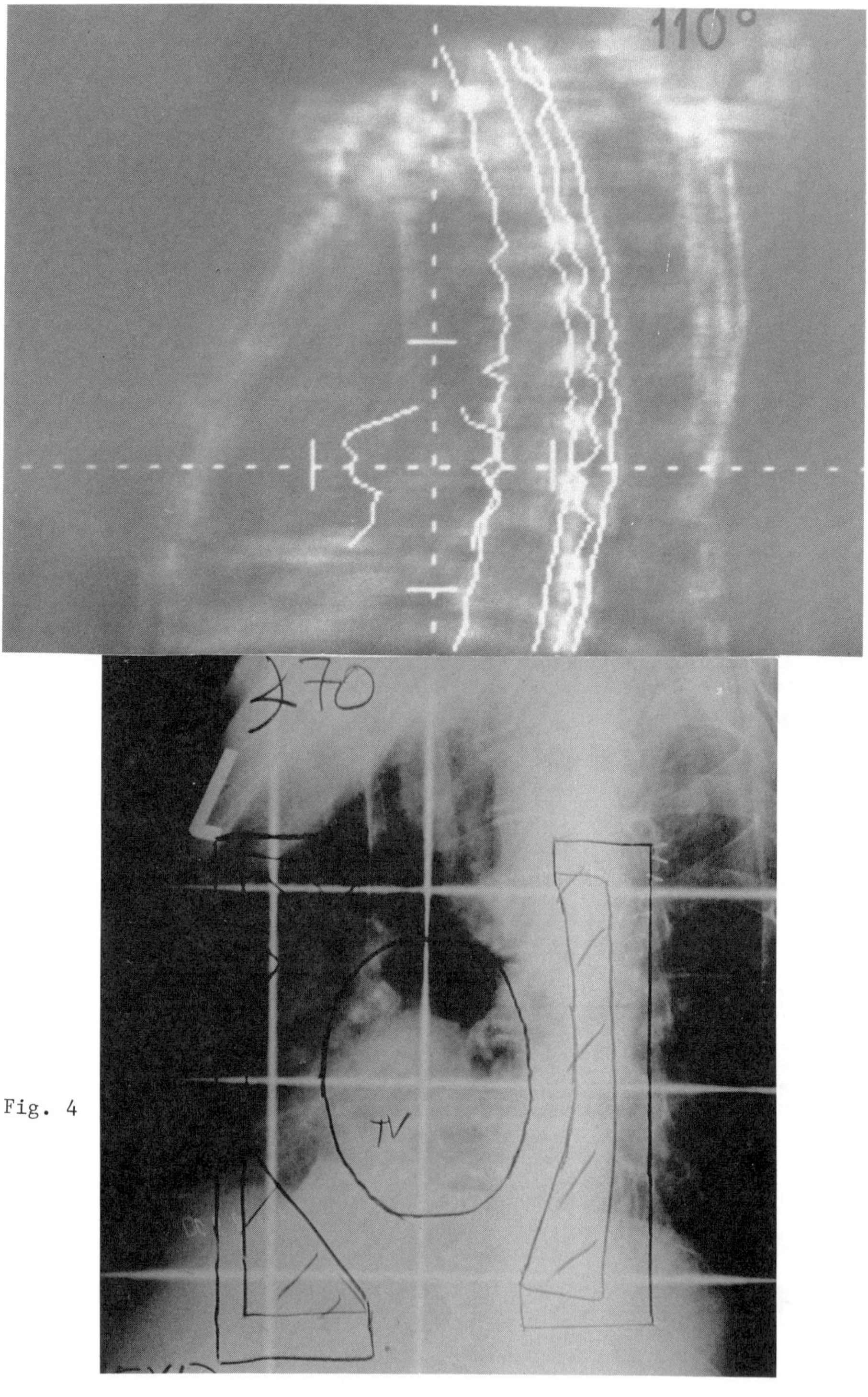

Fig. 4

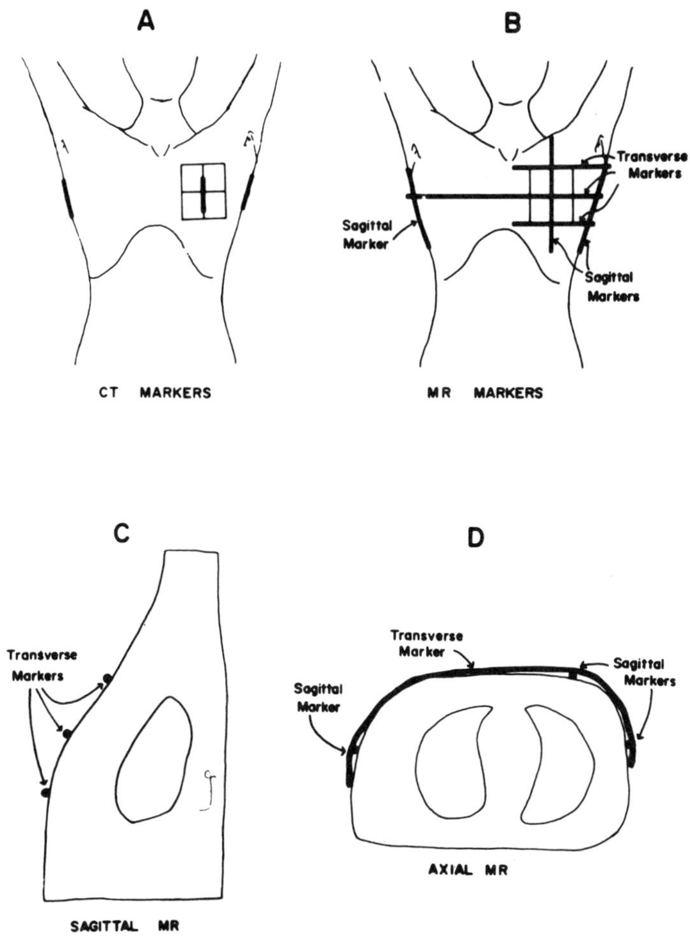

Fig. 5 Schematic illustrations of the differences in catheter placement.

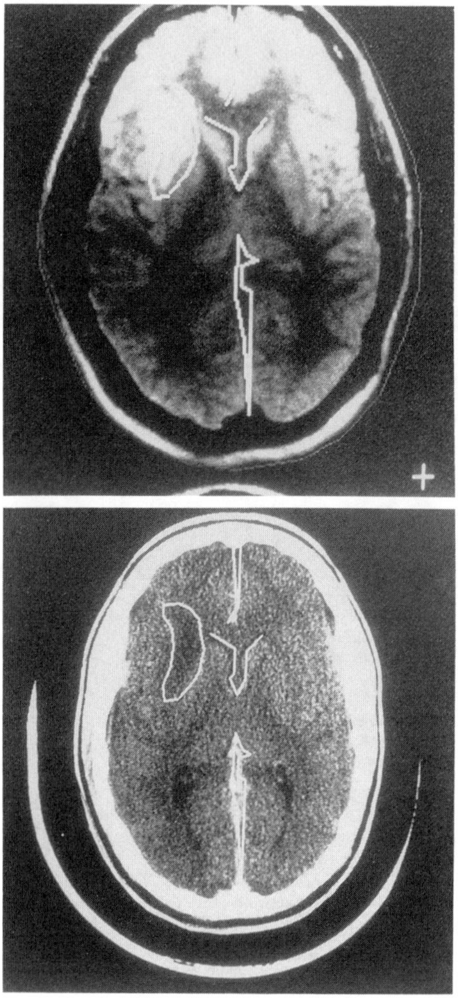

Fig. 6 Comparison of the CT and MR imaging for brain tumors. Target and normal structures outlined on the MR image (top) are transferred to the analogous CT slice (bottom).

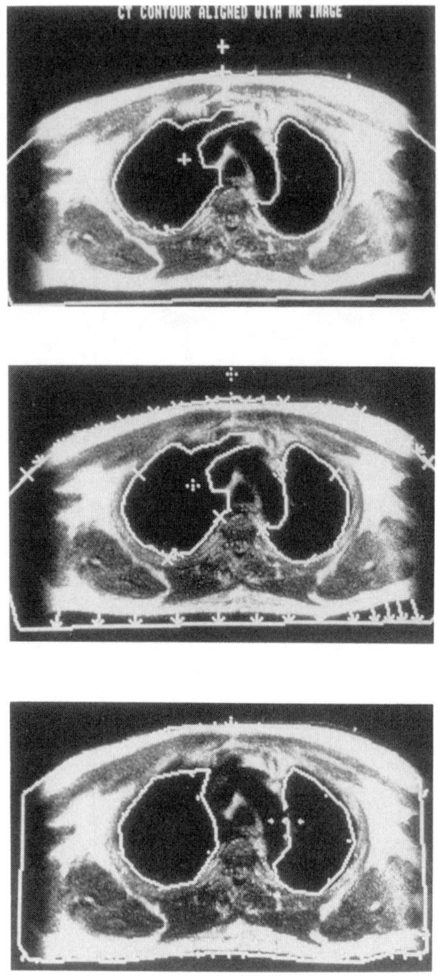

Fig. 7 2-D image unwarping procedure, see Reference 24 for details.

COMPUTERIZED BEAM SHAPING

B.A. Fraass, Ph.D., D.L.McShan, Ph.D., and K.J. Weeks, Ph.D.*
Department of Radiation Oncology
University of Michigan Medical Center
Ann Arbor, Michigan 48109

INTRODUCTION

The shapes of radiation fields used to treat patients have been individualized for many years, with lead cut-outs for orthovoltage beams and then hand-placed lead blocks once megavoltage machines came into use. With the introduction of low melting point lead alloy blocks(11) which are fabricated with a simple styrofoam molding technique (10), or with other types of field shaping systems (3), individualized beam shaping has become the standard of care in many radiotherapy centers. Typically, blocks have been designed by drawing the block shape on films, either portal films taken on the treatment machine, or diagnostic quality simulator films. The block shape drawn on the film is converted to a styrofoam mold with the use of a hotwire styrofoam cutter, and the mold is then filled with lead alloy. Once the alloy has cooled, the blocks are mounted on a tray and used for the patient's treatment.

In recent years, this routine has begun to be modified by the explosive growth in the capabilities of computerized treatment planning systems (see other articles in this volume). Treatment planning no longer consists of a few manual contours (or even a few CT scans), with hand-drawn target volumes identified from orthogonal simulator films.

State-of-the-art planning systems (5,9) now contain many capabilities which contribute information which is needed for block shape design:

- o Modern CT-based 3-D planning systems now make routine use of 30 to 60 CT slices per patient. Target volumes and critical normal structures can be outlined on each slice.
- o Beam's Eye View (BEV) displays (5,6,8,9) are often used (see Fig. 1). This computer graphics display shows the anatomy of the patient, as it would appear if one looked down the radiation beam from the source of radiation. Various contours represent target volume, normal structures, and other areas of interest.
- o Digitized portal and simulator films are used inside the planning system in a geometrically correct way (7).
- o Image enhancement and other image processing are used so that film images, particularly portal films, can be more useful in identifying anatomy (13).
- o Generalized 3-D treatment planning capabilities, including 3-D geometry and dose calculations, make the 3-D knowledge (and hence design) of field shaping more important than with 2-D systems.

The implementation and routine clinical use of the above-described features forces the development and use of computerized beam shaping. It is very difficult to view a BEV display of a target and a normal tissue without immediately requesting the ability to design the shape of the radiation field so that the target is fully irradiated and the normal tissue is adequately shielded. Our experience has been that use of 3-D planning capabilities, in particular the Beam's Eye View (BEV) display, lead directly to a need for computerized beam shape design.

BEAM SHAPE DESIGN

At the University of Michigan, we have used a fully three-dimensional treatment planning system ("U-U of M Plan") (4), which includes full BEV capabilities (5,7), for all clinical treatment planning for the last four years. Beam shape design has been included in the system since its implementation. Two main methods of block entry into the planning system are available. The first method makes use of an electromagnetic digitizer tablet to manually digitize the block shapes drawn on the simulator films into the computer. This method initially was used often, but is now relatively rare, since drawing the blocks on simulator films does not directly use the CT data which is available for most patients in our clinic. However, even if the blocks are designed on the simulator without CT information, the block shapes are digitized into the computer so that 1) the dose calculations reflect the actual block shapes used, and 2) the BEV display can be used to check the blocks against the other information contained in the planning system.

Block shapes can also be designed within the planning system. The most obvious method, and the most commonly used, is to generate a BEV display of each planned field, and to use a joystick and cursor to design the block shape using the color graphics BEV display. Referring to Fig. 1, various selected contours are displayed from the perspective of the source of the radiation. Usually presented with each structure color-coded, the BEV display gives the planner an excellent idea of the relationship of the various structures. Design of the block shapes using a joystick/trackball/mouse device can be done very quickly if the situation is not too complex. We have found that the use of a circular cursor (radius defined by the planner) helps the planner draw a block with a particular margin around the target volume (as seen on the BEV display).

A further enhancement involves the video digitization of the relevant simulator film for each field, and using it as a backdrop for the graphics BEV display (7). This type of image can help the physician make contact between the CT-defined (and relatively unfamiliar) BEV display of information, and the usual projective x-ray view of the anatomy.

As more familiarity and confidence in the 3-D planning system information became available, a third method for block shape definition became more important. This method, called "Autoblock" in the U of M Plan system (5), uses the 3-D description of the target volume to generate automatically the shape of the blocks which will leave a specified geometrical margin around the target volume. These computer-designed blocks thus conform to the shape of the target volume, as illustrated in Fig. 2. The ability to easily choose the margin used for each block, and the equally necessary ability to easily edit the computer-designed block shape, make the computerized design of the beam shape extremely simple.

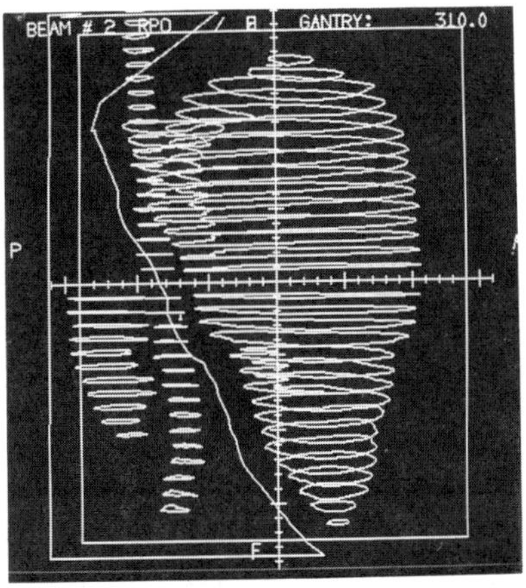

Fig. 1. Beam's Eye View (BEV) contour display for an oblique field used to treat the whole liver. Shown are contours for the liver (target volume), kidney, spinal cord. The block was drawn with the joystick using this display.

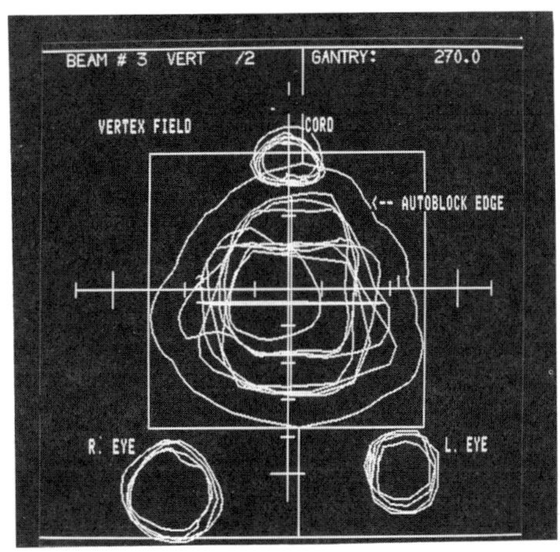

Fig. 2. Block shape designed for a vertex field used to treat a brain tumor. The target volume, eyes, spinal cord are shown. The "Autoblock" capability of the planning system was used to generate a computer-designed block with a 1 cm margin around the target volume.

AUTOMATED BLOCK FABRICATION

Computerized design of field shaping forces the development of ways to fabricate the computer-designed blocks. The simplest method is to use a hardcopy plot of the BEV display. If that plot is made with the correct magnification (the source film distance for the appropriate simulator film), then 1) it can be used directly by the block cutter just as if the block shapes were drawn on a film, and 2) the hardcopy BEV plot of the blocks (and CT-defined structures which have been outlined) can be underlaid below a simulator film. The block shapes and target volumes can be traced onto the film, if desired, for later use comparing portal to simulator films.

A second method for the generation of block molds is to use the block shape information as input into a computer-controlled styrofoam cutter (15). We have implemented this facility on a microprocessor-controlled hotwire milling machine manufactured by HEK GMBH in West Germany (marketed by Theratronics in the U.S.). There are several other similar devices currently on the market. As shown in Fig. 3, the HEK consists of a frame (containing the styrofoam mold) which swings about the fixed vertical hotwire. Two orthogonal linear drives swing the frame about a pivot point which mimics the radiation source of the treatment machine. The position, speed, and wire temperature are controlled by the microprocessor. Coordinates and other information are downloaded to the HEK processor from the treatment planning system. In our department, that link is by Ethernet from the VAX 8800 treatment planning computer to the microVAX II which controls the HEK, and then by serial link through a terminal server to the HEK.

The conversion of block coordinates from the planning system and their input into the HEK system requires several software routines. Conversion of the planning system block coordinates into those used by the HEK includes correction for the arc traveled by the frame when moved by the linear worm drive. The system then generates cuts along the principal axes of the field, to be used for mold alignment when the blocks are mounted to the tray. Coordinates are then reordered so that the cutting for each block can start at one of those alignment cuts if possible, thereby avoiding cuts which might cause the styrofoam to fall apart.

A summary of our experience (15) with the implementation of the automated block cutter is discussed below. Various problems with either the manual or automated cutter are identified.

- o The automated cutter was forced to set a high temperature on the cutter's wire to prevent dragging of the wire, and so the size of the cut in the styrofoam was larger than with the manual system. This resulted in the blocks being somewhat larger than desired. This is not a problem for the manual cutter because technicians learn to "cut along the inside edge of the line" so that the blocks will be the correct size. For the wire temperature and speed used in our automated system, we found that the block size had to be shrunk by 1.2 mm to account for the wire's cut. This shrinking is accomplished by a variation on the software which is used to design autoblocks.
- o Another problem that an experienced block cutter solves automatically is that of planning ahead so that there is adequate space to allow several mounting screws to be mounted to the blocks. Since we have a defined hole pattern in our block trays, we were forced to add a function to the treatment planning system that can overlay the block tray hole pattern over the BEV display so that the treatment planner who designs the blocks can also check that the block shape has adequate room for the mounting screws.
- o Large fields, for example mantle fields, were an area where the automated cutter was both more accurate and faster than the manual system. The difficulty and inaccuracy associated with taking two simulation films, taping them together, and cutting the block shapes from the combined films becomes larger than the basic accuracy of the automated cutter.
- o The automated cutter does not deal with irregular styrofoam as easily as a technician. When faced with irregularity in the cutting speed (or wire tension), the technician knows to change speed and try to maintain the

same cutting resistance so that the width of the line that is cut is maintained. The automated cutter does not recognize changes in wire tension, so if the speed stays constant, the width of the line cut in the styrofoam changes.

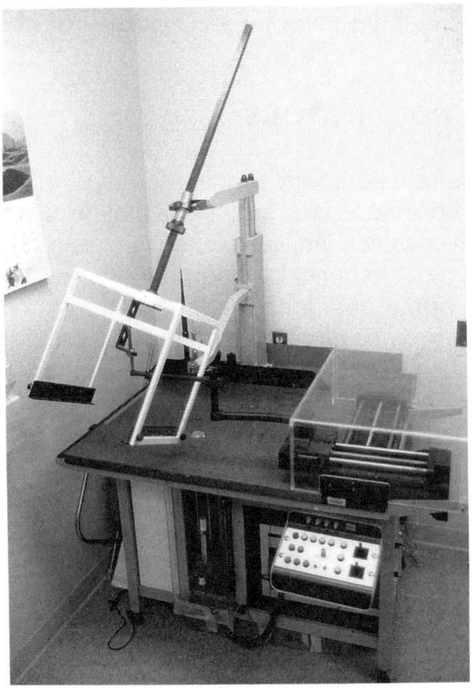

Fig. 3. Automated block cutter which uses linear drive mechanisms to drive a hinged frame (containing the styrofoam block mold) so that the hot wire in the center of the device cuts out the desired block shapes.

- o If correction to the linear coordinates input into the linear motions of the automated cutter is not performed, significant inaccuracies in position can occur.
- o The overall time spent on the automated system varied from much more than the manual system (for small fields with simple blocks) to much less (large complicated automatically-designed blocks).

Basic accuracy and usefulness of both the manual and automated systems were tested with both small and large blocks (15). Accuracy of both methods were comparable for small fields (< 0.5 mm) with the hint that the automated method was slightly less accurate. For large fields (a mantle field, for example), accuracy of both methods was again good. Few discrepancies were visible, and the precision of the block shapes was < 1 mm, although the automated system was generally better and more consistent. These results can be compared to the accuracy of a standard kind of sonic digitizer system (not used here) which is often used in radiotherapy planning: Rosen et al have shown that the digitization error can be as much as 2 mm (12).

Our studies with the automated block cutting system led to the following conclusions (15):

- o Automated block cutting is possible and can be time-efficient in some situations.

- o Accuracy does not degrade for large field blocks as it does for manual systems.
- o Overall time for block cutting is comparable to the manual methods, although the manual method is faster for small fields.
- o Accuracy of cutting (both manual and automatic systems) is reproducible to better than 1 mm, with the manual method showing an accuracy of about 0.5 mm for small fields.

MULTILEAF COLLIMATORS

Much of this discussion of computerized beam shaping leads directly to the multileaf collimator. A multileaf collimator is a beam collimation system that consists of a number of small leaves which individually move in and out of the field. Used as early as the 1950s by Takahashi in Japan (14), the multileaf collimator has been discussed by numerous authors (see ref 1). Only recently, however, have technology and computer-control capabilities developed and merged to the extent that multileaf collimator designs have begun to fulfill the clinical desires of the radiotherapy community. See, for example, the work of Brahme (1,2).

The use of a multileaf collimator for beam shaping is the logical extension of the computerized block design and fabrication techniques which are discussed in this paper. Instead of downloading coordinates to a block cutter, the field shapes will be downloaded directly to the computer control system of the treatment machine which is equipped with a multileaf collimator. The "Autoblock" facility of the planning system is easily modified to become the "Autoleaf" capability, in which the location of each leaf is computed so that a particular margin is maintained. More sophisticated and complicated algorithms are also possible (1). These capabilities then lead directly to the use of computer-controlled conformal therapy, as is discussed in another paper in this volume.

CONCLUSIONS

The routine use of 3-D treatment planning capabilities, and especially the Beam's Eye View (BEV) display, has directly led to the design, implementation and routine use of computerized beam shape design and block fabrication. Blocks are designed either using the joystick with a BEV display, or with an algorithm which generates the block shapes based on the shape of the target volume from the BEV point of view. The computerized design of a large percentage of the blocks used in the department then leads directly to implementation of an automated block cutting system.

The automated block cutting system was found to be approximately as fast and as accurate as the manual block cutter, once several corrections to the automated block cutter software were implemented. In general, the manual block cutter was somewhat faster and more accurate on small fields. For large and complex fields, however, the automated system's accuracy and speed did not degrade as it did for the manual system, and the automated system was slightly more accurate.

ACKNOWLEDGEMENTS

We would like to acknowledge the excellent assistance and interest of E. Hargreaves and S. Hardybala with the automated block cutter study and W. Roberts for interfacing the cutter to the VAX system.

* Present address: Radiation Oncology, Duke University, Durham NC

REFERENCES

1. A Brahme, Optimal setting of multileaf collimators in stationary beam radiation therapy. Strahlentherapie und Oncologie 164: 343-350, 1988.
2. A Brahme: Design principles and clinical possibilities with a new generation of radiation therapy equipment, Acta Oncol 26: 403-412, 1987
3. WFM Brouwer, I Kazem, CJHM van Mil: Four Years Experience with a Mercury Shielded Irregular Field system (MSIFS), Int. J. Rad. Oncol. Biol. Phys. 10: 225-229 (1984).
4. BA Fraass and DL McShan: 3-D Treatment Planning: I. Overview of a Clinical Planning System, in The Use of Computers in Radiation Therapy, ed. IAD Bruinvis et al, Elsevier Science Publishers B.V., (North-Holland), 1987, pp 273-276.
5. BA Fraass, DL McShan, and KJ Weeks: 3-D Treatment Planning: III. Complete Beam's-Eye-View Planning Capabilities, The Use of Computers in Radiation Therapy, ed. IAD Bruinvis et al, Elsevier Science Publishers B.V., (North-Holland), 1987, p 193-196.
6. M Goitein, M Abrams, D Rowell, H Pollari, J Wiles: Multi-dimensional Treatment Planning: II. Beam's Eye View Back Projection, and Projection Through CT Sections. Int. J. Rad. Oncol. Biol. Phys. 9: 789-797, 1983.
7. DL McShan, BA Fraass, and AS Lichter: Full Integration of the Beam's Eye View Concept in Clinical Treatment Planning, Int. J. Rad. Oncol. Biol. Phys., in press.
8. DL McShan, DR Hauman, LE Reinstien, and AS Glicksman: An Interactive Three-Dimensional Radiation Treatment Planning System. Br. J. Radiol. 15(Suppl.):144-146 (1979).
9. R Mohan, G Barest, LJ Brewster, CS Chui, GJ Kutcher, JS Laughlin, Z Fuks: A Comprehensive Three-Dimensional Radiation Treatment Planning System, Int. J. Rad. Oncol. Biol. Phys. 15: 481-495, 1988.
10. JA Purdy: Secondary Field Shaping, in Advances in Radiation Therapy Treatment Planning, ed. A.E. Wright and A.L. Boyer, Amer. Assoc. Phys. Med., New York, 1983, pp 456-476.
11. WE Powers, JJ Kinzie, AJ Demidecki, JS Bradfield, A Feldman: A New System for Field Shaping for External-Beam Radiation Therapy, Radiology 108: 407-411 (1973).

12. II Rosen, RG Lane, CA Kelsey: Accuracy of a Two-Sensor Sonic Digitizer, Med. Phys. 6: 536-538 (1979).

13. GW Sherhouse, J Rosenman, HL McMurry, SM Pizer, EL Chaney: Automatic Digital Contrast Enhancement of Radiotherapy Films, Int. J. Rad. Oncol. Biol. Phys. 13: 801-806, 1987.

14. S Takahashi, Conformation radiotherapy rotation techniques as applied to radiography and radiotherapy of cancer. Acta Radiol (suppl) 2: 27-59, 1962.

15. KJ Weeks, BA Fraass, DL McShan, SS Hardybala, EA Hargreaves, and AS Lichter: Comparison of Automated and Manual Shielding Block Fabrication, Int. J. Rad. Oncol. Biol. Phys., 16: 501-504 (1989).

CONFORMAL THERAPY

C. W. Cheng, Ph.D.
Joint Center for Radiation Therapy, Department of Radiation Therapy, Harvard Medical School, Boston, MA 02115

ABSTRACT

In radiotherapy normal tissue tolerance is very often the dose limiting factor. Conformal therapy with computer control permits simultaneous variation of several treatment parameters during irradiation, producing dose distributions which conform to the target shape. With reduced normal tissue exposure, higher dose can be delivered to the target. This may result in a higher local control rate. The objective of this presentation is to review the capabilities of conformal therapy in terms of dose distributions and normal tissue sparing. The procedures involved in conformal therapy will be described. The current status and the future development are also discussed.

INTRODUCTION

The principal goal in radiotherapy is to obtain a high level of local control while minimizing the risk of normal tissue damage. It has long been recognized that the dose that can be given to the target volume is often limited by the need to avoid complications. When both the patient anatomy and the target volume do not change significantly throughout the treatment field, conventional treatment techniques with shielding blocks are usually satisfactory. However, when the shape of the target volume is complex or when the target is close to critical structures, conventional treatment technique could result in a large treatment volume and the delivery of a tumoricidal dose usually leads to a high dose to the nearby structures. If radiation therapy is the only option, a suboptimal dose would usually be prescribed and lower control rates accepted for fear of complication. Conformal (or conformation) therapy, which can produce improved dose distributions with potentially reduced morbidity, provides the clinician with an alternative to plan a better treatment strategy.

The term conformation therapy was given by Takahashi[1] to describe irradiation techniques whereby a high dose region is made to fit the target volume. In conformal therapy, one or more treatment parameters can be varied during irradiation, depending on the sophistication of the treatment plan. The treatment parameters include the field size, the gantry angle, the collimator angle, the table position and the dose rate. The objective is to generate a dose distribution with a high dose region conforming as closely as possible to the target volume while at the same time a maximum sparing of the adjacent organs is achieved. A smaller volume of normal tissue irradiated means that the tolerance of the tissue is higher. This implies in turn that a higher dose can be given to the target volume.

The generation of dose distributions, however, is but one of the steps involved in conformal therapy. Localization of the target volume and the nearby structures, as well as the verification of the dose delivered, are equally important steps in the process. The techniques of using CT and MRI for accurate localization have been discussed in the chapter "CT and MRI treatment planning interface". The objective of this presentation is to review the capabilities of conformal therapy in terms of dose distributions and normal tissue sparing. An historical review of conformal therapy will be presented in the next section. Some techniques employed in conformal therapy will then be illustrated. Comparisons with conventional methods will be made. The simulation procedure, the quality assurance, and the safety aspects of the control system will be considered. Finally, the current status and the future development of conformal therapy will be discussed.

HISTORICAL REVIEW

While Takahashi was credited, and justifiably so, with developing various conformal techniques in the 60's, incorporating rotation, synchronous blocking, and multileaf collimators (interested readers can read Reference 1 for the theoretical development by Takahashi), the concept of conformal therapy had actually been used earlier by Wright[2] in the late 50's, and even earlier in the beginning of this century in what was then called the moving field technique. Wachsmann[3] pointed out in his review of moving field techniques that Kohl was the first to perform manually, in 1906, what we now recognize as the rotational arc technique.

Early conformal techniques utilized, in Japan and in the U.S., synchronous blocks for shielding and field shaping.[2,4-7] With the patient in the sitting position, specially designed shielding blocks were placed on a separate rotation table and made to rotate at the same speed as the patient by geared or servo mechanism. Thus, the blocks protected critical organs at all times during the treatment. A similar idea, using the gravity of the shielding block to maintain its orientation, was proposed by Takahashi and Iida.[8] The gravity block technique was originally used as eye blocks in rotation therapy of the head,[9,10] and later extended to other regions.[11] Other developments included the use of multileaf collimator in rotational therapy of the cervix,[12] and the moving table technique for treatments involving large oblong fields[13]. Mantel[14] described an automatic system using potentiometers to vary the field size and the dose rate for each gantry angle in rotational therapy. A full-fledged computer controlled linear accelerator in which up to 11 treatment parameters can be varied simultaneously was reported in 1978 by Levene.[15] With different combinations of the treatment parameters, a wide variety of conformal techniques can be carried out. Since then, development in conformal therapy has turned to computer control systems (both treatment planning and dose delivery).[16-19]

Somewhat different from what has happened in Japan and the U.S., much effort has been spent in the U.K. on the development of the tracking technique in conformal therapy. Tracking is a combination of arc rotation (or static field arrangement) and moving field technique, designed to treat long fields. The table can be moved in all three spatial directions by means of electromechanical devices as the gantry rotates (or stays at a fixed gantry angle). The original tracking technique was reported in 1960 by Green et al[20] using a modified 250 kV therapy unit. The system was modified and improved by Davy et al[21] of the Royal Free Hospital in London and used on a cobalt unit. A punch-card control system, installed in the University Women's Hospital in Erlangen in 1967, is perhaps the earliest automatic control system without the use of a computer. The system was used for automatic setting of several treatment parameters, such as gantry angle, field size, gantry rotation speed and treatment time.[22] A computer controlled tracking unit was reported in 1981 by Brace et al[23] at the Royal Free Hospital. The system was upgraded to include CT in 1982.[24] A fully automatic computer controlled system reported by Brace[25] was used to treat the entire spinal axis by Shentall and Brace.[26]

SOME EXAMPLES OF CONFORMAL THERAPY

Figure 1 shows the treatment parameters which can be varied in conformal therapy. By using different combinations of these parameters, a wide variety of conformal techniques can be obtained.

A simple application is the so-called dynamic wedge, in which the speed of motion of independently movable jaws is varied to simulate the function of a wedge.[27] This is done with collimators which can move past the central axis. Starting with the collimators closed, one of the jaws is slowly opened up until the final field width is reached. The radiation beam remains on throughout the collimator motion and the speed of the jaw is varied to produce

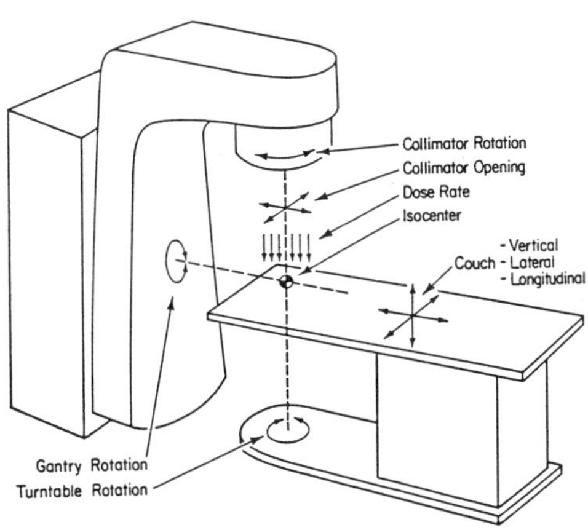

Fig. 1. The different treatment parameters which can be varied in conformal therapy.

the desired dose variation in the irradiated volume. The typical 'tilted' isodose distribution of a wedge filter can be generated. A schematic diagram of the dynamic wedge and the isodose distribution is shown in Fig. 2. Figure 3 shows the dosimetric verification of the isodose lines for a dynamic wedge. Similarly, by moving the collimator jaws along the longitudinal axis, the function of a compensator can be simulated. An example of dynamic compensators will be presented elsewhere.[28]

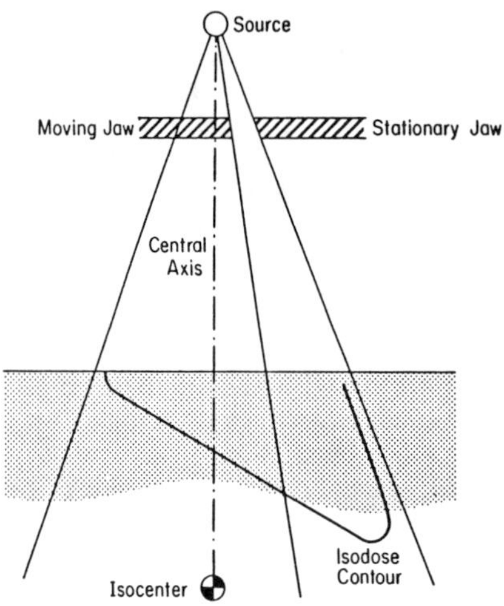

Fig. 2 A schematic diagram of a dynamic wedge.

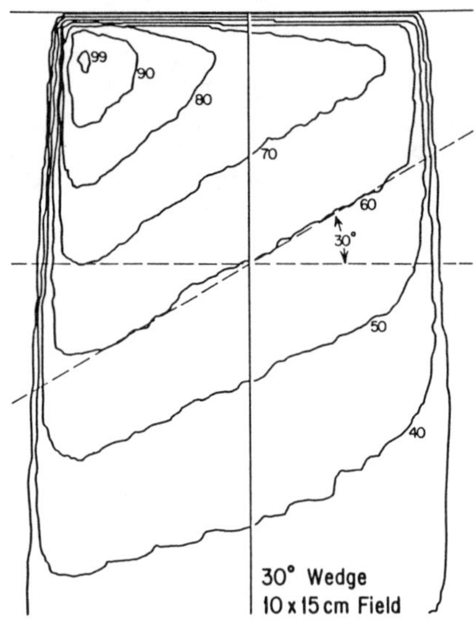

Fig. 3 Isodose lines of a dynamic wedge.

The technique of conformal rotation is also easily implemented. Fig. 4 shows a 360° rotation with collimator jaw motion for a large elliptical target volume[29] which remains unchanged along the longitudinal axis. The beam width is varied as the gantry is rotated so that the beam always encompasses the projection of the target volume with a suitable margin. The dose rate is also adjusted. As a result, a uniform dose distribution within the target volume is obtained which conforms to the target shape.

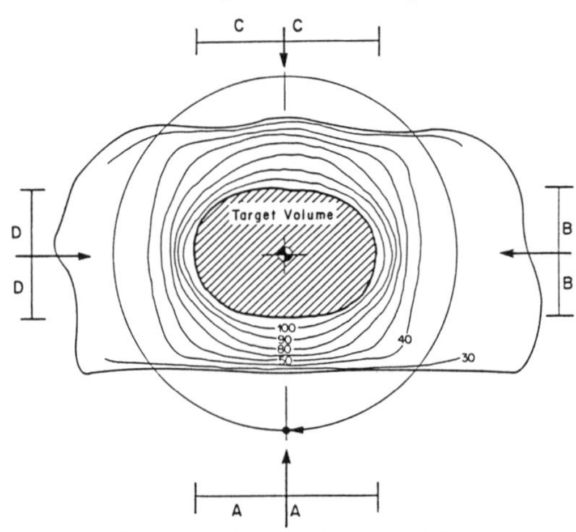

Fig. 4 Conformal rotation.

A slightly more complicated conformal technique is the dynamic wedge arcs.[15] This is accomplished by combining the dynamic wedge technique and arc rotation together with dose rate control. Fig. 5 is a schematic diagram of the dynamic wedge arc technique. A comparison of the dynamic wedge arc technique and a conventional plan in the treatment of a lung tumor[28] is shown in Fig. 6. The objective was to irradiate the target volume to a high dose while sparing the cord and the contralateral lung. The two distributions are very similar but the conformal treatment, which involves only one rotation arc, is simpler to execute than the conventional plan.

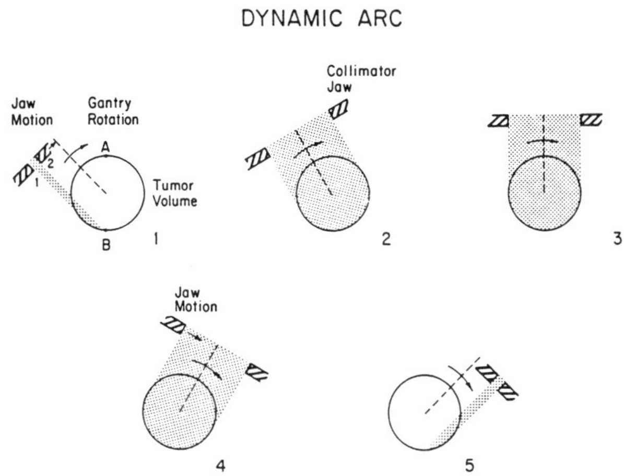

Fig.5 A schematic diagram of the dynamic wedge arc technique.

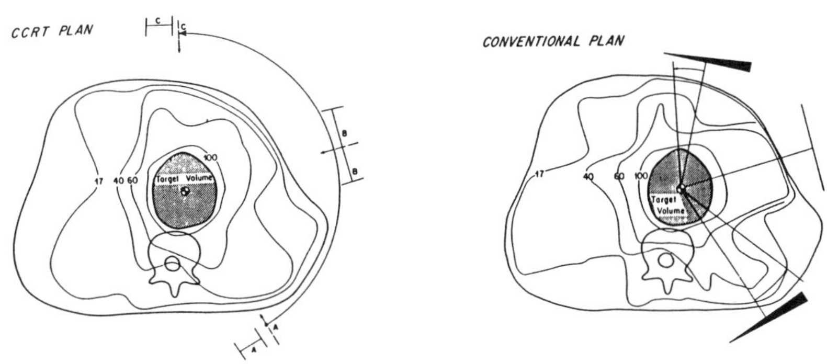

Fig. 6 A comparison of the dynamic wedge arc technique and the conventional plan. The conventional plan consists of two wedge arcs and one static field. The dynamic wedge arc technique consists of only one arc.

When the shape of the target volume is concave, the conventional technique is to use an arc rotation with a block, shielding more than half of the field. In conformal therapy, this is carried out by combining two rotation arcs with collimator jaw motion and dose rate control.[29] In each of the two arcs, a collimator jaw is moved past the central axis to simulate the block in the conventional plan. The advantage of conformal therapy is that the collimator jaw can be moved during rotation to allow shaping of the high dose region. The resulting cylindrical shell-like distributions are shown in Fig. 7. Using

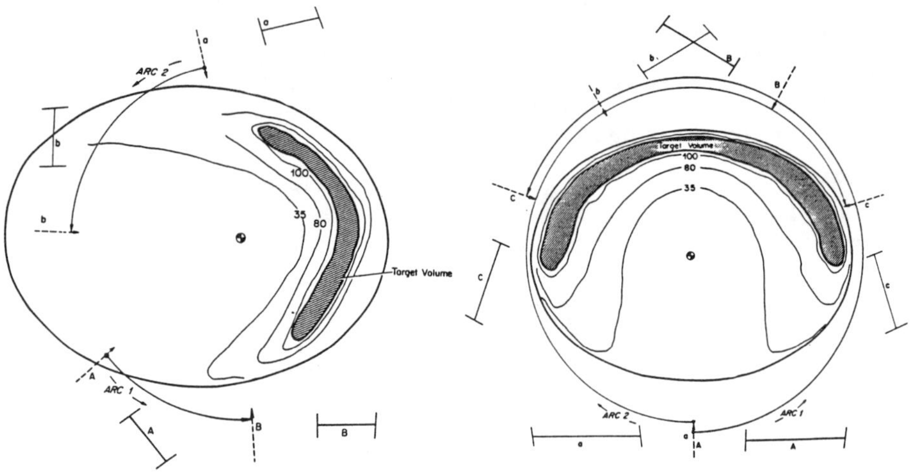

Fig. 7 Examples of cylindrical shell distributions by conformal techniques, consisting of arcs with eccentric, variable fields.

multiple arcs, the high dose region can be moved around the patient by choosing the range of the arcs. An example is shown in Fig. 8.[29]

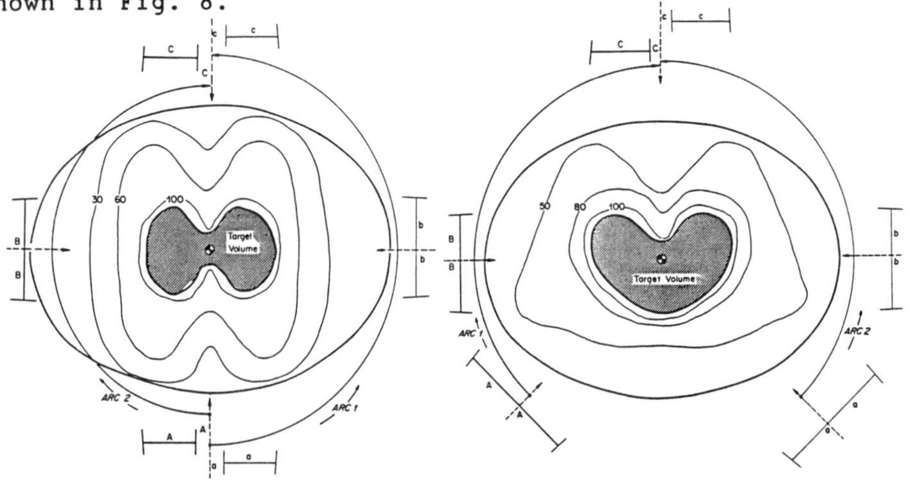

Fig. 8 Multiple arc technique in conformal therapy to shape irregular targets.

All the above conformal techniques are two-dimensional and used when the shape of the target volume remains unchanged in the longitudinal direction. An interesting clinical application of two dimensional techniques on a three-dimensional volume, one which also changes along the patient, is demonstrated in Fig. 9. This is a patient who required treatment of the residual soft tissue sarcoma along the left abdominal wall.[29] The target shape at the central section is shown in Fig. 9. No conventional technique could deliver a high dose to the irregular target and spare the bowel, the cord and the single kidney. By using two rotation arcs and a combination of the above techniques, an acceptable distribution could be generated. The area within the 100% isodose line is larger than the target in the central section in order to cover the target in the other sections. With a multileaf collimator, a much improved distribution can be generated as the leaves can be moved during the rotation to conform more closely to the target cross-section along the patient.

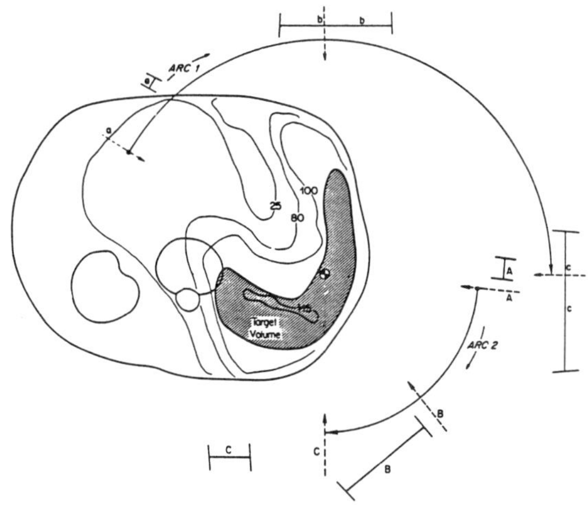

Fig. 9 The target shape and dose distribution for a patient with soft tissue sarcoma.

The two-dimensional techniques can be extended to three dimensions by incorporating motion of the treatment couch or collimator rotation. Table motion is suitable for long volumes. When the target volume is not too long but

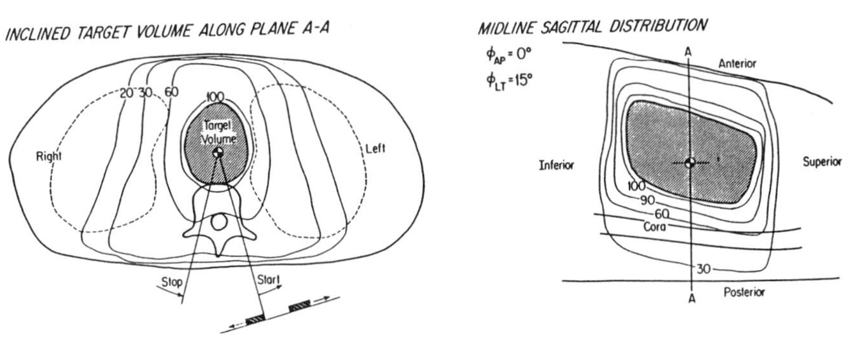

Fig. 10 Dose distribution for a 3 dimensional conformal technique, consisting of gantry rotation, collimator jaw motion and collimator rotation.

is inclined at an angle to the longitudinal axis, collimator rotation proves to be very useful.[30] Figure 10 shows one such example. The irradiation sequence consists of an almost full rotation, coupled with the rotation of the collimator assembly and the motion of the jaws to conform to the target.

For long target volumes, table motion can be combined with rotation to generate the desired distributions. This is illustrated in the treatment of a patient with Ewing's sarcoma. As shown in Fig. 11, the target volume varies both in shape and location along the chest wall. The distributions shown in Fig. 12 are obtained by two scans. During each scan, table translation is combined with gantry rotation, collimator jaw motion, and dose rate control. This results in a parallel-opposed tangential pair in dynamic motion. The spinal cord is spared and only a small fraction of the lung is in the high dose region. A conventional plan would result in a much larger treatment volume and a large fraction of the lung would receive an excessive dose.

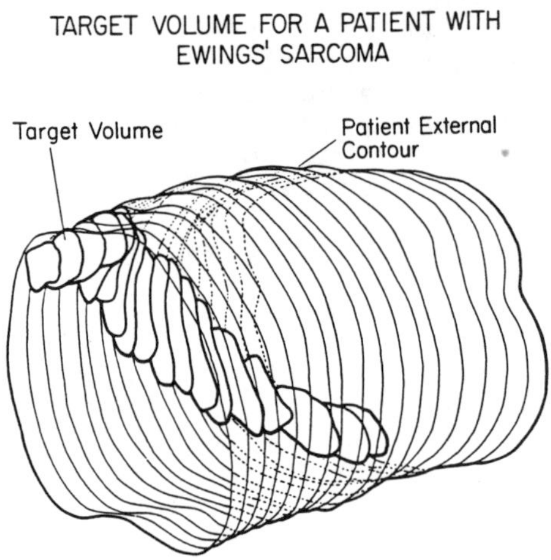

Fig. 11 Target volume of a Ewing's sarcoma.

The final example is the treatment of a primary pelvic tumor and the para-aortic nodes.[31] Because of the irregular shape and the length of the target(Fig. 13), the treatment was carried out with a series of scanned

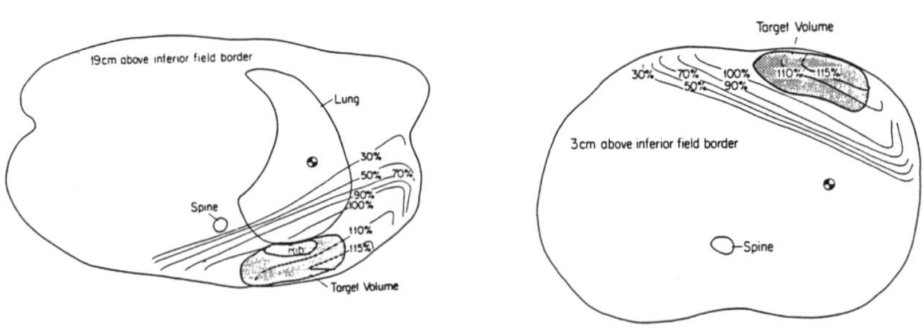

Fig. 12 Dose distributions for a 3 dimensional conformal technique for the treatment of Ewing's sarcoma.

fields. A detailed description of the treatment can be found in Reference 31. Fig. 14 is the dose distribution in the midline sagittal plane. The scanning technique is compared with two conventional methods in Fig. 15 for a transverse section through the target volume. A substantial decrease in dose to critical normal structures is achieved in the conformal technique. The field-matching problem involved in the conventional technique is also eliminated.

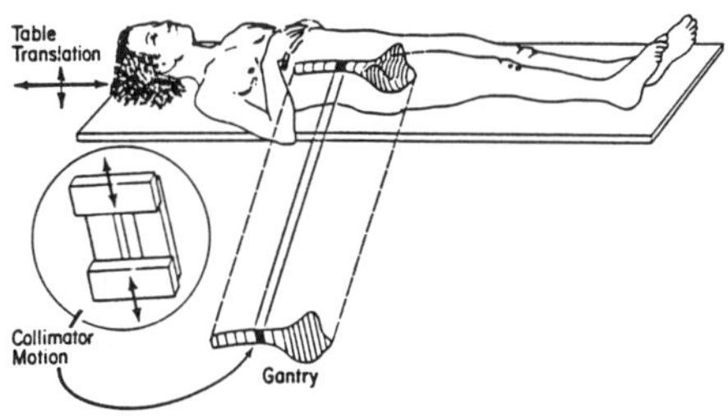

Fig. 13 3 dimensional representation of pelvic and para-aortic target volume.

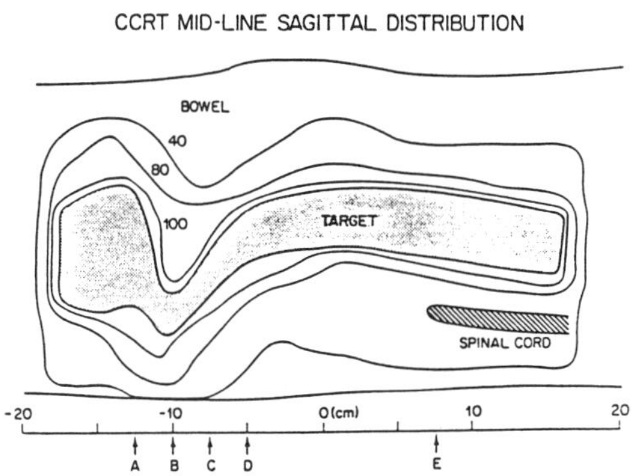

Fig. 14 A midline sagittal distribution of the conformal technique described in the text.

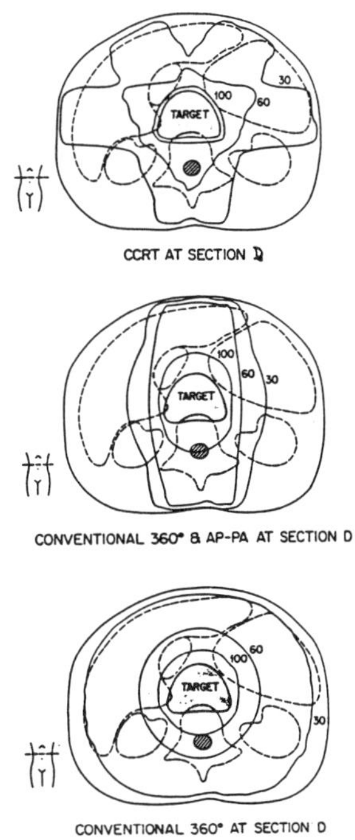

Fig. 15 A comparison of the conformal technique and conventional techniques through a section of the target volume.

SIMULATION, QUALITY ASSURANCE AND SAFETY

The simulation of a conformal treatment is the same as that of a CT based (and with MRI as well if available) simulation for a conventional technique. The procedure has been described in detail in the chapter " CT-MRI treatment planning interface " and will not be repeated. A conventional technique should also be simulated as the backup treatment. In case of unexpected delay in the conformation therapy, such as a technical malfunction, the patient can still be treated on schedule, using the conventional alternative.

To ensure that the dose distribution will be delivered to the patient as desired, it is necessary to carry out some quality assurance checks of the treatment machine and the associated computer control system. In our institution, a thorough check is carried out before the first treatment. The first step is to verify that all treatment parameters are entered correctly into the control system. A 'dry' run is then carried out to confirm that all machine operations involved in the treatment are as expected. In the case of a complex conformal treatment such as that of a scanning technique with multiple scanned fields, it is only necessary to carry out the 'dry run' on one of the fields as the other field is simply a mirror image of the field tested. The technologists and the physicist would observe closely throughout the test run. The final check consists of phantom measurements for dosimetric verifications. Films are used to verify the distribution at different levels of the phantom and ionization chamber measurements are carried out at selected locations within the phantom to verify the doses obtained from the calculations. On subsequent treatments, phantom measurement is not necessary unless there is a change in the treatment plan. Thus the checking procedure is simpler.

In conformal therapy, the positioning of patient on the treatment table is the same as in the conventional treatment. For 2D conformal treatment, the 'beam on' time is comparable to the treatment time in conventional therapy. For the 3D scanning technique it is substantially longer. However, while the conventional multiple field treatment have several 'breaks' in an irradiation session, it is usually a continuous run in conformal therapy. In other words, the patient must remain still on the table throughout the treatment session, which may be 1-2 minutes for simple conformal rotation and up to 6-20 minutes for scanning techniques. Patient immobilization is therefore an important issue in conformal therapy and safety precautions are important features in the computer control system.

There are several important safety aspects in the control system. (1) The system can detect position errors due to malfunction of the machine, turn the machine off if tolerances are exceeded, and prevent the machine from being turned on by the operator if a malfunction is detected. (2) Once the machine is turned off, it goes out of the computer control. The computer cannot change the system from the manual to the automatic mode. (3) The computer can never turn the machine on. Only the operator can do this. (4) The operator can always override the computer through the 'deadman' switch, which turns the radiation off and suspends computer control. The reasons for these safety features are obvious. The technologist also plays an important role in the safety aspect by monitoring, throughout the treatment, the position errors displayed on a CRT screen in the control room (e.g. in a simple bar pattern), and by watching for the patient motion on a closed-circuit TV. The technologist can turn the machine off if the patient moves or the machine malfunctions and the computer fails to respond (the latter never occurred in our experience). The treatment can be continued by the technologist after the interruption.

CONCLUSION

With computer control, conformal therapy permits simultaneous variation of multiple treatment parameters and produces improved dose distributions. Although its clinical benefits remain to be demonstrated, the potential of improving local control by conformal therapy should be explored.

It has been more than two decades since Takahashi developed the concept of conformal therapy. While there is a growing interest in conformal therapy and computer controlled linear accelerators have been manufactured and commissioned in the clinics,[32] the clinical use of conformal therapy is still very selective. The candidates for conformal therapy usually involve complex target shapes with adjacent critical structures. Compared to most conventional techniques, the entire treatment planning process involved in conformal therapy, from localization to generation of a dose distribution, is laborious and time consuming, not to mention the equally laborious quality assurance checks of the entire system. This is particularly true when multiple scanning fields are employed. Thus one of the obstacles to the implementation of conformal therapy is its inefficiency in the generation and the subsequent delivery of the distributions. On the other hand, it should not be overlooked that conformal therapy is definitely superior to conventional techniques in such complex cases as have been discussed earlier. This is a classical example of cost and benefit. When situation arises in which the patient would clearly benefit from conformal therapy because of the improved local control and the reduced normal tissue morbidity, the clinical decision is clear.

Since improved local control and reduced normal tissue exposure are the goals in radiotherapy, conformal techniques should not be restricted to selected cases but also extended to less complex demands. In my opinion, two approaches should be taken concurrently in the development of conformal therapy. One is to improve the current system to make it more efficient (a more in depth discussion on the long term research emphasis can be found elsewhere[33]). Another approach is to use current conformal techniques on simple routine cases to evaluate its benefits.

At the present time, there are still many time-consuming manual stages in the process, such as the outlining of critical structures, the setting-up of multiple beams with different field sizes in dose calculations, and the optimization of the three-dimensional distributions. These steps can be eliminated by developing efficient algorithms and using more powerful computers. There are ongoing studies at our institution and other centers aiming at automating the planning process.

The development of computer controlled linear accelerators with multileaf collimators and other special features are now undertaken by several manufacturers. Multileaf collimators are likely to be the most promising feature in computer controlled conformal therapy. Most of the irregular target volumes encountered in radiation therapy can be shaped by the motion of the leaves during the irradiation. The major advantage of using a multileaf collimator is its ability to achieve the desired conformation with a much reduced treatment time compared to that of the scanning technique. Undoubtedly, there would still be a small subset of target volumes which favor the use of scanning technique. However, it is probable that with improved algorithms for automatic contouring, dose optimization and multileaf collimators, much of the time currently involved in conformal therapy can be reduced.

At the present time, there are no published data comparing normal tissue sparing between conformal and conventional techniques. With the use of multileaf collimators, it is possible that conformation is superior to conventional methods, for at least some cases. This may prove an important point. The expensive conformal therapy is not just for the selectively few cases, but can be used as a routine treatment technique with superior normal tissue sparing. In our institution, we have used conformal therapy as boost to pelvis and other disease sites as well as in a full treatment course. With the existing system, the treatment planning of conformal therapy of a pelvis, for example, is simple and the treatment time is comparable to the conventional four-field plan. More work should be directed to the routine applications of simple conformal therapy in the future.

ACKNOWLEDGEMENTS

The author would like to thank Drs. Bjarngard, Chin, Kijewski, Langer, Siddon and Svensson for their critical criticisms on the manuscript.

Figures 4, 5, 6, 7, 8, 9 are reprinted with permission from Int. J. Rad. Oncol. Biol. Phys. vol. 9, L.M. Chin et al. "Dose optimization with computer-controlled gantry rotation, collimator motion and dose rate variation" 1983 Pergamon Press plc.

Figures 13, 14, 15 are reprinted with permission from Int. J. Rad. Oncol. Biol. Phys. vol. 7, L.M. Chin et al. "A computer controlled radiation therapy machine for pelvic and para-aortic nodal areas" 1981, Pergamon Press plc.

REFERENCES

1. S. Takahashi, Acta Radiol. Suppl. 242, 12 (1965).

2. K.A. Wright, B.S. Proimos, J.G. Trump, M.I. Smedal, D.O. Johnston, F.A. Salzman, Radiology 72, 101 (1959).

3. F. Wachsmann, G. Barth, L. Lanzl et al. Moving Field Radiation Therapy, Chicago, University of Chicago Press, 29 (1962).

4. B.S. Proimos, Radiology 74, 753 (1960).

5. B.S. Proimos, Radiology 77, 591 (1961).

6. J.G. Trump, K.A. Wright, M.I. Smedal, F.A. Salzman, Radiology 76, 275 (1961).

7. D.N. Ilfled, K.A. Wright, F.A. Salzamn, Am. J. Roent. Rad. Ther. Nucl. Med. 112, 792 (1971).

8. S. Takahashi, H. Iida, Nippon Acta Radiol. 21, 126 (1961).

9. B.S. Proimos, Radiology 87, 933 (1966).

10. T. Kitabatake, T. Ueda, Acta Medica et biologica 16, 63 (1968).

11. B.S. Proimos, A.L. Goldson, Int. J. Rad. Oncol. Biol. Phys. 7, 973 (1981).

12. T. Kitabatake, S. Takahashi, Tohoku J. exp. Med. 94, 37 (1968).

13. J. Mantel, A.C. Scheer, Radiology 90, 799 (1968).

14. J. Mantel, H. Perry, J.J. Weinkam, Int. J. Rad. Oncol. Biol. Phys. 2, 697 (1977).

15. M.B. Levene, P.K. Kijewski, L.M. Chin, B.E. Bjarngard, S. Hellman, Radiology 129, 769 (1978).

16. H. Perry, J. Mantel, D.M. Vieau, D.L. Higden, M.M. Lefkofsky, Proc. 8th I.C.C.R.T. Toronto, 474 (1984).

17. W. Schlegel, R. Boesecke, B. Bauer, K. Alandt, W. Lorenz, Proc. 9th I.C.C.R.T. Scheveningen, The Netherlands, 361 (1987).

18. J. Richter, P. Klemm, M. Neumann, G. Nowak, O. Sauer, Proc. 9th I.C.C.R.T. Scheveningen, The Netherlands, 545 (1987).

19. K. Nakagawa, Y. Aoki, K. Sakata et al., Proc. 9th I.C.C.R.T. Scheveningen, The Netherlands, 541 (187).

20. A. Green, W.A. Jennings, H.M. Christie, Transac. 9th Int. Cong. Radiol. Munich, 766 (1960).

21. T.J. Davy, P.H. Johnson, R. Redford, J.R. Williams, B.J.R. 48, 122 (1975).

22. J. Weishaar, W. Rummel, A. Reuss, Electromedica 3, 78 (1969).

23. J.A. Brace, T.J. Davy, D.B.L. Skeggs, H.S. Williams, B.J.R. 54, 1068 (1981).

24. T. Tait, J.A. Brace, T.J. Davy, H.M. Morgan, D.B. Skeggs, A.J. Tookman, Clin. Radiol. 36, 209 (1985).

25. J.A. Brace, Int. J. Rad. Oncol. Biol. Phys. 8, 2011, 1982.

26. G.S. Shentall, J.A. Brace, Proc. 9th I.C.C.R.T. Scheveningen, The Netherlands, 367 (1987).

27. P.K. Kijewski, L.M. Chin, B.E. Bjarngard, Med. Phys. 5, 426 (1978).

28. C.W. Cheng, R.L. Siddon, B.E. Bjarngard, World Cong. Med. Phys. Biomed. Eng., San Antonio, Paper MP24.3 (1988).

29. L.M. Chin, P.K. Kijewski, G.K. Svensson, B.E. Bjarngard, Int. J. Rad. Oncol. Biol. Phys. 9, 723, (1983).

30. R.L. Siddon, Med. Phys. 8, 766 (1981).

31. L.M. Chin, P.K. Kijewski, G.K. Svensson, J.T. Chaffey, M.B. Levene, B.E. Bjarngard, Int. J. Rad. Oncol. Biol. Phys. 7, 61 (1981).

32. T. Blake, P.A. Hickling, R.A. Lane, A.E. Nahum, I.I. Rosen, M.E. Rosenbloom, Proc. 9th I.C.C.R.T. Scheveningen, The Netherlands, 537 (1987).

33. B.E. Bjarngard, Am. J. Clin. Oncol. 11, 187 (1988).

ARTIFICIAL INTELLIGENCE

Ira J. Kalet, Ph.D.*
Radiation Oncology Department
University of Washington
Seattle, Washington

Abstract

The field of Artificial Intelligence concerns the design of systems that emulate human cognitive activities, such as reasoning, language understanding, problem solving, and learning. The heart of artificial intelligence programming is the ability of a computer program to do symbol processing, in which the symbols represent concepts, data and other entities that are non-numeric in nature. The methods and ideas that have evolved in this field have been applied to medical problems of diagnosis and therapy problem solving, and in so doing these medical projects have provided a ground for developing new techniques and insights into the fundamental issues of knowledge representation and reasoning methods. In radiation therapy and other medical physics areas the introduction of artificial intelligence programming techniques is new, and there is fertile ground for research and successful applications. This article summarizes some of the fundamental concepts and illustrates with an example from radiation therapy planning.

1 Introduction

Artificial intelligence refers to a collection of subfields of computer science. These subfields all are concerned with programs that attempt to emulate human thought processes or problem solving behavior. Early in the history of the field it was hoped that this emulation would shed light on how human thought processes work. More recently research in the field has established new programming methods, added to our understanding of data abstraction, and even produced some useful practical applications. An encyclopedic but readable survey of the field is the three volume work of Barr and Feigenbaum [1].

Some of these subfields are:

Games - programs that solve puzzles (e.g., the "Fifteen puzzle", the 4 by 4 array of numbers 1-15 with one blank, initially mixed up, where the object is to slide the numbers around and get them in order), play strategy games such as chess, and solve problems such as the "eight queens" problem (how to place eight queens on a chess board so that no queen attacks any other queen).

Logical reasoning - theorem proving, symbolic mathematics, programs that do algebra and calculus (not numerical integration but symbolic integration, i.e., indefinite integrals).

*This work was partially supported by NIH grant no. LM04174 from the National Library of Medicine and contract no. CM97282 from the National Cancer Institute

Natural language interpreters - story understanding programs as well as language translators. Of great interest are programs that accept commands in conversational language and translate them into data base queries or other formal command languages [2].

Automatic programming - automatic generation of a computer program in a recognized programming language from a higher level specification of the program's behavior.

Learning and inductive inference - programs that refine their procedures or knowledge base depending on the outcome of processing cases; also, programs that can make discoveries, such as Lenat's AM program that generates new mathematical theorems [3].

Robotics, vision, pattern recognition - includes many applications from industrial use of robots to automated analysis of satellite photographs and medical images.

Expert systems - programs that capture in a functional way the knowledge of an expert in some area, such as medical diagnosis, machine troubleshooting, planning of experiments, analysis of scientific data beyond traditional numerical algorithms.

In this article we will focus on expert systems since this area is the one most actively applied to medical problems. However, computer vision, pattern recognition, and robotics are rapidly becoming important for some medical applications also. These topics are beyond the scope of this article; however, the reader is referred to [1] for examples and further discussion.

2 Artificial Intelligence in Medicine

Most medical AI applications are expert systems. The most well known, the MYCIN program [4], which provides advice for diagnosis and therapy for blood infections, is a "rule-based" system, and has become the paradigm for design and application of rule-based systems not only in medicine, but for hundreds of applications in industry and other areas. Because of this, we will discuss in some detail how a rule-based system works and include an example of a research project under way using rule-based programming for planning of radiation therapy treatments [5,6]. A good survey of the early work in medical artificial intelligence is given in [7].

2.1 Medical objectives

It is useful to distinguish between the medical issues that these systems address and the design principles they follow. The medical objectives include such categories of medical problems as:

- solving diagnostic problems
- recommending therapy
- critique of management plans
- training in clinical medicine

These medical objectives may be achieved by a variety of methods, or a new method may be developed for them. In defining the medical problem we have to answer questions such as:

- What is already known? You can not build an expert system program without an established body of expertise. This may not be in a form suitable for direct translation into a computer program. In some cases it is not even available in written form, but may be the collective knowledge of identifiable expert practitioners.
- How is the knowledge used? What kinds of reasoning processes do experts utilize? This may be hard to evoke.
- Is the knowledge divisible into many small relatively independent pieces, or is it in large chunks? The former is easily adaptable to a knowledge programming system, while the latter is more challenging.

It is also important to consider what the purpose of the computer program is. Do we wish to make high level expertise more widely available? Do we intend to provide a supplemental reference medium for practitioners?

2.2 Medical AI examples

Medical applications of artificial intelligence techniques include examples from many areas of clinical medicine as well as basic medical science. There have also been applications of AI ideas to medical education [8]. Some interesting examples of medical artificial intelligence projects are listed here. This is not an exhaustive or balanced list but a sample representing a variety of methods and applications.

- MYCIN - an infectious disease consultant [4]. The MYCIN program pioneered the use of rules and uncertainty measures in reasoning systems.
- INTERNIST - an internal medicine diagnostic tool [9]. The INTERNIST program attempted to cover a large comprehensive subject area by organizing diseases into a taxonomy or disease tree, and to model reasoning that clinicians use to solve diagnostic problems.
- CENTAUR - pulmonary function test analysis [10]. This program derives from the earlier PUFF [11] program, one of the few in routine clinical use. It demonstrated the use of a data representation scheme called "frames", and the organization of diseases into prototypical descriptions as a way to focus the application of rules and gain efficiency.
- HODGKINS - decision analysis for diagnostic workup of Hodgkin's Disease [12]. This program differs from the others in that it uses statistical calculations to select between alternatives, rather than rules.
- ONCOCIN - management of protocols for cancer chemotherapy [13]. The ONCOCIN program uses a division of labor, in that there are two separate programs, an INTERVIEWER and a REASONER, which communicate data back and forth. The ONCOCIN project also has resulted in many offshoot investigations, including projects on time-oriented reasoning, tools for clinical protocol design, and integration of statistical decision theory and rule-based strategies.

- RADEK - planning of radiation therapy treatments for cancer [5]. This project is a pioneering effort to systematize knowledge about radiation therapy treatment strategies. It also uses two separate programs, a reasoner and a radiation treatment simulator. The program uses frames, heirarchies of anatomical sites and prototype plan building blocks to reduce a numerically intractable search to a design procedure.

2.3 Design issues

From the variety of examples above, it is clear that there are many ways to approach the problem of design of an expert system to solve a given medical problem. It should be kept in mind that these systems do not necessarily "solve" clinical problems but may function as tools that aid in solving clinical problems, much like instruments or textbooks.

Although the field of expert systems has made significant advances and many small to medium sized rule-based expert systems are in use in industry, few medical systems are in clinical use. Some of the difficulties are in the design of the user interface, but also there is some question about what problems these systems address. Do they assist the physician and the other medical professionals or attempt to replicate their function? The latter is of little use but the former would find wide acceptance now that computers are a commonplace phenomenon.

2.3.1 Knowledge representation

A fundamental part of the expert system design process is the choice of method(s) for knowledge representation. In writing any computer program we are taking what we know about some problem and representing that knowledge in some abstract language that a computer can process. Some considerations in this process are:

- How to integrate factual and judgmental knowledge. Factual knowledge may be derived from large patient databases by statistical methods. Judgmental knowledge is a distilled collection of principles modified by judgment and experience.

- Is the knowledge formalized already? If so, does this formalism have a straightforward counterpart in a computer program or programming paradigm?

- Knowledge acquisition - how do we carry out the translation of knowledge into computerized form? Must this be done by a hybrid expert at both the subject matter and computer science or can it be automated in some way?

- Knowledge vs. representation language - the choice of representation language will be influenced by the kind of knowledge. When a researcher says that the knowledge base in a program to do radiation therapy planning contains descriptions of prototype plan fragments and directions for when to use them, he is describing the knowledge in the system, but when he says that plans and rules for modifying plans are represented as frames with details organized into slots, he is describing the representation language.

- Meta-knowledge - sometimes it is possible to describe a higher level of know-how than the details of the medical phenomenon itself, a kind of more general problem solving methodology (see, for example, Mark Stefik's MOLGEN program [14]).

Then there may be different classes of knowledge and methods of representing and using each.

The distinction we make here between knowledge and data is that knowledge represents a uniform principle or property of the data that may be succinctly summarized and generally applied. Knowledge has predictive and deductive power.

2.3.2 Reasoning strategies

The use of medical knowledge concerns reasoning strategies. Because of the wide availability of so-called expert system shells, the novice does not consider in detail the reasoning strategies appropriate to the problem, but may simply use the default method provided by the shell.

Methods of reasoning include classification, pattern matching, planning, and design. Of these, classification is widely applied to medical problems. Classification involves the selection of (or accumulating evidence for) a solution to a problem, when there are a relatively few candidate solutions. Search strategies explore a network or tree of nodes and links between nodes, to find a solution. Systems in which probabilities are available either from "training sets" or from clinical databases can use Bayes' Theorem to evaluate each branch of a decision tree to decide which node to examine next. Rule-based systems may use other methods to select rules from the knowledge base. Planning and design problems present new difficulties, most particularly the problems of satisfying constraints.

Another approach to medical problem solving uses case based reasoning. A case based system will contain a large database of actual clinical cases, not as a source of statistical probability data, but as examples, which can be searched for a case similar to the one under consideration. The known properties of that case can then become candidates for conclusions about the case at hand.

Some systems use knowledge in the form of models, from which the system may draw conclusions or derive predictions or other data. Several kinds of modeling techniques are used in expert systems. They include:

- models of prognosis [15]
- clinical algorithms (decision trees) [16]
- mathematical models
- statistical (e.g., Bayesian) models [12]
- qualitative models [17]

More recently, qualitative modeling and mathematical modeling have been bridged by semi-quantitative techniques something like coarse numerical integration [18]. In the radiation therapy example we will present, the model is a computation of radiation dose for a proposed treatment plan, and is used to determine if the plan meets the constraints that were set by the system.

2.4 What makes a system useful?

Although we have focused on the design of a knowledge base and its use in problem solving, the usefulness of an artificial intelligence program will depend on other factors as well as the accuracy of its reasoning system. These factors include:

- dialogue, other interaction techniques
- explanation mechanisms
- efficiency

Examples of issues in dialogue management and user interaction are given in [19,20].

3 Components of an Expert System

Having discussed the background, purposes and overall organization of medical expert systems, we now turn to a discussion of the detailed workings of a rule-based expert system. In order to illustrate these ideas in a familiar example, we then describe an expert system for design of radiation therapy plans, which is under development at the University of Washington [5,6].

All rule-based expert systems have three basic components:

- A Global Data Base of Facts
- A Collection of Rules (Productions), the so-called knowledge base
- A Control Strategy

Facts are assertions about the case at hand. They are organized in one or another formal way. One way is to represent facts as lists of three elements, consisting of an identifier of an object, an attribute of the object, and a value for that attribute. For example, the object-attribute-value triple `(PATIENT-1 TUMOR-SIZE (7.5 CM))` asserts that the size of the tumor found in the patient represented by the symbol `PATIENT-1` is 7.5 cm. Another way to represent these facts is to create an outline-like structure called a "frame", which has "slots" or attribute labels, so that all the facts for a given object can be organized together. These facts can be organized into more complex structures, but in any case, their role is to define the specific problem that needs to be solved. Thus the facts may collectively describe a patient, or a specific episode of illness for a patient.

Rules represent generic knowledge that may be applied to the problem. By generic, we mean that they apply to any case that may satisfy the prerequisites for the rule, not just the specific case under consideration. Rules are in `IF-THEN` form, e.g., there is an 'if' part, which may be a set of assertions (facts) or a logical expression, and a 'then' part, which may be additional assertions or an operation to perform. When the facts of the 'if' part are present, or the logical expression is true, the facts in the 'then' part may be added to the database (if they are not already present), or the operation in the 'then' part is performed. See Figure 1 for an example rule for radiation therapy planning. Note that in this example, facts are object-attribute-value triples, and the if-part is a logical expression requiring that all the listed facts be present in order to use the conclusion.

```
(IF (PATIENT-1 PRIMARY RESECTED)
    AND (PATIENT-1 NODES POSITIVE)
    AND (NODES TREATMENT (COMBINED RESECTION AND RADIATION))

(THEN (NODES DOSE (6,000 RAD))
```

Figure 1: An example rule for radiation therapy planning

A control strategy is a procedure for applying the rules to the data until you have obtained a solution. This requires that there be a termination condition, a criterion for knowing when you have either obtained a solution to the problem or have exhausted all means for doing so. Figure 2 illustrates the general form of the control procedure. The important details are of course in how the procedure selects a rule to examine, and how it determines success or failure.

```
procedure PRODUCTION

    set DATA to initial database;

    until DATA satisfies the termination condition, do:
    begin
        select some rule, R, in the set of rules
           that can be applied to DATA

        set DATA to the result of applying R to DATA
    end
```

Figure 2: The basic procedure for a production system

There are several classes of control strategies, depending on how the search for a solution proceeds. An *irrevocable* strategy applies a rule without any possibility for reconsidering it later, in the sense of undoing its conclusion (it may be used again, but not revoked). On the other hand, a *tentative* strategy allows reconsideration of steps or decisions made earlier. If one wishes to apply rules and then reconsider, there must be some record kept of how rules were applied, so they may be reversed. One such tentative strategy is called *backtracking*, in which one may apply a rule, and establish a backtracking point, a record of the state of the database at that point. Then if the process reaches a dead end later, one may resume from the backtracking point, but applying a different sequence of rules than before. Another method of tentative control involves maintaining a search graph, in which several sequences of steps or decision alternatives may be maintained and followed simultaneously.

It is also possible to classify control strategies according to how they approach the goal or solution. A *forward chaining* strategy applies rules to the given information, working towards a goal by deducing as many useful conclusions as possible. As each new set of facts is added, more rules may be applicable, leading forward (hopefully) to a solution. A *backward chaining* strategy picks a hypothetical solution and then works backwards to find evidence to support it. In this latter case, having picked a tentative conclusion, you look first at the 'then' part of a rule, to see if it contains data or conclusions you need, then if so, you examine the 'if' part to see if those facts are in the database. One may extend the process (chaining rules together) by taking the (as yet not valid) 'if' part facts as the new tentative conclusions and looking for another rule that enables you to conclude them. Thus you work backwards until you reach known facts.

4 Artificial Intelligence Programming Languages

The two programming languages that have most been the basis for artificial intelligence programming are LISP and Prolog. Until recently, LISP has been predominant. In this section we give a brief summary of what characteristics of these languages have made them so effective in this area. Both are general purpose programming languages and there is no reason one can't write any kinds of programs using them. In fact, LISP is used for many other kinds of programming besides artificial intelligence research and applications.

Much artificial intelligence research is devoted to developing languages at a higher level than LISP, so-called knowledge representation languages and others. Expert system "shells" are also examples of higher level languages. The early expert system shells such as EMYCIN [21] were written in LISP, but more recently, expert system shells have been written in C or other languages, to gain greater efficiency, where the full underpinnings of a powerful LISP environment are not needed.

4.1 LISP

The LISP programming language is almost as old as FORTRAN, having been invented by John McCarthy in 1958 [22]. It is particularly noted for being able to do *symbol processing* and *recursion*. Although the original version of the language was obscure and slow in its implementation, modern LISP systems are remarkably powerful and find application in many areas, some rather unexpected (for example, in the Congressional Record, testimony by MIT physicist Kosta Tsipis, on the vulnerability of MX missiles in "dense pack" basing mode, includes a LISP program for some of the computations). A good textbook on LISP, for people already familiar with computer programming, is the book by Winston and Horn [23]. Here we can only give a flavor of programming in LISP.

LISP is usually interpreted; the interpreter prompts the user for input, then evaluates the expression the user typed. A LISP expression is either an atom or a list. A list consists of an ordered collection of LISP objects, which can each themselves be an atom or a list.

An atom is the smallest unit of structure in LISP. The three common kinds of atoms are: symbols, numbers, and strings. Atoms have print-names, which serve to label them and enable them to be typed by a user or printed by a program. Some examples of atoms are: `6734`, `45.9`, `i-am-an-atom`, `ANOTHER`, and `"a string atom"`.

Lists begin with a left parenthesis and end with a matching right parenthesis. An example of a list is `(this is a list)`. In this list there are four elements, the four atomic symbols, `this`, `is`, `a`, and `list`. Note that this is NOT the same as the character string `"this is a list"`. Another example is the list, `(this is (a (very nested) list))`. In that example, there are three elements, and the third element is a list of three elements, the second element of which is a list of two elements.

Symbols may have values attached to them (we say "bound") or they may be unbound. Unlike conventional languages like FORTRAN, the symbol does not have some binary words of memory allocated for a predefined type of value, such as integer or floating point or string. Instead, a symbol may have any LISP expression bound to it as its value. A symbol may also have a function definition.

If you type the name of an atom which has a value bound to it, the interpreter prints that value. If you type in a list expression the interpreter tries to find a function definition for the symbol that is the first element of the list. A function takes certain inputs and computes some result (the value). Functions may have side effects as well, which means that other symbols' bindings may change, and there may be input or output at a terminal or graphic display. Side effects are important in order to get anything done. They are not always undesirable. There are functions to manipulate any kind of LISP expressions, not just numbers.

The remaining elements of the list are considered the arguments or inputs to the function. The function is evaluated and the interpreter prints the result on your terminal. For example, the `+` symbol has a function definition; it adds all the arguments and returns the result. So if you typed the expression `(+ 4 7)` the interpreter will print `11`. Another example is the function `FIRST` which takes a list as input and returns the first element of the list. So if the symbol `MYLIST` is bound to the list `(A B C D)` as its value, and you type `(FIRST MYLIST)` the interpreter prints the value `A`.

To create functions you use the LISP function `DEFUN` which creates a function definition for a symbol as a side effect. So to *define* the function `FACTORIAL` (which takes as input an integer, N, and returns the value of N!) you would type:

```
(DEFUN FACTORIAL (N)
   (IF (EQ N 0) 1
        (* N (FACTORIAL (- N 1)))))
```

Then you could *use* this LISP function by typing, for example, `(FACTORIAL 24)` and the interpreter will respond with `620448401733239439360000`. The `FACTORIAL` function first checks if the argument (input) is 0, since we know that 0! is 1. If not, it computes N! by first subtracting 1 from N, then uses itself to compute (N-1)!, then multiplies that by N. This illustrates the idea of recursion, in which a procedure or function is defined in terms of itself. LISP explicitly supports recursion, as does Pascal. However, FORTRAN and BASIC do not provide for this type of programming.

By the way, there is no restriction on the number of digits in most LISP implementations of integer arithmetic. This same function would just as easily compute 1000! down to the last digit, providing sufficient dynamic (scratch-pad) memory was available. The result is about 3200 digits long.

LISP functions that manipulate symbols are just as easy. Many excellent examples are discussed in the Winston text [23].

4.2 Prolog

Prolog is a language designed to make it easy to write production systems and logic programs. It is a declarative language, in that you write expressions declaring facts and relations, and let the interpreter try to satisfy the requirements you have specified. The most well known reference on Prolog is the book by Clocksin and Mellish [24].

In Prolog, the primitive concepts are much closer to the expert system components discussed above. Facts in Prolog are written in a standard form consisting of a relationship between objects, something like the object-attribute-value idea above. For example, the classic assertion, "Socrates is a man", would in Prolog be expressed as `man(socrates)`. Another example, `likes(ira,soccer)` would be a way of expressing, "Ira likes soccer". The names of objects in these facts all must begin with lowercase letters. Questions begin with a question mark and a hyphen, followed by the same format as a fact.

```
?- likes(ira,hiking)
```

This expression asks the question, "Does Ira like hiking?". If you type this in to the Prolog interpreter, it will search the list of facts and try to find a match for that fact. If a match is found, the response is `yes` but if no match is found the interpreter will respond with `no`.

Variables are represented by names that begin with uppercase letters. Variables may be used in questions, as in:

```
?- likes(X,soccer)
```

If a match is found, the interpreter prints `X=ira` but if no match is found, the interpreter prints `no`.

Rules are easy to express in Prolog. Suppose you want to express the classic rule, "all men are mortals", or, X is mortal if X is a man. In Prolog you use the special symbol `:-` (for "if"). The rule then looks like this:

```
mortal(X) :- man(X)
```

Then we could ask

```
?- mortal(socrates)
```

The interpreter will search the database of facts to find a match. Not having found a match but having found a rule that matches with X as socrates, the interpreter then searches for a match with the antecedent, man(socrates). Having found it, the interpreter answers, `yes`.

Another way in which a rule may be used is to find instances of a relation or assertion. For example, if you type the expression

```
?- mortal(X)
```

the interpreter will respond

```
X=socrates
```

This is only the barest introduction to Prolog. For more examples, and a more thorough discussion of principles, see the text mentioned above. As you might surmise, the greatest power of Prolog is to express and solve problems involving logic and predicate calculus.

4.3 Expert system shells and other languages

There are now available many expert system development tools. These tools provide facilities for creating a rule-based expert system, including a built-in rule interpreter, a means for entering rules into the knowledge base, and a user interface of some sort. They may provide more or less flexibility in the choice of control strategy, and knowledge representation techniques. They are especially good for rapid prototyping.

One of the earliest expert system shells was the EMYCIN program, which consisted of the control code, user interface, rule editing programs and other superstructure of the MYCIN expert system, but without the rule database of knowledge about bacteriology and blood infections. The EMYCIN program therefore provided a framework in which someone could put in rules about some other subject and thereby generate an expert system for a different subject without writing all the rule interpretation and other code. Some of the commercial expert system shells available today are derived from EMYCIN and have a lot of attractive user interface features added.

5 Radiation Treatment Planning: An Expert System Example

In this section we describe an expert system that generates radiation therapy plans from clinical and anatomical data on a patient. It is designed as a "dosimetrist's assistant" and the dosimetrist is expected to fine tune the plan or plans that the system generates. It can generate more than one candidate plan in some cases; it is then up to the clinical team to evaluate the plans and select the best one to treat the patient.

In planning complex cases, the clinician does not do elaborate optimization calculations. Rather, one considers factors that suggest a general or prototypical approach. This first level of decision making involves the choice of treatment modality, the determination of target volume(s) and doses desired (we include here considerations of areas at risk), and the identification of critical organs whose maximum tolerance dose will limit treatment options. Next, one must consider the types of radiation (photons, electrons, neutrons, other) that may alone or in combination achieve the goals. This problem involves beam geometry considerations, the dosimetric properties of the radiation type and machinery, and practical matters of treatment technique (reproducibility and setup time, for example). There are no uniform or standard procedures for solving this problem. Instead, physicists, dosimetrists, physicians, and clinicians distill from years of experience some heuristics, ideas that may apply in one or another case. The initial formulation of a treatment plan relies on selecting prototypical treatment geometries such as "parallel-opposed lateral fields", or "antero-lateral wedged pair", or other such combination fields, and adapting these to the case at hand. The plan may be modified by adding components or by altering parameters (or both).

A project is under way at the University of Washington to build a plan generating program that accomplishes the goals described above, of creating and refining candidate plans based on clinical information as well as patient geometry. This "reasoning" program is written in COMMON LISP on a Digital Equipment Corporation VAX computer running the VMS operating system. We chose this environment because we had available a sophisticated graphic simulation and dose computation program already running on this system [25].

The program involves the following concepts in its design:

- Constraint checking by use of a simulation system coupled to the expert system. Constraints include required tumor dose, maximum tolerance of normal tissues, extent of area(s) of treatment, and possibly others.
- Frames as the basis for an object-oriented design.
- Prototypes - "standard" techniques or building blocks that are used to create rough approximations, to be refined by plan improvement rules.

5.1 Coupled systems

Traditional radiation treatment planning program systems compute doses for specified treatments for a given patient. They therefore can provide data from which constraint satisfaction can be checked. In our system the dose calculation is run as a parallel computation by another program. The expert system provides the plan data and gets back the dose values, as organ specific dose information as well as target dose data. The plan is evaluated not with a single figure of merit but is analysed in detail for its weak points so the system can decide how best to improve it.

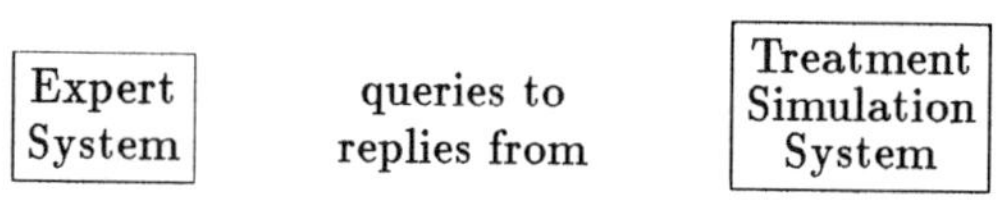

Figure 3: Coupled Systems for Constraint Checking

5.2 Frames - the patient and the knowledge base

5.2.1 The patient (case description)

The patient can be represented by a frame with slots for each of the items of information needed to generate candidate plans, and a slot for the list of plans themselves. Plans are also frames, since much detailed information is needed to specify a plan, e.g., multiple treatment fields. In Figure 4, a part of the patient frame is shown. The slot `SELF` identifies the type of object this frame represents, and the slot `NAME` contains a unique name for the particular instance of the object. The other slots contain symbolic data for the case, organized by keywords. The data symbols are not strings but labels referring to objects. For example, the value for `SITE` in slot `DIAGNOSIS` is used to find rules that can be used for tumors in that area (hard palate). While we could have invented and used some kind of obscure numeric coding scheme for all these symbols, a virtue in using LISP is that the LISP system does it automatically for you and you only need to make up names for the concepts and phenomena you are working on. The LISP code then refers to and manipulates this symbolic data, without the programmer (much less the user) needing to know anything about numeric coding schemes.

```
(Object P1)

(Slot SELF) --->  PATIENT
(Slot NAME)--->  P1
(Slot TNM-STAGE)
    (Keyword T) --->  1
    (Keyword N) --->  0
    (Keyword M) --->  0

(Slot DIAGNOSIS)
    (Keyword PRIMARY-CANCER) ---> SQUAMOUS-C-C
    (Keyword MICROSCOPIC-VARIATION) ---> KER-W-D
    (Keyword LESION-SIZE) ---> 1.5
    (Keyword SITE) ---> HARD-PALATE

(Slot TREATMENT-STAGE)
    (Keyword PRIMARY) ---> INTACT
    (Keyword NODES) ---> INTACT

(Slot TREATMENT-PLAN)
    (Keyword RADIATION) ---> (R0034)
    (Keyword SURGERY) ---> (S0035)
```

Figure 4: Part of a PATIENT frame

5.2.2 Treatment planning rules

Treatment planning rules can be represented similarly as frames. Our implementation includes an English-like readable form of the rule, which is ignored by the program. This readable text part includes an index number, an abbreviation for the source of the rule (e.g., `C06` is the ACS textbook edited by Rubin [26]), and an abbreviated text version of the rule. The remainder of the rule is used by the program, and contains a logical expression to evaluate for the 'if' part and a stylized form for the 'then' part (a list consisting of object, action, slot, and value for that slot).

5.3 Control strategy

In this system we use an agenda-based control strategy. There are a variety of subtasks in the planning process, and they are not always done in the same order. So we have a small upper level expert system with rules for deciding what tasks need to be done. The system generates an agenda of tasks from these rules, then runs each task, using only the rules pertaining to that task. This focuses the search for a solution, and makes looking up and trying rules much more efficient than simply searching the rule database for a rule that can be applied.

```
(2 GEL ((Massive (=T4 and large T3) but localized
   (NO and exophytic) disease (no nodes involved))
   (Inclined to surgery (=MULTIPLE (POSTOPERATIVE) plan)))
(AND (EQ 4 (TSTAGE CURRENT-PATIENT))
     (IS-ITEM CURRENT-PATIENT 'DIAGNOSIS 'LESION-TYPE 'EXOPHYTIC)
     (IS-ITEM CURRENT-PATIENT 'TNM-STAGE 'N 0 ))
(CURRENT-PATIENT
   ESTABLISH
   TREATMENT-PLAN
   (MAKE-NEW-PLAN 'MULTIPLE '(PRIMARY-SEQUENCE POSTOPERATIVE))))

(4 C06 ((The primary is resected (= at all) and
     nodes are involved and node treatment is MULTIPLE)
   (6,000 rad for nodes))
(AND (MEMBER (GETPROP CURRENT-PLAN 'PRIMARY-MODALITY)
             '(SURGERY MULTIPLE))
     (IS-SLOT CURRENT-PLAN 'NODES-MODALITY 'MULTIPLE))
(CURRENT-PLAN
   ESTABLISH
   NODES-DOSE
   6000))
```

Figure 5: Some Radiation Treatment Planning Rules

5.4 Prototypes for treatment plans

In creating a radiation treatment plan, we must start with some values for a set of numeric parameters that can assume values in a continuous range, e.g., radiation beam gantry angle, from 0 to 360 degrees. Certain "standard" geometric arrangements of radiation beams have evolved as basic approaches to treatment in various disease sites, so we use these geometric arrangements as starting points, or prototypes. A prototype is not a complete plan or model plan, but a building block for a plan, so a plan may be constructed from one or more prototypes in combination. Thus prototypes represent useful starting points and abstractions of treatment strategies.

Examples of prototypes commonly used for head and neck cancer include:

- Parallel opposed lateral fields
- Anterior-lateral wedged pair
- Posterior oblique wedged pair with anterior field

For disease sites other than head and neck, there may be additional prototypes that are useful there but not in the head or neck, e.g., the four field box arrangement for bladder cancer.

The knowledge base therefore includes rules for selecting prototypes, based on disease site and other factors. Planning proceeds by selecting one or more prototypes, creating a plan, computing the dose distribution and then analysing the results to see what improvements can be made.

5.5 Plan improvement rules

These are examples of rules to use when trying to improve a plan that is not satisfactory or perhaps not optimal (that is, it may be possible to obtain the same target dose and lower critical organ dose by modifying the treatment parameters somewhat).

1. If there is a cold spot in a deep area and a hot spot in a shallow area, then switch that beam to higher energy.
2. If there is a cold spot just outside a beam's edge, then increase the beam aperture in that direction.
3. If an organ is overdosed by one beam, then
   (a) reduce that beams's contribution and increase the others, or
   (b) change the gantry angle of that beam, or
   (c) include a shielding block in that beam's path.

Which rules are selected depends primarily on the outcome of analysis of the dose computation results. Some rules involve qualitative changes in the plan (even the addition or removal of a prototype component); others involve quantitative changes. Some problem conditions suggest more than one kind of change to make. In these cases we need to be able to keep track of what has been tried so we do not needlessly explore dead end combinations of parameter ranges.

Implementation of the rules is somewhat difficult, since we are imposing a kind of qualitative procedure in an area where the objects are described by continuously varying numeric parameters.

5.6 Work in progress

The radiation therapy planning project is far from completed but there are some definite areas in which we believe progress can be made. These include:

- Refinement of the evaluation scheme for treatment plans. It is crucial to the plan refinement step that the evaluation of a treatment plan provides useful information about a plan's strengths and weaknesses, and the possible causes of them. This will likely involve qualitative modeling of the properties of radiation beams in various treatment geometries.

- Efficient control strategy for coupling between expert system and dose computation. It is important to know when to compute the dose distribution and when qualitative reasoning can provide what is needed for the next step. Dose calculations are computationally costly. A few seconds to several minutes is the time required for plans in a typical range of complexity. Many iterations of dose calculation would not be practical.

- How to handle failures ("dead ends") with continuously varying parameters. We would like to know not each and every plan that was tried but whether a whole region of parameter values seems fruitless to explore further. A technique called "dependency directed backtracking" [27] can be extended to numerical (continuous) parameters in order to solve this problem.

- Evaluation of the working system will be important to determine if it can generate clinically acceptable or useful plans.

- Redesign of user interface - if the system appears to be a useful tool, it will need to be integrated into the clinical workflow. In this, the coupling with the treatment simulation system will be particularly important. A smooth integration here could make the system available in an already well established context of computer applications.

- Extension to other areas of cancer would be relatively straightforward if the basic ideas are proven sound.

5.7 Clinical application

How do we envision use of this work? There are a number of ways in which an expert system program for radiation treatment planning could be used.

- To assist a busy clinic with work load. At larger radiotherapy clinics, as many as four dosimetrists are employed to handle this work. Qualified dosimetrists are very scarce and this shows little sign of changing.

- To provide a tool (assistance) for smaller clinics where it is just not financially feasible to hire more dosimetrists.

- To teach clinical residents. While the application of expert systems to teaching is still a developing field, it is clear that clinical simulations can play an important role in teaching clinical medicine, and this in particular is an extension of a simulation system.

6 Conclusions

The field of artificial intelligence in general (and medical artificial intelligence in particular) is maturing rapidly. While there are significant research problems to solve, we are also entering an era where attention to practical matters will yield useful computer programs and systems that augment the practice of clinical medicine. Here we briefly list some of these research area and practical issues.

6.1 Research areas

The following research topics represent ongoing work in various medical artificial intelligence projects.

Represent common-sense knowledge - medical practicioners bring to their practice an enormous store of general knowledge about the world, and about their patients as human beings. Is it possible or appropriate or necessary to incorporate this vast content into an expert system?

Reason temporally - much work is now in progress to define ways of representing temporal relationships as well as time-varying data. This is particularly important in chemotherapy but also important in other fields of medicine as well.

Represent anatomic knowledge - spatial reasoning is important to radiation therapy planning where the geometry or the relation between the surrounding normal tissue and the disease site is central to the process of designing a good treatment. Other areas such as diagnosis of brain lesions [28] also address this issue. The teaching of anatomy may well be revolutionized by development of large anatomical knowledge databases coupled with high performance three dimensional graphic systems.

Construct hybrid systems - it is now recognized that rule-based programming is appropriate for some things but a procedural language like Pascal or FORTRAN is better for some others. Radiation treatment planning is a good example, in which there is already much work on dose calculations, written in a somewhat modular way so that programs written in these different languages can communicate.

Provide access to medical literature - computerized databases are available that provide a reference service, such as MEDLINE and PDQ. Research is under way to provide an analytical "front end", such as the ROUNDSMAN program [29].

Develop constructive problem solving methods - many problems that have been addressed so far in medical AI have been solvable by classification methods, but some (particularly radiation treatment planning) are constructive, or design problems.

6.2 Practical issues

These are the practical issues that we believe need to be addressed in order for medical artificial intelligence systems to become useful in clinical practice.

Medically sound explanations - while it has always been characteristic of AI systems that an explanation mechanism is readily available or built in, the explanation is often in terms of the system's internal techniques, not something that would make sense to a physician or nurse.

Integrate into clinical workflow - it is critical that use of AI systems should not require extra work or time involving data entry or unusual procedures.

Derive patient information from existing sources - As progress is made in hospital information systems design, we will need to give attention to integrating the AI systems into the hospital data systems. The duplication of data entry is a problem even in present hospital systems, where the laboratory, the pharmacy, the radiology department and the admit/discharge computers are all separate and contain redundant (and uncorrelated) information.

Validation - of course any system used in practice must be validated to verify that it performs the tasks it was intended to perform, as required, and correctly. This is not simply a matter of coming up with the same diagnosis as the expert physician but will require the writing of explicit performance specifications that say just what input the program needs, and what it is expected to do.

Improvement of human-computer interface - this is a subject in its own right, but it is clear that modern workstation equipment makes this a reasonable task, whereas earlier the limitations of terminal equipment and the performance of time-shared minicomputers made it impossible to create a really powerful user interface.

If one can avoid the pitfalls of overselling this interesting area of computing, it should be a valuable addition to the use of information systems in the practice of medicine. As radiation therapy was one of the first medical fields to make extensive use of computers in clinical practice, it should also be a fruitful one for applications of artificial intelligence programming.

7 Acknowledgements

It is a pleasure to acknowledge advice and friendly critique from Dr. Jonathan Jacky.

References

[1] Avron Barr and Edward A. Feigenbaum, editors. *The Handbook of Artificial Intelligence.* William Kaufmann, Inc., 1981.

[2] Mary Woodyard and Baruch Hamel. A natural language interface to a clinical data base management system. *Computers and Biomedical Research*, 14:41–62, 1981.

[3] Douglas Lenat. AM: an AI approach to discovery in mathematics. In Randall Davis and Douglas Lenat, editors, *Knowledge-Based Systems in Artificial Intelligence*, McGraw-Hill, New York, 1980.

[4] Bruce Buchanan and Edward H. Shortliffe, editors. *Rule-Based Expert Systems: The MYCIN Experiments of the Stanford Heuristic Programming Project.* Addison-Wesley, 1984.

[5] Ira J. Kalet and Witold Paluszyński. A production expert system for radiation therapy planning. In Allan H. Levy and Ben T. Williams, editors, *AAMSI Congress 1985*, pages 315–319, American Association for Medical Systems and Informatics, Washington, D.C., 1985.

[6] Witold Paluszyński and Ira J. Kalet. Applying expert system techniques to design problems. In *Proceedings of the Second Western Expert Systems Conference*, pages 169–176, Anaheim, California, 1987.

[7] William J. Clancey and Edward H. Shortliffe, editors. *Readings in Medical Artificial Intelligence: The First Decade.* Addison-Wesley, Inc., Reading, Massachusetts, 1984.

[8] William J. Clancy. Dialogue management for rule-based tutorials. In *Proceedings of the Sixth International Joint Conference on Artificial Intelligence*, pages 155–161, William Kaufmann, 1979.

[9] Randolph A. Miller, Harry E. Pople, Jr., and Jack D. Myers. INTERNIST-1, an experimental computer-based diagnostic consultant for general internal medicine. *New England Journal of Medicine*, 307:468–476, 1982.

[10] Janice S. Aikins. Prototypical knowledge for expert systems. *Artificial Intelligence*, 20:163–210, 1983.

[11] Janice S. Aikins, John C. Kunz, Edward H. Shortliffe, and Robert J. Fallat. PUFF: an expert system for interpretation of pulmonary function data. *Computers in Biomedical Research*, 16:199–208, 1983.

[12] Charles Safran et al. Diagnostic planning using computer assisted decision-making for patients with hodgkin's disease. *Cancer*, 39:2426–2434, 1977.

[13] Edward H. Shortliffe et al. ONCOCIN: an expert system for oncology protocol management. In *Proceedings of the Seventh International Joint Conference on Artificial Intelligence*, pages 876–881, William Kaufmann, Inc., Los Altos, California, 1981.

[14] Mark Stefik. Planning with constraints (MOLGEN: part 1). *Artificial Intelligence*, 16:111–140, 1981.

[15] Thomas W. Griffin et al. Predicting the response of head and neck cancers to radiation therapy with a multivariate modelling system: an analysis of the RTOG head and neck registry. presented at the 24th Annual Scientific Meeting of the American Society of Therapeutic Radiologists, October 1982.

[16] Theodore J. Brickner. Carcinoma of the breast. Patterns of Care Study Newsletter, July 1978. Published by the American College of Radiology, Philadelphia, PA.

[17] Benjamin J. Kuipers. Qualitative simulation. *Artificial Intelligence*, 29:289–338, 1986.

[18] Lawrence E. Widman. Expert system reasoning about dynamic systems by semi-quantitative simulation. July 1988. Submitted to Biomedical Measurement, Informatics and Control.

[19] James D. Hollan, Edwin L. Hutchins, and Louis Weitzman. STEAMER: an interactive inspectable simulation-based training system. *The AI Magazine*, 5:15–27, 1984.

[20] C. D. Lane, Joan D. Walton, and Edward H. Shortliffe. Graphical access to medical expert systems: II. design of an interface for physicians. *Methods of Information in Medicine*, 25:143–150, 1986.

[21] William van Melle. A domain-independent production-rule system for consultation programs. In *Proceedings of the Sixth International Joint Conference on Artificial Intelligence*, pages 923–925, William Kaufmann, 1979.

[22] John McCarthy. Recursive functions of symbolic expressions and their computation by machine, part I. *Communications of the ACM*, 3, 1960.

[23] Patrick Henry Winston and Berthold Paul Klaus Horn. *LISP*. Addison-Wesley, second edition, 1984.

[24] William F. Clocksin and Christopher S. Mellish. *Programming in PROLOG*. Springer-Verlag, Berlin, 1981.

[25] Ira J. Kalet and Jonathan P. Jacky. A research-oriented treatment planning program system. *Computer Programs in Biomedicine*, 14:85–98, 1982.

[26] Philip Rubin, editor. *Clinical Oncology for Medical Students and Physicians, a Multidisciplinary Approach*. American Cancer Society, sixth edition, 1983.

[27] R. Dechter. Learning while searching in constraint satisfaction problems. In *Proceedings of the Fifth National Conference on Artificial Intelligence*, pages 178–183, William Kaufmann, 1986.

[28] Gordon Banks, John K. Vries, and Sean McLinden. Radiologic automated diagnosis (RAD). In Helmuth F. Orthner, editor, *Proceedings of the Tenth Annual Symposium on Computer Applications in Medical Care*, pages 228–239, IEEE Computer Society Press, 1986.

[29] Glenn D. Rennels. A computational model of reasoning from the clinical literature. In Helmuth F. Orthner, editor, *Proceedings of the Tenth Annual Symposium on Computer Applications in Medical Care*, pages 373–380, IEEE Computer Society Press, 1986.

IV. APPLICATIONS IN EDUCATION AND MANAGEMENT

COMPUTER-BASED EDUCATION

James C. Carlson, MS, FACR
Current Computer Concepts
3976 Woodmore Drive, Greenwood, IN 46142

ABSTRACT

Computer-based education (CBE) techniques include not only computer aided instruction (CAI) but computer-managed instruction (CMI). Seven principal techniques - drill and practice, simulation, problem solving, testing, dialogue, games, discovery learning - are used in CAI. Hardware and software costs for creating or operating CBE courses range from a few hundred dollars to many thousands of dollars. While the enhanced options and opportunities available with larger systems enable creation of complex, complete courseware, they also dictate greatly increased courseware development time and funding. Newly developed, inexpensive CD ROM systems are expected to enhance development of superior courseware and the universal application of CBE techniques.

INTRODUCTION

Computer-based education (CBE) is frequently viewed as a modern version of the programmed learning techniques that were popular in the 1960's. To the extent this belief survives CBE suffers. CBE can be much more and failure to recognize and utilize its full capacity contributes to its lack of growth.

What started as computer aided instruction (CAI) developed into what is now called CBE. CBE includes CAI and record management techniques, labeled CMI. CBE is the sum of CAI and CMI. Various other names for CAI exist. They are: CBI - computer-based instruction; CAL- computer-aided learning; CBT - computer-based training.

Programmed instruction techniques require the student to respond to a question(s) relating to the paragraph of information just presented. After confirming the correct response by looking on another page, the student proceeds to the next paragraph. Simple forms of CBE follow this same format, but the tremendous capabilities of the computer enable much more complex presentations and reactions to the student's response -- especially in relation to image presentation and course modification. CBE offers many innovative opportunities.

It is assumed that the reader will be a medical physicist interested in using CBE techniques in his/her department for training residents, technologists and other hospital personnel. The self directed and operated format of CBE is especially appropriate for these students who can now intermix their physics training needs at different times and frequency with their many other required activities. Although the effort required to develop a CBE course is extensive, the reduction in repetitive, routine teaching chores may warrant it.

TECHNIQUES OF CBE

CBE terminology and techniques are presented before discussing the hardware and software required to reach the goals of CBE.

Computer based teaching strategies range from simple linear drill and practice (D & P) to complex simulations. Between these extremes lie 5 other categories: problem solving, testing, dialogues, games, and discovery learning.

Drill and practice is the easiest to implement and therefore, the most common. In D & P the student is expected to memorize the association of 2 elements. In physics they learn to associate 3.1415 with the Greek letter PI. If paired associated material is presented sequentially, it's called linear D & P. If it's randomly selected, it's called random D & P.

Some of the tedium of repeating linear D & P associations can be eliminated by using the random D & P form where the material to be presented is presented in a random form each time it is called up. Carried one step further, the pool of 50 or 60 associated items to be presented in a session may be randomly drawn from a pool of 300 - 400 items each time the student reviews the material.

Studying from a unit that is linear is inefficient because the known material must be repeated to review the unknown material. This fault exists even in random D & P; known material is repeated. It can be corrected by programming the computer to delete known material from further review.

In either the linear or random form the associations in a unit should be limited to a level that permits completion of the unit in 15 to 20 minutes.

In a more complex version of drill and practice where math skills are involved, the computer's capabilities can be more fully utilized, e.g., to teach any simple mathematical operation, the computer randomly selects the parameters while making its own calculation of the correct answer.

The dialogue approach has been frequently used despite its more difficult development. Dialogue techniques lead the student to new material according to the type of response obtained. For example, a wrong choice in multi-choice answers will lead to additional frames explaining why the choice is wrong. Additional frames related to the original question may follow or the student may rejoin the mainstream of the lesson. On the other hand, correct choices on several closely related questions can result in elimination of this topic from further review.

Dialogues are of two basic types: tutorial and inquiries. In tutorials the lesson plan is completely predetermined by the course while with inquiries the student selects the topics of study, proceeding in whatever direction and pace desired.

Computer simulation techniques are particularly appropriate in teaching the use of complex, expensive and dangerous equipment. Computer simulation of real word situations sometimes exceeds the capabilities of human teachers. A clear cut example is pilot training in flight simulators; but simulation courses need not be this complex. One appropriate for teaching radiation safety could show what happens to radiation levels on the outside of a linear accelerator room as the student changes the beam direction. When enhanced by allowing the student to vary scattering media or room wall thickness, the dynamic presentations will give the student a better "feel" of what's happening.

Games can be very instructive, particularly with students who have little motivation. In single player games the player strives to reach a goal without getting zapped; in multiple players games they compete against each other. Games may be based on simple fact recital or on model building where reaching the goal is achieved only when appropriate model parameters have been chosen.

Testing is a new kind of CAT - computer assisted testing. Testing has many advantages, especially in the environment of the medical physics department. CAT allows:

* the student to take the exam at his/her convenience
* differently worded questions to each student thereby helping to eliminate answer exchanges
* evaluation of areas of weakness - not only for individual students but the entire class
* build up of a large bank of questions
* elimination of much typing
* elimination of tedious exam grading
* immediate feedback to the student
* item analysis

In problem solving techniques the computer can be used to lead the student through complex problems. To do so the problem must be broken into small steps where incorrect input is corrected by the computer with appropriate explanations. Properly designed, the student is relieved of complex calculations but is required to analyze the problem and provide direction.

It is frequently said, "experience is the best teacher". In discovery learning the student is given the opportunity to select different parameters and topics in the area of study. A discovery learning course allows the student to jump to different topics after every frame. A frame on decay schemes could be followed by one on the operation of a survey meter and then followed by one on the definition of the rad. A comparison here is the student opening his/her book to a particular chapter or page after checking the index for the topic.

HARDWARE

These operations require appropriate hardware. Simple activities require minimum hardware - desirable from an economic viewpoint but unacceptable from that of effective utilization of CBE techniques. A dilemma exists. Although funding, time and experience parameters dictate a minimal system, future plans and needs should be incorporated in the initial hardware specifications.

Hardware items to consider are computers, disks, display systems, and input devices. This discussion of hardware will tend to stress IBM PC compatible equipment ignoring the fact that much existing CBE is designed for mini- and mainframe computers (e.g., DEC's authoring language and course authoring system will be ignored) and Apple systems have a wide following with many products in the education field.

What are the hardware and software costs for CBE? They can range from under $700 to over $50,000. In the under $700 category you may create simple courseware that runs on an IBM PC clone with 640K bytes of memory and a floppy disk system. At the high end you're buying a very complete system that will accommodate many students and complex presentations.

Single user IBMs (and clones) are appropriate for simple cost delimited systems. At the low end, at about $600, is the computer with 640K of memory, a single floppy disk drive and a monochrome screen. Dedicating this inexpensive system to students would enable independent study as their schedule permits.

At the high end are the 80386 based systems employing a multiuser operating system such as UNIX or OS/2. A minimum price is about $4,000. Add to this the cost of extra terminals, extra memory and an expensive operating system.

Memory - the more the merrier. Don't skimp on memory in a large system. Multiuser operating systems require 2 or 3 megabytes of memory without any application programs; and multiusers means you will need another 2 or 3 megabytes to accommodate their programs.

Although a system containing only floppy disk drives may be adequate for student use on simple courses, any development system should include at least one hard disk of at least 30 megabytes capacity. When considering a larger drive - basically, the bigger the better - give considerable weight to the need for increased retrieval speed.

Optical disks are of interest as a distribution medium and as a source of images. They may contain all of the course (software and images) or only the images associated with it. Optical disks are preeminent for the distribution of huge amounts of data (and images) to large groups and are the systems of the future.

Some fundamental optical disk characteristics are: 1) film based; 2) reflective; 3) constant angular or constant linear velocity; 4) write once read many (WORM); and 5) read only.

Film based optical systems are rare. To my knowledge the only supplier of film based systems is McDonnell Douglas. These systems have features rivaling those of reflective systems and have the advantage of very low cost of mastering and duplication. Their disadvantages are only 18 minutes of motion video (or 32,440 still frames) and system cost is relatively high.

Reflective systems are CLV - constant linear velocity - or CAV-constant angular velocity. Worst time access is 2 - 3 seconds. A 12 inch CAV disk will store 30 minutes of video per side, i.e., 54,000 frames. A 12 inch CLV disk contains 60 minutes of video and audio on one side.

Standard features on playback system hardware are:

* stop on single frame
* frame or time display
* scan forward/reverse
* fast scan forward/reverse
* step forward/reverse
* multi-speed forward/reverse
* dual or single audio channels
* frame/time search
* instant jump

Playback hardware may be purchased at 3 different levels. All have the characteristics of level 1 which are those above. Level 2 playback equipment has a built-in microprocessor which allows the system to be used as a stand-alone interactive system. Any optical disc player interfaced to a computer is a level 3 playback system. Level 4 systems with additional features are being developed.

Optical systems manufactured by Hitachi, Philips, Pioneer and Sony range in price from $550 to $1700 for a level 1 system. Level 2 systems range from $1600 to $1900 and level 3 systems range from $1600 to $3500.

The display system will always be CRTs -- except LCDs may replace them in the future. The term CBE conjures the image of a student in front of a screen. After all CBE is intended to be interactive and it's difficult for the computer to be interactive with a large group. Yet it can be done. In these cases, large 25 or 45 inch screens will be useful.

To follow fiscal constraints you may upgrade from a simple monochrome monitor to a high resolution color monitor when your courseware advances require it. In the long run a high resolution color monitor for compatibility to the high quality images obtained from optical disks is desired. In this situation you may use the monochrome monitor for questions and responses and the color monitor for high resolution images.

What's the best interactive input device? The answer is highly dependent on the type of user you have. The keyboard is not too complex for your users but, obviously, for children a touch screen is superior.

The mouse, digital graphics board and light pen all work on the same principle - a device highlights your choice of screen presentations. It's a menu selection device - a feature many people appreciate. Menu selection eliminates the common keyboard problem of accidentally hitting wrong keys. (Incidentally, if you are programming in a standard language like Fortran, C, Pascal, etc. and are programming for keyboard response, you must always include error routines that prevent the user from creating havoc with erroneous keyboard activity. Essentially, you have a hidden menu that must match typed material.)

What's the saying: "A picture is worth a thousand words"? Image displays and even video presentation can add much interest and capability to the course. An additional item to include in a complete system is a scanner or video camera that creates digitized images to add images to your course. Images can also be purchased by the thousands on digital disks. For example, NEC Home Electronics now sells a CD-ROM player (with interface) for $1200 and 2 disks for $399 each. One disk contains 2500 clip art images; the other is a library of more than 4,000 color photos.

SOFTWARE

Special programs (authoring software) designed to ease the burden of creating CBE courses (courseware) are available. Examine these carefully before proceeding. Addresses for many companies supplying authoring software are found in appendix H of The Educators Handbook To Interactive Videodisc.

Authoring software is not a new idea. In the late 60's IBM developed Coursewriter for the systems 1500 machine. Also during the 60's the U. of Illinois developed PLATO which was later distributed commercially by Control Data Corp. It used multiple terminals on a large central computer. Brigham Young and the U. of Texas jointly developed a CBE system running on a minicomputer. The product TICCIT is currently distributed by Hazeltine Corp.

Why use software designed to produce courseware? It's known that being able to create a program with Fortran, C, or C++ to solve your specific needs is very convenient; and whenever you want something additional, you can do it because you have the source listing. Why buy authoring software? Answer: because anything other than a very simple program is very time consuming.

Authoring software prices range from $80 to $5500. Maybe some of the inexpensive ones contain everything you need - but that may be a computerized form of 3X5 cards. A simple drill and practice course is relatively easy to produce. However, even at this level you probably want the management capabilities obtained in authoring software.

Even in a simple course, as illustrated in figures 1-3, you should use branching according to the student's response. The first screen, figure 1, requires a response after describing a formula for absorption calculations. If the response is correct, the lesson proceeds to the next paragraph, screen 3, further developing the topic. If the response is not correct, an explanation is provided, screen 2.

$I=I_0e^{-kt}$ describes the absorption of a monoenergetic beam of radiation (photons).
where: I - transmitted radiation level
I_0 - radiation level with no absorber
k - linear absorption coefficient
t - thickness of absorber

I/I_0 * 100 represents the _____ transmisssion of the absorber.
choose: a. loss of b. increase in
c. percentage d. ideal

Fig. 1. Screen 1. A simple CBE course follows the techniques of programmed learning. After a short presentation the student is required to answer a question about it.

I/I_0 is the ratio of transmitted radiation level to the original level. When multiplied by 100 the ratio becomes the percentage transmitted.

(press any key to continue)

Fig. 2. Screen 2. If the student chooses the wrong answer, follow up information is provided.

This formula is only true for monoenergetic beams because the linear absorption coefficient varies with photon energy. Plotted on semi-log paper this formula produces a straight line for monoenergetic beams and a curved one for heterogeneous ones.

for the following choose: T/F
For a heterogeneous beam you would expect the first part of the curve to be steep representing absorption of soft radiation. Later the curve flattens because the beam is now composed mostly of harder radiation.

Fig. 3. Screen 3. Rejoin the primary lesson path after completing all remedial comments and subsidiary lessons.

Although a course at this level may accomplish the desired goals, the super system described below would be better. Do you have that much time and money? It should have at least the following characteristics:

* distributed on optical disk
* disk contains all the software, images and sound
* operating on a network, several students can use it simultaneously
* extremely fast response times because each student loads his/her lesson into his/her workstation with 8 megabytes of memory
* every statement and response has an image or images associated with it
* the workstation has 2 color screens, one for text and one with high resolution for images
* each item within a topic has 5 questions of different wording
* at least 3 course techniques are used
* wrong answers lead to remedial presentations
* automatic suppression of each student's known material
* complete user data management
    + student review of his/her study report
    + reports show correct and incorrect response frequency for each item and topic
    + report lists the specific topics needing additional study
    + report maintains individual student utilization records
    + report contains complete class summary records and displays them graphically
* video materials are incorporated into the lessons and on request a 1 to 2 minute video presentation on any of the topics in the course is obtained
* complete index

Having created this super course, how do you get it onto optical disk? One method involves writing it onto a WORM (write once read many) drive. Most frequently, in today's environment, TV tape editing techniques are used to get everything onto the tape in the exact sequence it is to go onto the optical disk. A CD-ROM service organization then transfers the video tape to optical disk. This mastering process including the set up charge is about $2-4,000. But duplicates of the master cost only $10-20.

THE FUTURE REQUIRES OPTICAL DISK

It appears optical disks are essential. If asked whether it is best to buy a bicycle or a Rolls Royce to travel from Boston to San Francisco, the choice is easy - given enough money. There's a lot in between and if you don't have the money for the Rolls, you must settle for less. CBE has been around for more than 20 years, yet its success rate is very small. Why? The bicycle products are grossly inadequate, and the Rolls is too expensive. In addition, the unknown, developing technology and pricing of the essential component, optical disk, discouraged their use. Changes in this field enabling the use of optical disks in moderately priced systems should unleash the true capabilities of CBE.

Keep in mind what the optical disk does for you. Your courseware essentially uses the computer to deliver selections from thousands of images on the disk. If you like, you can use a short sequence of images, which is actually a short video presentation including sound. The student's responses to questions determines which of these many images or video series are appropriate and it varies for every student. Follow up material can be based on the student's responses; consequently, the course becomes tailored to each and every student.

To date, the field of CBE has mimicked the early stages of the movie industry. Originally movies were made by simply putting on film what was already happening on stage; consequently, there was little difference between the stage and movies. Today it's impossible to do on stage what is done in the movies. The enormous potential of CBE has not been released (except in a few outstanding examples) and will only be released when the time, money and effort required to utilize the available techniques and equipment are expended.

BIBLIOGRAPHY

Ed Schwartz. The Educator's Handbook To Interactive Videodisc. Washington, DC: Association for Educatiional Communications & Technology, 1987

Introduction to Computer-Based Education, Marlborough, MA: Digital Equipment Corporation, 1984

DEPARTMENT MANAGEMENT

Don P. Ragan, Ph.D.
Wayne State University
Detroit, Michigan 48201

ABSTRACT

Hospital department management systems have been under development and in use for two decades.[1] Especially suitable for automation have been departments such as clinical labs, pharmacy, and radiology. These systems provide improved productivity, charge recovery, patient flow, and quality assurance. This is important because medicine is a service industry dependent on information for improved efficiency. Data processing provides to medicine benefits analogous to the benefits robotics provides to manufacturing.

INTRODUCTION

The lack of management and clinical information in medicine has been identified by others. Dr. Otten pointed out that "one can obtain more information about the content of a hot dog one buys than the quality of one's medical care".[2] Variation exists in clinical results; documented differences in end results in several Illinois hospitals have been reported where the 5 year survival varied between 7% and 46% in the least to most successful institution.[3]

The amount of clinical data to be processed in medicine is well within mini or micro computer capabilities. Typical departments process hundreds or thousands of procedures per day. Compared to banking or point of sale applications, this is trivial. For a sample clinical study of 386 cervix patients, 112 data elements were recorded per patient resulting in approximately 43,000 possible data items.[4] Searching or sorting this information by hand is onerous, but computer processing is simple.

Someone once defined medical judgement as the willingness to make medical decisions based on inadequate information. This author believes medicine needs to spend more on information systems. The U.S. spends 20% of the Gross National Product on information processing and only 10% on health care. Another point of comparison: the banking industry spends 5% of gross revenue on data processing and only 1.3% is spent in medicine.

Department management systems can enhance scheduling, communication, patient flow, charge capture, and clinical research in a medical setting. There are approximately 14 companies offering such systems in radiology and 2 companies in radiation oncology. It's most illustrative to review the capabilities and characteristics of these systems by example. This article will use the Radiotherapy Analysis and Management System (RAMS) as an example, not to endorse it, but because it is the one the author is most familiar with.

RAMS DESCRIPTION

RAMS is divided into two modules, Analysis and Management. The original motivation for system development was the accumulation of clinical research data. The Analysis module provides this tool. It has the capability of adding special study information to conveniently augment the tumor registry and follow-up data. This data base may be searched quickly by entering any arbitrary set of such criteria. Statistical analysis is available.

The remainder of this paper is devoted to a discussion of department management systems which are exemplified by the Management module of RAMS. The objective of this module is to provide information on communication, patient flow, facility utilization, charge capture, and quality assurance within the department.

RAMS structure is menu driven with sequential question and answer (Q/A) presentation to the user. Other systems may be frame or screen, rather than Q/A, driven. Interaction with the system is predominantly through interactive CRTs scattered throughout the department. This is supplemented by printed output which is available on demand or through a batch routine that runs nightly producing selected reports. These are available for each person when they arrive in the morning. Printouts are usually available in a short (abbreviated) form on a single page and an expanded form with more data and space to provide a "worksheet" style tool.

This is exemplified by traveling down the Physician area of the menu and selecting a printout of the "Short Physician Schedule" (Figures 1 and 2). This provides a list of patients sorted by function with a four-day window into scheduled procedures. The last column provides scheduled procedures falling outside this four-day window and protocol number (if applicable). The printout lists <u>all</u> patients for whom this physician is responsible; many of these are not yet scheduled for treatment and are kept in file as "Treatment Decision Pending" or "Treatment Delay" patients. It is important for systems to accommodate "suspense files" like these to reflect accurately patient flow through the diagnosis and treatment process. RAMS provides five patient statuses: Consult (Pre-Treatment), Treatment Decision Pending, Treatment Delay, Treatment (including Hold) and ready for Billing.

Radiotherapy care is centered around patient treatment and RAMS reflects this process. Machine schedules (Figure 3) are set up on ten-minute default intervals so that empty slots can be easily identified. This can be overridden by assigning slots of arbitrary length. The system provides a scheduling aid for technologists; it does not attempt to select an optimal time. Treatments are scheduled in perpetuity (block scheduled) as opposed to consults, simulation, follow-ups, etc., which are scheduled as single events. RAMS allows restricted double booking of patients for treatment through the scheduling of patients "to-start" and on "hold". It is important to warn the user of conflicts or over-commitment of staff and/or facilities; this is particularly important in radiology systems where multiple equivalent resources may be available or where two tests may be incompatible typically due to use of contrast media.

Significant effort has been invested in RAMS to integrate the system into the normal operation of the department. There are no individuals dedicated to data entry; the clerical/scheduling/records staff enter demographic and insurance (registration) information in

ENTER RAMS OPTION: ?

RAMS is menu driven; Entry of "?" at any prompt lists the available option for the current page of the menu.

ANALYSIS
CHARGE CAPTURE
FOLLOWUP
LIBRARY
MANAGEMENT
NURSING & QUEUE
PHYSICS
PRE-TREATMENT
SCHEDULES & LISTS
TREATMENT

SPECIAL COMMANDS: KILL, INDEX, DETAIL

ENTER RAMS OPTION: SC

ENTER SCHEDULES & LISTS OPTION: ?

SPECIAL COMMANDS: (KILL, INDEX, DETAIL) provide audit trail, index and usage information at any level.

BATCHED OUTPUT
GENERAL LISTS
PATIENT INFORMATION
POST TREATMENT LISTS
PRETREATMENT LISTS
SUMMARIES
TREATMENT LISTS

SPECIAL COMMANDS: KILL, INDEX, DETAIL

ENTER SCHEDULES & LISTS OPTION: GE

ENTER GENERAL LISTS OPTION: DO

One could have gotten to this point through the shorthand notation SC,GE, DO at any option prompt.

DOCTORS SCHEDULES ARE AVAILABLE IN:
1) LONG OR
2) SHORT FORM
SELECT 1 OR 2: 2

SELECT OUTPUT DEVICE:

WHICH PHYSICIAN? WEP

FIG. 1. Example Menu Traverse.

```
******** DOCTOR'S SCHEDULE  FOR TUESDAY  11/01/88   TUESDAY  TUESDAY  TUESDAY  TUESDAY  TUESDAY  TUESDAY  ********

< GCO >==========[ PBL ]=========================================== 11/01/88=== 11/02/88=== 11/03/88=== 11/04/88====OTHER=====
   BELL,VERNITA                 BREAST;POORLY DIFFERENTIATED  2:00 16/28 |          |          |          |          |
 * BROOKS,ALBERTIN              MET RENAL;;                  10:30 7/20  |          |          |          |          |
   CADDELL,W                    BLADDER;TRANSITIONAL CELL;GR 11:20 29/36 |          |          |          |          |
   CAPERS,HENRY                 PROSTATE;;                   10:30 22/15 |          |          |          |          |
   CAYCE,GLORIA                 LUNG;POORLY DIFFERENTIATED;  10:50 36/38 |          |          |          |          |
 H CHRISTIAN,TANYA   221-1      LUNG;;                       10:30 4/4   |          |          |          |          |
   COVA,RUTH                    COLON;ADENO;                  1:50 20/28 |          |          |          |          |
   CROUCH,GEORGE                RECTUM;ADENO;                 8:40 15/26 |          |          |          |          |
   FARKAS,WALTER                KIDNEY;;                      8:40 15/16 |          |          |          |          |
   FERGUSON,JANE                RECTUM;;                      2:00 4/28  |          |          |          |          |
   FRYE,MARION                  PAROTID;UNDIFF NEOPLASM;      9:20 12/25 |          |          |          |          |
   GILES,JOSEPHINE   323-1      ESOPHAGUS;;                  11:00 12/15 |          |          |          |          |
   GLENN,FRANK       411-1      PROSTATE;;                    1:40 7/12  |          |          |          |          |
   GLOVER,LILLIE                GALLBLADDER;;                 9:40 12/25 |          |          |          |          |
   GOREN,LESLIE                 BREAST;;                      7:30 21/28 |          |          |          |          |
   JOHNSON,BESSIE               LUNG;;                        8:00 34/38 |          |          |          |          |
   LACKO,ROSE                   LUNG;OAT CELL;                1:00 18/25 |          |          |          |          |
   LEWIS,SHIRLEY     327-1      ESOPHAGUS;;                   1:40 6/15  |          |          |          |          |
 * MACK,BEN                     PROSTATE;ADENO;              11:20 0/0   |          |          |          |          |
 S MACUMBER,AMELIA              MALIGNANT LYMPHOMA;;          9:00 0/0   |          |          |          |          |
   MATYJANOWSKI,TI              PROSTATE;ADENO;GRD.III        1:00 24/38 |          |          |          |          |
   MCMAHON,ABBAS                BREAST;;                      8:00 10/14 |          |          |          |          |
   PARKER,SOPHIE                LUNG;;                        3:10 6/12  |          |          |          |          |
   POTTER,GEORGIA               LARYNX;;                     11:00 20/35 |          |          |          |          |
   RAYMOND,FRANCIS              PROSTATE;;                    7:40 20/38 |          |          |          |          |
   ROSENBURG,CONRA   PCU-1      LUNG;OAT CELL;               11:15 3/7   |          |          |          |          |
   RUDDY,JAMES       517-2      BRAIN;;                      10:40 2/0   |          |          |          |          |
   TWOREK,JAMES                 ENDOMETRIUM;;                 9:20 18/28 |          |          |          |          |
 * WILLIAMS,ABRIZA   338-2      PROSTATE;;                   12:45 8/15  |          |          |          |          |
   ZIMMERMAN,BERTH   316-1      LYMPHOMA;;                    1:10 7/10  |          |          |          |          |
< 20 MEV >===========================================================================================================
 * WILLIAMS,ABRIZA   338-2      PROSTATE;;                    8:30 0/0   |          |          |          |          |
< BCO >==============================================================================================================
 * BROOKS.ALBERTIN              MET RENAL;;                   9:10 0/0   |          |          |          |          |
   JONES,LEO                    COLON;;                      10:00 0/0   |          |          |          |          |
 * MACK,BEN                     PROSTATE;ADENO;               9:30 0/0   |          |          |          |          |
< BLOCK >============================================================================================================
   TSAO,WALTER       422-1      BLADDER;ADENO;                           |          |          |          |          |
< CONSULT >==========================================================================================================
   KANGAS,ROBERT                PROSTATE;;                               |          |          |          |          |
< DELAY >============================================================================================================
   LANGFORD,BENJAM   436-2      BREAST;INFILTRATING DUCTAL C             |          |          |          |          |
< SIM 1 >============================================================================================================
   COATS,IAN                    RECTUM;;                                 |          |          |          |          |
   GOODSELL,ANNA     437-2      PANCREAS;;                               |          |          |          |          |
   JOLIFF,EDWARD     432-1      KIDNEY;;                                 |          |          |          |          |
   SIEGEL,JOSEPH     406-1      RECTUM;ADENO;                            |          |          |          |          |
   SKELLEY,CARL      426-1      COLON;;                                  |          |          |          |          |
< TX DEC >===========================================================================================================
   ALLEN,MICKEY      624-2      LUNG;SMALL CELL;                         |          |          |          |          |
   ENGLUND,ESTHER    421-1      BREAST;INFILTRATING DUCTAL C             |          |          |          |          |
   LASWELL,STELLA               LYMPHOMA;;                               |          |          |          |          |
   MULARONI,JOAN     401-1      BREAST;INFILTRATING DUCTAL C             |          |          |          |          |
   PEREZ,BERNADINE              BREAST;INFILTRATING DUCTAL C             |          |          |          |          |
   TRAYLOR,CARL                 RECTUM;;                                 |          |          |          |          |
< TX PL >============================================================================================================
```

FIG. 2. The Doctor's schedule provides a four-day window into future scheduled events with patient name, room no., diagram, treatment time, treatment progress and scheduled events.

```
GCO  SCHEDULE   10/24/88

TIME      ROOM     ID #     NAME              DR.  TFC/TFP
==============================================================================================
7:00                                                      2:00   4P  OUT    ..870018..FERGUSON,JANE ..PBL..4 /28
7:10                                                      2:10
7:20                                                      2:20   2P  WS8888..870075..ANDERSON,MARGA..WEP..2 /25
7:30   3T  OUT    ..870009..GOREN,LESLIE  ..PBL..21/28    2:30
7:35H  2P  OUT    ..870047..PICKARD,BELLE ..WEP..23/30    2:35
7:40   4P  OUT    ..870033..RAYMOND,FRANCI..PBL..20/38    2:40
7:50                                                      2:50
8:00   1P  OUT    ..870038..MCMAHON,ABBAS ..PBL..10/14    3:00
8:10   2E  OUT    ..870042..PARKER,SOPHIE ..PBL..6 /12    3:10   1P  OUT    ..870042..PARKER,SOPHIE ..PBL..6 /12
8:20                870042..PARKER,SOPHIE                 3:20
H8:30                                                     3:30
8:40   2P  OUT    ..870062..FARKAS,WALTER ..PBL..15/16    3:40
8:50                                                      3:50
*9:00                                                     4:00
9:10                                                      4:10
9:20   4P  OUT    ..870050..FRYE,MARION   ..PBL..12/25    4:20
9:30                                                      4:30
9:40   2P  OUT    ..870059..VINIC,RODNEY  ..YHK..7 /25    4:40
9:50                                                      4:50
10:00                                                     5:00
10:10  2P  OUT    ..840010..ENGLISH,JOE   ..AMH..11/12    5:10
10:20       .                                             5:20
H10:30 2P  OUT    ..870036..CAPERS,HENRY  ..PBL..22/15    5:30
10:40  2P  517-2  ..870069..RUDDY,JAMES   ..PBL..2 /0     5:40
10:50  2P  OUT    ..870007..CAYCE,GLORIA  ..PBL..36/38    5:50
11:00  2P  OUT    ..870055..POTTER,GEORGIA..PBL..20/35    6:00
11:10                                                     6:10
11:15  4P  PCU-1  ..870076..ROSENBURG,CONR..PBL..3 /7     6:15
11:20  4P  OUT    ..870044..MACK,BEN      ..PBL..0 /0     6:20
11:30                                                     6:30
11:40                                                     6:40
11:50                                                     6:50
12:00                                                     7:00
12:10                                                     7:10
12:20                                                     7:20
12:30                                                     7:30
12:40                                                     7:40
12:45  4P  338-2  ..812668..WILLIAMS,ABRIZ..PBL..8 /15    7:45
12:50                                                     7:50
1:00   4P  OUT    ..870048..MATYJANOWSKI,T..PBL..24/38    8:00
1:10   2P  316-1  ..860178..ZIMMERMAN,BERT..PBL..7 /10    8:10
1:20   3P  OUT    ...870052..APPIE,LINDA  ..WEP..19/25    8:20
1:30                                                      8:30
1:40   4P  411-1  ..870067..GLENN,FRANK   ..PBL..7 /12    8:40
1:50   2P  OUT    ..870014..COVA,RUTH     ..PBL..20/28    8:50

*  TO-START PATIENTS                                          H  HOLD PATIENTS
==============================================================================================
9:00S      OUT    ..870079..MACUMBER,AMELI..PBL..0 /0 ..04/24/87     8:30H    OUT    ..870047..PICKARD,BELLE ..WEP..23/30
                                                                     10:30H   221-1  ..870071..CHRISTIAN,TANY..PBL..4 /4

=PY-
   -IC1000-IS00204-PN-
```

FIG. 3. The short machine schedule lists all patients scheduled on a therapy machine with time, ports, room number, ID, name, physician and treatment progress.

the process of scheduling consults, simulation, and the like. In return for their participation, the clerks are rewarded with such aids as alphabetical list of patients. The strength of such systems is in appropriate selection and presentation of information and communication of this information to the staff who need it.

Patient flow through the department is organized through scheduled events. When a patient enters the department, reception, nursing, or patient transport staff note this in RAMS. The patient's name is displayed on a queue (Figure 4) which is displayed everywhere in the department. This queue is organized by function; the patient name will appear in a time ordered list under each function for which he/she is scheduled. For weekly checkup, the patient is automatically moved from the treatment to nursing queue when treatment is complete. Blood request slips can be automatically generated. Implants scheduled, diagnostic films needed, and patient transport lists are available. Further, as patients move through the scheduled procedures, every event is time stamped and saved for patient flow management purposes. A graph of treatment times (Figure 5) is typical of the reports available for this data rich resource.

Dosimetry and Physics activities are scheduled and in other ways supported. Treatment planning and treatment aid construction are requested (normally) by the simulator technologist. In order to capture the patient prescription, RAMS will perform irregular field calculations for prescription and points of interest. Output is in dose and monitor units (linac) or dose and timer (cobalt) setting. As a quality assurance tool, the technologists are required to enter the dose and setting daily for each port. This dose entry provides a quality assurance check on treatment prescription, clinical research data, dose summaries during and after treatment, and automatically drives the charge reporting. Techs also enter machine down time from which notes, lists, and graphic summaries flow.

As a result of supporting all aspects of department operation, management reports are readily available. Daily summaries (Figure 6) present yesterday's activities broken down by physician, therapy machine, in/out status, diagnosis, or patient status. Charge capture information is presented daily, monthly, or annually, and may be contrasted with the previous year's activity (Figure 7). Summaries exist for procedures grouped by department area (Figure 8). Patient waiting times are available for a specific patient visit, in tabular form, or in graphic form.

CONCLUSIONS

RAMS has been in clinical use for over a decade at numerous institutions. It is mutable (can be tailored to the needs of a specific institution) and is cost-justified based on charge recovery alone. The improvement in charge recovery at Harper Hospital was a whopping 340%.

The cost of a departmental system can be quite reasonable. Hardware power has increased phenomenally over the last two decades. System costs range from $10,000 to $500,000. Software development and maintenance costs are prohibitive, but many commercial systems exist and they may be tailored to one's specific needs. These systems easily pay for themselves and offer additional benefits in terms of improved patient flow, communication and quality assurance.

REFERENCES

1. Lodwick, G.S., Wickizer, Robert C., Tully, R.J., Hakimi, B.R., The Missouri Automated Radiology System (Ap Rad, Sept-Oct 1980), p. 73.

2. Otten, J.W., Report to the ICC Site Visit. Am Can Soc, Illinois Division, 1976.

3. Am Can Soc, Colon Cancer - Differences in Survival, Illinois Division, 1976.

4. Ragan, D., Lee, J.Y., Lee, Juhlin, K.D., Statistics - An Interactive Survival Analysis Package, Proceedings of the Fifth Conference on Computer Applications in Radiology (Chicago, Illinois, ACR, June 1977), Pp 389-400.

```
:------ 20  M E V ------:--- GR  C O B A L T ---:--T R A N S P O R T-:
:Brooks, A      0910 PBL:Frye, H       0900 WEP:Zimmerman*1008 PBL  :
:                       :Brooks, A     0920 PBL:                    :
:                       :Apple, L     *1030 WEP:                    :
:                       :English, J    1010 AMH:                    :
:                       :Vinic, R       940 YHK:--- N U R S I N G --:
:                       :                      :Washington5 1000 WEP:
:                       :                      :Fox, A      1008 AMH:
:                       :                      :                    :
:--- BR  C O B A L T ---:----- 12  M E V ------:                    :
:                       :Anderson, M  *0810 WEP :                   :
:                       :Applel        1010 WEP :----- SIMULATION ---:
:                       :                      :Anderson, A 1016 WEP:
:                       :                      :                    :
:                       :                      :                    :
:                       :                      :-- FLOOR COVERAGE --:
:                       :                      :WEP  ON CALL        :
:                       :                      :YHK  ON CALL        :
:                       :                      :                    :
:.......................:......................:...10:24 . 11/02/88 :
```

FIG 4. The Queue is displayed throughout the facility and presents all patients currently in the facility sorted by function and arrival times.

Radiation Oncology - Harper Hospital
SUMMARY OF PATIENT QUEUE PARAMETERS/STATISTICS DATE: 11/01/88 PAGE 1

DATES OF SAMPLE: 10/27/88
PATIENTS SERVICED: 207 MEAN TIME: 29 (MINUTES)
PATIENTS SAMPLED: 122 MEDIAN TIME: 23 (MINUTES)
ALL TREATMENT MACHINE QUEUES

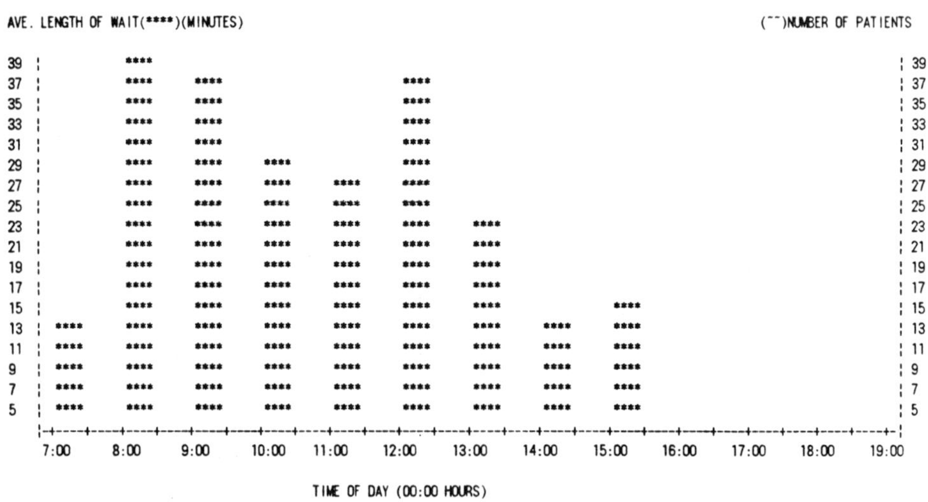

FIG. 5. Average length of wait is available in order to provide management information to shift scheduled patient arrivals.

RADIATION ONCOLOGY DAILY SUMMARY OF ACTIVITY FOR THURSDAY 10/13/88
PAGE 1

TREATMENT SUMMARY BY MACHINE

	PATIENTS						PORTS		
	SCH	T ST	HLD	TRT	NS	PST	PF	SCH	TR
GCO	21	5	6	19	2	2	2	49	43
SUPER	1	0	0	1	0	0	0	1	1
12 MEV	39	4	11	33	6	2	8	84	70
20 MEV	48	9	7	48	0	1	13	123	124
BCO	13	3	8	11	0	1	7	29	26
HDR	0	4	2	0	0	0	0	0	0
TOTAL	122	25	34	112	8	6	30	286	264

TREATMENT SUMMARY BY RADIOTHERAPIST OR TEAM

	PATIENTS						PORTS		
	SCH	T ST	HLD	TRT	NS	PST	PF	SCH	TR
BLU	23	7	11	24	0	2	9	49	52
GRN	38	12	10	35	3	3	10	112	105
RED	0	0	0	0	0	0	0	0	0
YEL	54	6	13	46	5	0	11	125	107
TOTAL	115	25	34	105	8	5	30	286	264

**** SUMMARY OF IN/OUT TREATED PATIENTS ****

IN	20	HARPER HOSPITAL	16
OUT	85	CHILDRENS HOSPITAL	0
		DETROIT RECEIVING HOSPITAL	1
		HUTZEL HOSPITAL	1
		REHABILITATION INSTITUTE	2
		VA HOSPITAL	0
TOTAL	105		20

PRETREATMENT SUMMARY BY RADIOTHERAPIST OR TEAM

	SCHEDULED								
	CONSULTS			PHYSICS					
	NEW	RET	TOT	CBM	RXP	MOL	SIM	IMP	FUP
BLU	0	0	0	0	0	0	0	0	1
GRN	3	0	3	5	3	0	2	1	1
RED	0	0	0	0	0	0	0	0	0
YEL	3	0	3	0	0	0	5	0	17
TOT	6	0	6	5	3	0	7	1	19

FIG. 6. Daily summaries provide machine treatment scheduled, to start, on hold, treatment, no shows, ports started, port films, ports scheduled and treated by machine or therapist.

In and outpatient information, pretreatment, medical service and patient status information is also presented.

RADIATION ONCOLOGY   DAILY SUMMARY OF ACTIVITY FOR THURSDAY 10/13/88
PAGE 2

SERVICED

	CONSULTS NEW	RET	TOT	PHYSICS CBM	RXP	MOL	SIM	IMP	FUP
BLU	0	0	0	0	0	0	1	0	0
GRN	0	0	0	3	5	1	1	0	0
RED	0	0	0	0	0	0	0	0	0
YEL	1	0	1	1	0	0	3	0	12
TOT	1	0	1	4	5	1	5	0	12

MEDICAL SERVICE BY PHYSICIAN OR TEAM

	BEN	BRS	CNS	END	GI	GU	GYN	H&N	HEM	LNG	OTH	PED	SAR	SKN	UNK	TOTAL
BLU	0	33	0	0	0	20	0	1	0	0	0	0	0	0	0	54
GRN	0	0	8	0	1	0	15	47	8	0	1	0	0	3	1	84
RED	0	0	0	0	0	0	0	0	0	0	0	0	0	0	0	0
YEL	0	0	0	0	28	0	0	0	0	37	1	10	4	0	1	81
TTL	0	33	8	0	29	20	15	48	8	37	2	10	4	3	2	219

STATUS SUMMARY

	TXD	DEL	PATIENTS STARTED NEW	RET	TOT	COT COM	CHARGING PEN	COM	ACTIVE PATIENTS IN	OUT	TOT
BLU	8	4	1	1	2	0	12	0	6	48	54
GRN	12	0	1	1	2	0	16	0	25	59	84
RED	0	0	0	0	0	0	0	0	0	0	0
YEL	2	0	0	2	2	0	25	1	11	70	81
TOT	22	4	2	4	6	0	53	1	42	177	219

PATIENT STATUS

	TXT	PTX	TXD	DEL	HLD	TOT
BLU	30	6	8	4	6	54
GRN	51	18	12	0	3	84
RED	0	0	0	0	0	0
YEL	65	7	2	0	7	81
TOT	146	31	22	4	16	219

THE ABOVE SUMMARIES MAY BE INACCURATE.
QUITS ARE NOT RECORDED FOR THESE MACHINES:

HDR

FIG. 6. (cont'd)

GRACE HOSPITAL
CHARGE CAPTURE SUMMARY FOR 10/1/88 TO 10/31/88

DATE: 11/01/88 PAGE 1

CODE	SERVICE DESCRIPTION	CURRENT MONTH SERVICES FOR MONTH IN	OUT	TOT	CHARGES FOR MONTH IN	OUT	TOTAL	YEAR TO DATE SERVICES FOR YEAR IN	OUT	TOT	CHARGES FOR YEAR IN	OUT	TOTAL
****	PATIENTS VISITS ****												
101	CONSULTATION - LIMITED	0	0	0	$ 0	$ 0	$ 0	4	1	5	$ 0	$ 0	$ 0
102	CONSULTATIONS - COMPREHEN	9	11	20	$ 0	$ 0	$ 0	133	85	218	$ 0	$ 0	$ 0
109	SUBSEQUENT HOSPITAL CARE	0	0	0	$ 0	$ 0	$ 0	4	0	4	$ 0	$ 0	$ 0
201	LIMITED EXAM (FOLLOW-UP)	4	26	30	$ 0	$ 0	$ 0	32	347	379	$ 0	$ 0	$ 0
	CHARGES				$ 0	$ 0	$ 0				$ 0	$ 0	$ 0
****	BRACHYTHERAPY ****												
	CHARGES				$ 0	$ 0	$ [illegible]				$ 0	$ 0	$ 0
****	TREATMENT AIDS ****												
401	WEDGE FILTER DESIGN AND F	1	1	2	$ 118	$ 118	$ 237	2	8	10	$ 237	$ 950	$ 1188
402	BOLUS DESIGN AND FABRICAT	0	0	0	$ 0	$ 0	$ 0	3	4	7	$ 356	$ 475	$ 831
403	FIELD BLOCK DESIGN AND FA	14	28	42	$ 2762	$ 5524	$ 8286	166	362	528	$ 32750	$ 71418	$ 104169
404	COMPENSATION FILTER DESIG	0	5	5	$ 0	$ 986	$ 986	14	44	58	$ 2762	$ 8680	$ 11442
405	MOLDS OR CASTS FOR IMMOBL	0	0	0	$ 0	$ 0	$ 0	3	8	11	$ 709	$ 1892	$ 2601
	CHARGES				$ 2880	$ 6629	$ 9510				$ 36815	$ 83417	$ 120233
****	DOSIMETRY ****												
501	CENTRAL AXIS DEPTH DOSE C	14	26	40	$ 1098	$ 2040	$ 3139	171	270	441	$ 13420	$ 21189	$ 34609
502	ISODOSE PLAN-SIMPLE (1 OR	1	6	7	$ 203	$ 1222	$ 1426	36	44	80	$ 7337	$ 8968	$ 16306
503	ISODOSE PLAN-INTERMEDIATE	1	0	1	$ 277	$ 0	$ 277	5	17	22	$ 1389	$ 4725	$ 6114
504	ISODOSE PLAN COMPLEX (4 O	1	5	6	$ 427	$ 2136	$ 2563	6	44	50	$ 2563	$ 18800	$ 21364
	CHARGES				$ 2007	$ 5399	$ 7407				$ 24711	$ 53683	$ 78394
***	PRETREATMENT SET-UP ***												
701	SIMULATION SIMPLE	3	3	6	$ 163	$ 163	$ 327	56	37	93	$ 3052	$ 2016	$ 5068
702	SIMULATION INTERMEDIATE	2	8	10	$ 220	$ 880	$ 1100	39	29	68	$ 4293	$ 3192	$ 7486
703	SIMULATION COMPLEX	1	4	5	$ 171	$ 684	$ 855	20	64	84	$ 3422	$ 10952	$ 14374
710	TREATMENT PLANNING SIMPLE	1	0	1	$ 578	$ 0	$ 578	40	23	63	$ 23151	$ 13312	$ 36463
711	TREAT PLAN INTERMEDIATE	0	8	8	$ 0	$ 6173	$ 6173	32	25	57	$ 24695	$ 19293	$ 43988
712	TREATMENT PLAN COMPLEX	1	4	5	$ 963	$ 3854	$ 4817	17	57	74	$ 16380	$ 54922	$ 71303
	CHARGES				$ 2097	$ 11756	$ 13853				$ 74995	$ 103689	$ 178684
****	TREATMENT ****												
810	TREATMENT-PORT FILMS	23	71	94	$ 2130	$ 6578	$ 8709	237	679	916	$ 21958	$ 62909	$ 84867
821	TREATMENT 4 MEV SIMPLE	16	12	28	$ 1866	$ 1399	$ 3265	276	104	380	$ 32189	$ 12129	$ 44319
822	TREATMENT 4 MEV INTERMEDI	66	117	183	$ 13452	$ 23848	$ 37300	476	1217	1693	$ 97023	$ 248061	$ 345084
823	TREATMENT 4 MEV COMPLEX	7	198	205	$ 2090	$ 59134	$ 61225	119	1594	1713	$ 35540	$ 476064	$ 511604
	CHARGES				$ 19540	$ 80960	$ 110500				$ 186711	$ 799164	$ 985875
****	GRACE SERVICES ****												
	CHARGES				$ 0	$ 0	$ 0				$ 0	$ 0	$ 0
	TOTAL SERVICES FOR PERIOD	165	533	698				1891	5063	6954			
	TOTAL CHARGES FOR PERIOD				$ 26526	$ 114746	$ 141272				$ 323234	$ 1039954	$ 1363188

TOTAL NO. OF PATIENTS SERVED ON INPATIENT BASIS 22
TOTAL NO. OF PATIENTS SERVED ON OUTPATIENT BASIS 66

TOTAL NUMBER OF PATIENTS CHARGED DURING PERIOD 82

FIG. 7. Charge Capture information is summarized monthly and contrasted with the previous year's information.

HARPER HOSPITAL
END OF MONTH TREATMENT SUMMARY
FOR AUG 1988 DATE: 10/05/88

TREATMENT SUMMARY

MACHINE	VISITS IN	VISITS OUT	VISITS TOTAL	TREATMENTS IN	TREATMENTS OUT	TREATMENTS TOTAL
*UNK	0	0	0	0	0	0
GCO	62	198	260	157	455	612
SUPER	1	10	11	1	10	11
12 MEV	70	407	477	160	877	1037
20 MEV	40	399	439	67	1118	1185
BCO	27	177	204	44	439	483
HDR	2	13	15	2	13	15
TOTAL	228	1475	1703	431	2912	3343

PRETREATMENT ACTIVITIES	IN	OUT	TOTAL	NEW	RET	TOTAL
CONSULTS & HOSPITAL VISITS	50	57	107	64	29	93
SIMULATIONS	44	140	184			
TREATMENT PLANS	27	69	96			

PHYSICS ACTIVITIES	IN	OUT	TOTAL
TREATMENT AIDS	43	199	242
CENTRAL AXIS COMPUTATIONS	51	209	260
ISODOSE PLANS	25	104	129
ISODOSE SURCHARGES	0	0	0
BRACHYTHERAPY PLANS	1	0	1

INTERNAL PROCEDURES	IN	OUT	TOTAL
BRACHYTHERAPY	1	1	2
HANDLING & LOADING	2	0	2

POST TREATMENT ACTIVITIES	IN	OUT	TOTAL
FOLLOW UPS	0	135	135

FIG. 8. Monthly treatment summary information is available in logical groups.

SPREADSHEETS AND DATA BASE MANAGEMENT SYSTEMS

Don P. Ragan, Ph.D.
Wayne State University
Detroit, Michigan 48201

ABSTRACT

The personal computer (PC) revolution has been made possible by increasingly fast, inexpensive, and small computer hardware. Equally critical to the popularity of these systems is application software of general interest. Word processing, spreadsheets, and data base management are the most popular applications on PCs. This paper will discuss spreadsheets and data base applications.

INTRODUCTION

The availability of PCs comes at a time when changes have made personal productivity important and, therefore, aids such as spreadsheets and data base management systems (DBMS) valuable. There has been a national change in priorities which resulted in a shift of workforce and capital expenditure from the industrial to the service sector of society. In 1984, IBM revenue exceeded that of the total U.S. steel industry. This results in a shift in cost from factory to office workers with a resultant need to increase white-collar rather than blue-collar productivity.

Seventy-five percent of the costs in industry are now associated with management. These managers spend 85% of their time communicating. This is equally true for medicine. Despite the cost of CTs, MRs, linear accelerators, and the like, medicine remains a service and labor intensive industry. Computers, and their associated personal productivity tools, offer efficiencies to medicine analogous to the improvements robotics offer to manufacturing.

Spreadsheets and data base management systems are the most ubiquitous and popular personnel productivity tools on microcomputers. It is well recognized that cost per worker has increased while computer costs have decreased. Computers offer the possibility of increasing productivity while decreasing costs. More importantly, _you_ can personally increase _your_ productivity by utilizing spreadsheets and DBMS.

These tools are important for the organization and communication of information rather than mathematical computation. DBMS provides storage and retrieval of information. Spreadsheets organize and process data. Word processing and graphics concentrate on presentation and communication.

SPREADSHEETS

Spreadsheets are very simple structured data base systems with special interface and compute capability. Data is organized in "worksheets" which are organized by "rows" and "columns" (Figure 1). This results in a two-dimensional array of "cells." Later in this article we'll return to extensions which allow three-dimensional and hierarchical arrays. These 2-D arrays are displayed on the CRT with a "border" and an "active cell" highlighted.

```
                ______The status line shows the
               |      contents of the active cell.
               |               ------------------------The prompt line
               |              |                        shows what options
               |              |                        are available.
 ______________|______________|_____________________
|B6:       Total Sales:       |          READY      |  }Program status and
|RANGE COPY MOVE FILE QUIT                          |  }help information
|                                                   |  }(control panel).
|        A             B          C          D  <-- |------Border
|1                     Units      Price      Sales  |  }
|2  Shirts             12,250     $6.90     84,525  |  }
|3  Pants               6,700     $9.30     62,310  |  }  Window into the
|4  Dresses             9,000    $12.50    112,500  |  }  worksheet.
|5                                                  |  }
|6                     Total Sales:        259,335  |  }
|_____________________|_____________________________|     The active cell
                      |___________________________________________|
```

FIG. 1. A typical (reduced-size) spreadsheet screen display. Examples are taken from "Using Computers Today".

Control of these arrays is governed by the "program status" area. The "status line" displays the information in the cell which has produced the value displayed. The "prompt" line allows data to be added or changed or commands to be entered.

Three types of data can be entered into cells. Literal data includes such things as labels and comments; these can extend past a cell boundary. Numeric information is recognized by the system as a number or if preceded by a "?". Cells can also have references to other cells such as "+B6" (Figure 2), formulas such as "+B6*3.1416", and functions such as "+B6*3.1416-COS(+B6)"; formulas are, in general, calculated on the fly, and any changes in dependent cells are adjusted automatically. Two important exceptions to this are circular references and references to external spreadsheets.

```
C5: (C2) +(1.+$C$1)*B5    READY
      A      B       C      D
1      Sales Mark-up:    .55
2
3  Item   Cost   Price  Margin
4  -----  -----  -----  ------
5  Skis  $140.00 $217.00 $77.00
6  Poles  $44.50
7  Boots  $52.80
```

-->

```
C6: (C2) +(1.+$C$1)*B6  READY
    A        B       C        D
1     Sales Mark-up:         .55
2
3 Item    Cost     Price   Margin
4 ------  -----   ------   ------
5 Skis    $140.00 $217.00  $77.00
6 Poles    $44.50  $68.98  $24.48
7 Boots    $52.80  $81.84  $29.04
```

FIG. 2. Copying the formulas in C5..D5 to C6..D7 using both relative and absolute cell references.

Operations can be performed on the spreadsheet to control the organization and presentation. Data may be right, center, or left justified. Numeric data can be displayed with a specified number of decimal places or "$", "%", etc. Rows and columns can be inserted or deleted.

It is important to note that addressing can be either relative or absolute. Relative addressing is assumed (default) with absolute addressing indicated with an "@" symbol preceding the value. This is best seen in the example in Figure 2. Absolute addressing can be associated with rows, columns, or both. When rows are added, deleted, or moved, the system will try to extend relative addresses to the new environment. The effects of address changes may result in error indications or, even more likely, incorrect results. Note, by way of warning, that various spreadsheets operate differently.

Once one gets past basic spreadsheet functions, many features exist to extend the spreadsheet's utility or improve user interaction. It may become useful to protect cells in a spreadsheet from being altered. This is most useful for a template that will be turned over to a user to fill in prescribed data; such might be the case for monthly generated ^{60}Co decay tables for radiotherapy. When spreadsheets exceed a screen or two, scrolling with headers or windows becomes useful.

Circular references occur when the result of a calculation affects the data used to make that calculation. Such calculations can usually be made to quickly reach an asymptotic result by use of the "recalculate" function.

Worksheets can generally draw from other worksheets; this can fill an individual cell or group of cells. If the information exists in another area of the existing worksheet, changes will be reflected as changes are made. On the other hand, changes in an external worksheet will only be incorporated if the "recalculate" function is exercised. This external reference to another spreadsheet is a special case of a three-dimensional spreadsheet. Spreadsheets that specialize in three-dimensional data may be appropriate if you have frequent need for a hierarchical data structure.

With this brief introduction, you'll probably wish to rush out and purchase a spreadsheet. The first and, by far, the most important thing to weigh in selecting a software product is what are your colleagues using. The differences between packages are small, and your need to exchange problems and "tricks" of use will be ongoing and important. Look at the printer interface and the graphics capability.

Package integration may be extremely important. This can be anything from an ability to produce a data interchange format (DIF) file to a single package that provides word processing, spreadsheets, DBMS, and graphics.

Special spreadsheet systems exist for scientific, business, statistics, etc.[1,2]. These are differentiated most prominently by the type of functions provided. An example of a minimal function set is given in Table 1.

Table I ref. Calc 7 Users Manual

Statistical Functions

CORR	Correlation coefficient
DEPD	Dependent value of linear regression
INDEPD	Independent value of linear regression
SLOPE	Slope of linear regression
SDEV	Standard deviation
VAR	Variance
GROWTH	Growth rate for list of values
TREND	Trend of a list of values

Other Functions

COL	Returns column number of cell
DB7	Value from DB7 data base file
ROW	Returns row number of cell
SS	Returns value of field "ref" in worksheet file with name "filename"

User Definable Functions

FILE	User definable file interface
CALC	User definable calc function
DB	User definable data base value

Trigonometric Functions

ACOS	Angle (radians) with given cosine
ASIN	Angle (radians) with given sine
COS	Cosine of angle

COSH	Hyperbolic cosine
DEG	Convert radians to degrees
RAD	Converts degrees to radians
SIN	Sine of angle
SINH	Hyperbolic sine
TAN	Tangent of angle

Financial Functions

PMT	Loan payment amount
RATE	Loan interest rate
TERM	Loan term (periods)
PRIN	Loan principal amount
FV	Future value of investment
FVA	Future value of annuity
PV	Present value of investment
PVA	Present value of annuity
IRR	Internal rate of return
NPV	Net present value

Mathematical Functions

ABS	Absolute value of a number
CEIL	Smallest whole number equal to or greater than "ref"
COUNT	Count of the number of items
E	The constant 2.718281828
EXP	Calculate the exponent to the base e
FLOOR	Largest whole number equal to or less than the value of "ref"
LOG	Log to the base 10 of a number
LN	Natural log of a number
MEAN	Mean of fields and/or constants
MAX	Maximum of fields and/or constants
MEDIAN	Median value of fields and/or constants
MIN	Minimum of fields and/or constants
PI	The constant 3.141592654
RAND	Random number
ROUND	Round number to "n" decimal places
SUM	Sum of cells and/or constants
SQR	Square
SQRT	Square root
SUMSQ	Sum of squares of numbers
TRUNC	Truncate number to "n" decimal places

There are many reasons for utilizing spreadsheets. They offer clear and graphic presentation of the results. Special functions, macros, or the ability to write your own functions may enhance the power of these tools. The most important and most immediate benefit will be reliability (consistency) resulting from a lack of arithmetical errors or even mundane transcription errors.

DATA BASE MANAGEMENT SYSTEMS

As the volume of data increases, the number of data elements increases, and the display format becomes more variable, spreadsheets become more and more constraining. Under these circumstances data base management systems allow more data elements, more flexible data structures and better searching, sorting, and presentation capability. Applications such as the AAPM mailing list, a tumor registry, or an equipment inventory are appropriate candidate applications for DBMS.

Performance of DBMS is a function of storage architecture. This is seen in speed of retrieval and in ease of user interface. For our purposes we will review three classes: sequential, hierarchical, and relational. Sequential or unstructured is neither relevant nor interesting and is useful for very small or batch-oriented systems. An example might be a file of all AAPM members that could be used to produce mailing labels, phone directories, etc.

Hierarchical data is the most straightforward. It is naturally well suited for patient files where one might have a patient who has multiple encounter records each with multiple interventions within an encounter (Figure 3). This is referred to as a tree with branches, sub-branches, etc. Hierarchical data structures provide the fastest access to individual patient or entity information and the slowest access when searching for data across entities deep into the tree. A good example here might be a tumor registry where each patient may have multiple primaries, with multiple courses of therapy and multiple follow-ups. A problem with hierarchical structures exists if one has a logical connection between elements within two different branches of a tree (the join operation which is generally not accommodated).

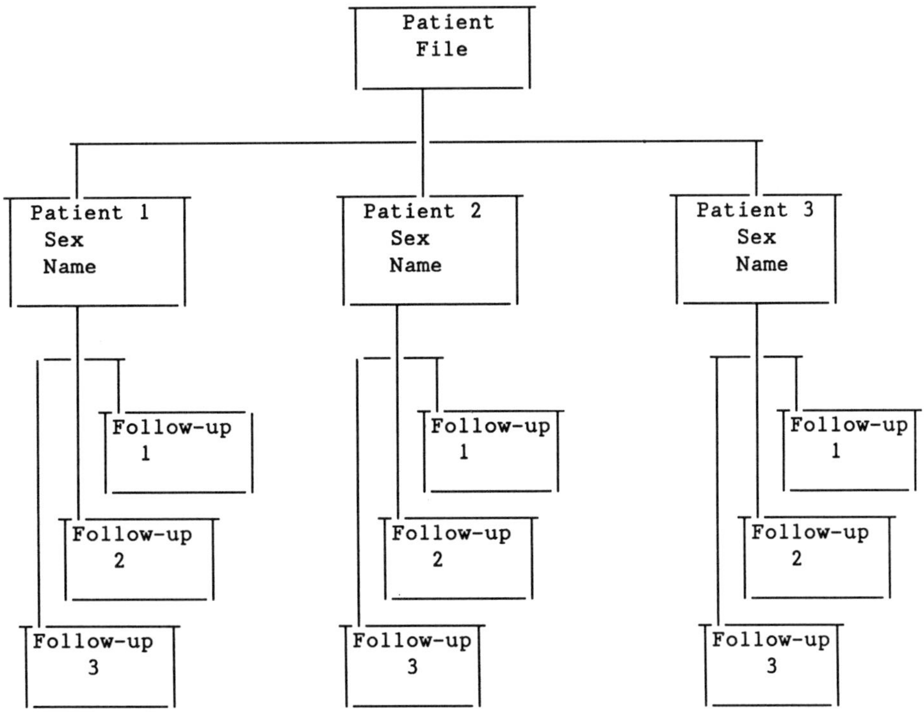

FIG. 3. A hierarchial data base

Relational DBMS are the most popular. This allows simple tables (like spreadsheets) with the ability to construct relations between these tables (Figure 4). Systems tend to be flexible. Complex applications tend to run very slowly, use large amounts of system resources (especially disk space and access) and demand a great deal of knowledge of the data base and sophistication on the part of the user. An example here might be an equipment inventory where one would want an equipment file, a manufacturer file, a down-time file, a repair file, etc.

SCHEDULE Relation

CLASS	INSTRUCTOR
CS211	Jones
CS215	Franklin
CS317	Lee
CS201	Feldstein
.	
.	
.	

REGISTRATION Relation

CLASS	INSTRUCTOR
CS212	Adams
CS317	Schmidt
CS211	Wilson
CS212	Schmidt
.	
.	
.	

FIG. 4. A relational data base stores data in tables called relations.

Many combinations, permutations, or hybrid systems exist. Inverted files can be constructed for index sequential or hierarchical systems. These can allow direct access to entries in the data base with a specific characteristic or comparisons of criteria without access to the primary record.

All DBMS are controlled by a data dictionary. This file contains a map to the location of each data element. It should also provide data typing (i.e. numeric, literal, date, ...), field size, error checking criteria, relationships to other data elements, etc. The ability to recognize data such as date and time is important if operations are to be allowed on this data type (e.g., patients born before 1950 or patients treated after 1:00 p.m.).

When retrieving information, data elements need to be identified and criteria on relations requested. Here the ease of interaction separates the good from the poor system. The ability to search on dates or time for "first occurrence of," "dates of most serious complications," "treatment following recurrence," etc. can be critical with time oriented data. Combinations of criteria must, of course, be possible. Intermediate search results are extremely useful with the ability to combine intermediate searches. Finally, free text searches are often useful.

The most recent piece of jargon, SQL, requires a brief explanation. SQL is a standard query language that allows data bases to be accessed through a standard programming interface. This is a language standard for ease of programming; it is not a user interface.

In selecting a DBMS, like a spreadsheet, select the one that others around you are using. If special requirements exist or there is no clear institutional preference, remember the KISS motto (Keep It Simple, Stupid!); sophisticated systems tend to be hard to learn. Often you will trade performance for ease of use. It may be useful to choose something like SAS if you anticipate doing statistical analyses. A system like Oracle will allow you to use micros to start and migrate to minis or main frames if the data dictates.

CONCLUSIONS

Microcomputers plus appropriate software are personal management tools that will allow you to become both more productive and effective. They are affordable, generally bug free, and easily learned.

REFERENCES

1. Cooney, P.J., What Spreadsheets Can Solve for Scientists, Comp. in Phy., pp. 99-10, May/June 1988.

2. Dory, R.A., Spreadsheets for Physics, Comp. in Physics, pp. 70-74, May/June 1988.

V. THE FUTURE?

FUTURE TRENDS IN HARDWARE AND SOFTWARE

Trevor D. Cradduck,
Victoria Hospital, University of Western Ontario
London, Ontario, Canada N6A 4G5

ABSTRACT

The future developments of hardware and software will depend on the applications to which computers are put and in each of diagnostic radiology, radiotherapy and nuclear medicine, the applications differ. Diagnostic radiology requires fast real-time acquisition of huge amounts of data with immediate data manipulation and subsequent image analysis of a fairly routine nature. Radiotherapy tends to have much less stringent speed requirements, significant "number crunching" requirements and display systems which can store large amounts of data for real-time image analysis. Nuclear Medicine requires fast real-time data acquisition with significantly smaller data storage requirements but sophisticated data analysis and image display requirements.

It is true that the application of computers is limited primarily by the imagination of the medical physicists using them. We must, however, be careful to avoid allowing those imaginations to act in isolation of the real problems and in ignorance of the fact that the patients' interests should be served first.

INTRODUCTION

The act of "crystal balling" the future of the computer industry is one that is fraught with difficulties. On the one hand it is possible to review the uses to which computers are presently being applied and to try to predict where those applications might lead us. On the other hand there must surely be some applications for which even today's sophisticated technology is insufficiently developed and when new methods, greater speed or large storage capacity is developed these new applications will spring to life. The history of the CT scanner is a superb example of a technique which had to await the development of sufficient computing power before it could finally be put into production.

The history of the CT scanner also provides some other significant lessons. The first scanners were extremely expensive - even by radiology standards - yet aggressive marketing was able to overcome that financial barrier and it is now possible to discuss the huge cost of "hi-tech" diagnostic equipment without causing immediate cardiac distress among hospital administrators.

Yet another lesson to be learned from the experiences with CT scanners, digital radiography and MRI is that the future trends of these technologies will be dictated largely by the profit motive of the relatively few large companies which dominate the North American market. It is frequently stated by these companies that they do "listen" to their customers. This activity is, of course, only useful providing that their market share continues to grow. If it does not, then presumably they are listening to the wrong people and will then change their market thrust.

One superb example of a changed market philosophy is the case of a nuclear medicine computer company that found that its targeted market place of large, primarily university based institutions was saturated. It therefore, quite sensibly, decided to target the smaller, 200 bed and up community hospitals that did not possess computers. This caused significant problems because the larger institutions had been capable of (indeed preferred) providing their own internal support. The smaller hospitals lacked that expertise and the company found that it had to fill the gap; this was an expensive venture and, as a consequence, the product did not receive the continuing attention to development that is required to stay ahead of the competition with the result that the market base with the larger institutions has been significantly eroded.

It may be regarded as a somewhat heretical statement, but I suspect that future trends will be predicated largely on what activity puts more money into the physicians' pockets. It is noticeable that all "hi-tech" developments become accepted as "good medical practice" once they have been approved by third-party insurers, regardless of whether they really do have a significant impact on the actual quality of care, diagnostic or therapeutic outcome.

In an effort to chart the future of computer applications in medical physics let us divide the activities into the three major areas of interest - diagnostic radiology, radiotherapy and nuclear medicine. I am sure my fellow contributors will be able to give far more details concerning radiology and radiotherapy than I. I will attempt, however, to look at each in broad perspective and note those areas where I think we might expect significant changes to take place.

DIAGNOSTIC RADIOLOGY

Depending largely upon where the divisions are made, there are three primary applications of computers in diagnostic radiology. These may be identified as 1) Administrative, patient records; 2) Picture Archival and Communications Systems (PACS); and 3) Image data acquisition and manipulation.

Administrative

As hospitals move more towards fully integrated Hospital Information Systems (HIS) it is inevitable that the radiology department will follow that trend. Patient scheduling, film envelope tracking, record keeping and all the other administrative functions have already been introduced in most large departments. Some attempts have also been made to automate film reporting from touch tablets and mark-sense cards. These have not proven to be particularly acceptable for a variety of reasons. Perhaps if we can eventually overcome the barrier of speech recognition it will be possible to make the direct transition from the radiologists' interpretation to the printed report. Further development in the area of computer aided diagnosis may impact this area significantly.

Of one thing we can be quite certain - radiology systems will not become universally acceptable until they can be integrated into a complete HIS or Medical Information System (MIS). Progress in this direction has been slow. Many hospitals have attempted to become "computerized", but the results to date have not been representative of overwhelming success.

PACS

Several programs for PACS have received considerable attention in recent years. The 1987 AAPM Summer School was devoted to this very topic. The overriding conclusion to be derived from that school was that PACS is very, very expensive and cannot be justified in the immediate future. Reports of limited success are made for systems that do not yet include fully integrated departments. This topic has been discussed elsewhere in these proceedings.

It is of some interest to note that the two major projects for the installation of total department-wide systems are both funded by the US military. These are at George Washington University, Washington, DC and University of Washington, Seattle. The motivation behind the US military's support of these developments is that when (please note, not IF) a third world war erupts the logistics of getting radiological film and chemicals to the front will surpass the capabilities of the transportation systems as well as the stocks of those items. Therefore, the logic is that everything should be recorded digitally. There does not seem to be an answer to the overriding question of who will interpret this huge number of radiographs or how the power will be provided to maintain all the digital storage and display components (assuming they, too, survive the hostilities).

There are many sound and perfectly logical reasons for development of PACS, but so long as the friendly piece of x-ray film can survive the price of silver it is unlikely that this medium will suffer a major demise.

PACS requires very large storage capacity and, to be useful, it must have rapid access to old records. The workstations must have very high resolution and must be capable of displaying not just several images from one imaging modality, but several images from different imaging modalities at once. Some display systems with 1024 X 1024 resolution are available. It is generally accepted, however, that in the case of chest films, resolution of the order 2048 X 2048 is required. This imposes a bandwidth demand that is in excess of 120MHz and probably closer to 1GHz.

Digital Image Processing

The impact of digital image processing on angiography has been substantial. Nevertheless, much of the early promise has not been totally fulfilled. A great deal of research continues to be directed towards development of effective semiconductor radiation detectors (charge coupled devices) as replacements for image intensifiers. The development of pattern or texture recognition and edge detection algorithms for diagnostic radiology will be steps towards some elementary forms of computer aided diagnosis as knowledge based or artificial intelligence systems become more readily available.

Computer requirements in this area are for fast real-time data acquisition with large storage arrays allowing rapid on-line processing.

RADIOTHERAPY

Two major applications of computers in radiotherapy are treatment machine control and treatment planning. These two applications may conceivably be linked if the treatment planning unit can transmit the necessary data to the treatment unit for position and dosage control.

Machine Control

The tragic results of several incidents involving computer control of treatment machines cannot be ignored. The results of these accidents is that we can expect more stringent control by the FDA on software as well as hardware used in this type of situation. There will undoubtedly be a much more rigorous routine for performing quality control of these machines. This watch-dog activity on the part of government agencies can be expected to spill over into other areas where we use computers in medical physics. We may find development of some of the knowledge based software systems for computer aided diagnosis that might have been expected in the next decade to be delayed by new legislation as a result of the accelerator incidents.

Despite this setback there will undoubtedly be greater computer control of treatment machines. It is most likely that these systems will receive their treatment prescription directly from the treatment planning system after some form of security to ensure human verification has taken place. Although it is self-evident that any control system must be extremely reliable and that both hardware and software shall act in a fail-safe fashion, there is no other stringent demand on the system. Compared to the data acquisition systems of radiology and nuclear medicine these systems can be relatively slow and do not have any demands for large memory capacity or sophisticated displays.

Treatment Planning

As the data and algorithms upon which treatment planning is based have progressed, so has the level of sophistication of treatment planning. Some systems are capable of displaying changes in the dose maps in real-time as the treatment port conditions are changed. This requires very fast and powerful real-time image display processors.

Further sophistication includes the 3-D display of the treatment volume with the treatment beams super-imposed. In these instances, the demand for even greater display memory becomes paramount.

From the scientific perspective it remains to be demonstrated that the ability to determine radiotherapy treatment patterns to within a few percent really does result in significant change in the therapeutic outcome. From a physicist's point of view there is a desire to be able to make a calculation as accurately as possible given the error involved in the input data. From the patient's perspective a satisfactory outcome from therapy is the measure of success. From the point of view of the health care provider it is the latter objective, not the former, that must be obtained for the least possible cost.

NUCLEAR MEDICINE

As in the case of diagnostic radiology, the two major applications in nuclear medicine are administrative and image processing. The requirements for these applications are similar though those for image processing tend to be less stringent in terms of the memory and storage capacities.

Administrative

Unless a nuclear medicine department is part of a radiology department it is unlikely to have its own dedicated patient scheduling, tracking and recording system. In most cases, the workload cannot justify computerization of this activity. However, there is an enormous amount of paperwork associated with radiopharmacy - inventory control, quality assurance and records of patient administration. This provides an excellent target for the application of computer technology and systems based on PC's are presently being marketed. A single PC with a large storage capacity will probably suffice in most situations. A large department may find that a networked system which can have access to a central file server is better suited to their needs. This is the sort of application that can be put onto a PC that is primarily used for such activities as word processing. It is important that such a system be capable of integration into a full Hospital Information System which is not a trivial requirement.

Image Analysis

The activities to which nuclear medicine imaging systems have been applied has grown considerably. The very first applications were for the enhancement of static images. From there, the collection of dynamic studies allowed more sophisticated and detailed analysis of function studies to which nuclear medicine is more properly suited.

Gated cardiac studies would not be possible without a computer to collect and format the data. SPECT would not be possible without a computer to reconstruct the data into tomographic slices.

The general requirement is for a camera/computer interface that can accommodate relatively high data rates of the order of up to 200k counts per second, and a display capable of displaying 256X256 matrices. When SPECT is contemplated then the system should include an array processor.

The early nuclear medicine computers were stand-alone units based on general purpose computers. As scintillation cameras have become more sophisticated, the image correction modules have assumed greater importance so that digitization of the image has moved closer to the detector itself. As a natural consequence of this, and their desire to command a larger market share, the manufacturers of scintillation cameras have incorporated the cameras and computers as single entities.

It is difficult to purchase a camera that does not incorporate a computer system and this leaves the potential buyer facing the dilemma of buying a good camera, but poor computer, or vice versa. It is likely that this trend will continue, certainly into the foreseeable future, although some camera manufacturers do provide analog signal outputs to interface to third party computers.

There seems to be two, somewhat diverging, technologies being applied to nuclear medicine imaging computers. On the one hand we have multi-workstation environments networked to a central file server. On the other hand we see the development of PC based systems.

The first development pathway is motivated by the logistics of wishing to provide access to all the acquired data from any suitable workstation. These network environments tend to be complex and demand considerable in-house systems support. They also tend to be slower than the stand-alone systems, but this disadvantage is offset by the other advantages that can result.

PC based systems are frequently regarded as extremely cost effective systems, but a strict analysis of the situation reveals this not to be the case. It is not possible to merely purchase a cheap PC clone to which simple A/D and display boards can be added. The speed and linearity requirements for the ADC are such that special camera interfaces are necessary. The speed and memory capacity of the computer requires that a computer with 80286 - or 80386 - equivalent CPU and with large memory capacity must be used. Benchmark programs have demonstrated that, even when equipped with an arithmetic co-processor, these systems are still slower than their mini-computer competitors. The requirements for the display are such that sophisticated display interfaces and display monitors are necessary. This only addresses the hardware issues and the development of software can make considerable demands on resources causing that component to assume a major portion of the cost of a system.

CONCLUSIONS

Hardware

There seems to be a general consensus that most future computer systems will be based on 32-bit CPU's. This may be appropriate for the larger number-crunching systems used in radiotherapy treatment planning systems where it is important to maintain high accuracy in mathematical operations. In nuclear medicine systems where most of the analysis uses integer arithmetic, a shift to 32-bit machines cannot be so readily justified.

There is also a considerable interest in networking distributed systems. This does have considerable virtue, but, at the same time, it is important to keep the confidentiality of patient records very much in mind. The security of files becomes more difficult to maintain as access to those files is provided to a wider audience.

The development of high resolution, large capacity display processors for military applications and for the entertainment industry will provide cheaper and more sophisticated image analysis systems for applications such as computer aided design and manufacturing (CAD/CAM) and, by extension, for the medical market.

Software

The programs developed for diagnostic radiology tend to be relatively closed in the sense that they are of the so-called push button variety. By way of contrast, those developed for nuclear medicine tend to be fairly open-ended. It seems likely that these two relatively contrasting approaches will tend to merge and that there will be a more common approach to the field of diagnostic imaging. In this context, it is necessary that some form of standard be adopted to facilitate image transfer. Both the ACR and the AAPM have published suggested protocols for image file descriptors and formats, but these do not seem to have been adopted by any of the manufacturers who do not find it in their best interests to do so.

If any prediction can be made it is that software can be expected to become more user-friendly and, at the same time, applications programs will become more sophisticated. In diagnostic situations it is likely that we shall have development of knowledge based (or artificial intelligence) systems capable of providing a tentative differential diagnosis. Some research into the area of computer aided diagnosis based on such features as edges and texture is already being done in diagnostic radiology. This development, will, however, be tempered by the need to exert far more quality assurance over software than has existed in the past in order to satisfy the FDA requirements regarding software classified as medical devices.

The ultimate goal expressed to me by one nuclear medicine physician is to have one unified system (a local area network) that is capable of integrating all of the activities in the department. This may be regarded as extending from the registering and scheduling of a patient study through the acquisition and analysis of the data to the final output of a printed report. This, I suspect, is probably too much to ask of any single system and almost certainly embodies as many disadvantages as it does advantages. It does, however, suggest that greater integration of our computer systems is a desirable goal and represents the ideal to which many radiology departments are directing their efforts.

The Patient

In this world of hi-tech diagnostic modalities, it is, I think, necessary to reflect on who is supposed to benefit from these activities and where our interests lie. It is too easy to be carried away by the glamour of applying computers to medicine and to marvel at the ability to display images in a number of different ways. What must be kept in mind is that the patient is at the centre of this activity and it is the diagnostic outcome of some test that is the important factor. The ultimate question surely has to be: Does this imaging modality or method of display really provide information that significantly aids in the diagnosis of the patient's disease? Further, if it does, at what cost is it being done, and can that cost be justified in terms of the incremental benefit to the patient?

As an example of this application of technology I would cite the recent interest in displaying nuclear medicine images in 3-D. These are very exciting images and certainly most impressive. It must, however, be reiterated that nuclear medicine is inherently functional in nature - not anatomic, and that 3-D imaging requires that a binary image be created in order to provide for surface display. Thus, if one is displaying perfusion, for example, a decision must be made about the level of perfusion that is deemed appropriate for the surface to be displayed. All voxels above that level of perfusion will be displayed, all those below will not. The operator is given the impression from such a display that only those parts of the organ which are displayed exist and that a gaping hole exists below that threshold. This is not the case, and it may well be that the poorly perfused portions of the organ are still viable and are capable of recovery. Care must be exericsed when emerging technologies are applied to be sure that the interests of the patient are placed before those of the developer - whether or not the latter's interests are scientific or financial in nature.

INDEX

A

B

C

D

E

G

I

L

M

N

O

P

Q

R

S

T

U